M000211348

Stephen S. Mack
March 3, 2022

MELTING SUN

THE HISTORY OF NUCLEAR POWER IN JAPAN AND THE DISASTER AT FUKUSHIMA DAIICHI

Andrew Leatherbarrow

What will happen if an earthquake greater than
magnitude 8.5 strikes a Japanese nuclear plant?

*"Then it will be the same as Chernobyl....
But that kind of case will never happen."*

•

Yuji Kurotani, senior examiner,
Nuclear Power Safety Examination Division
(Agency for Natural Resources and Energy), 1999

For Nicole, Jo and Noah, whom I neglected far too much during the years I spent working on this book.

MELTING SUN

ISBN 978-0-9935975-7-2

CONTENTS

"What must be admitted – very painfully – is that this was a disaster 'Made in Japan.' Its fundamental causes are to be found in the ingrained conventions of Japanese culture: our reflexive obedience; our reluctance to question authority; our devotion to 'sticking with the programme'; our groupism; and our insularity. Had other Japanese been in the shoes of those who bear responsibility for this accident, the result may well have been the same."

•

**Kiyoshi Kurokawa, chairman
National Diet of Japan Fukushima Nuclear Accident
Independent Investigation Commission**

ACRONYMS

Companies and Organisations

AEC	Atomic Energy Commission (United States)
AEG	Allgemeine Elektrizitäts Gesellschaft (Germany)
AFC	Atomic Fuel Corporation
ANRE	Agency for Natural Resources and Energy
BNFL	British Nuclear Fuels Limited (United Kingdom)
CRIEPI	Central Research Institute of the Electric Power Industry
DPJ	Democratic Party of Japan (left-wing)
EPDC	Electric Power Development Company
FEPC	Federation of Electric Power Companies
GEC	General Electric Company (United Kingdom)
GE	General Electric (United States)
IAEA	International Atomic Energy Agency
INSS	Institute of Nuclear Safety System [sic]
JAEC	Japan Atomic Energy Commission
JAERI	Japan Atomic Energy Research Institute
JAPC	Japan Atomic Power Company
JCO	Japan Nuclear Fuel Conversion Company
JCP	Japan Communist Party
JMA	Japan Meteorological Agency
JNC	Japan Nuclear Cycle Development Institute
JNFL	Japan Nuclear Fuel Limited
JNSDA	Japan Nuclear Ship Development Agency
JSCCS	Japan Society for Corporate Communication Studies
JSDF	Japan Self-Defense Forces
KEPCO	Kansai Electric Power Company
LDP	Liberal Democratic Party (right-wing)

METI   Ministry of Economy, Trade and Industry (2001–present)

MEXT   Ministry of Education, Culture, Sports, Science and Technology (2001–present)

NAIIC   National Diet of Japan Fukushima Nuclear Accident Independent Investigation Commission

MHLW   Ministry of Health, Labour and Welfare

MITI   Ministry of International Trade and Industry (1949–2001)

NIRS   National Institute of Radiological Sciences

NISA   Nuclear and Industrial Safety Agency

NRA   Nuclear Regulation Authority

NRC   Nuclear Regulatory Commission (United States)

NSC   Nuclear Safety Commission

OELCO   Osaka Electric Light Company

OPEC   Organization of Petroleum Exporting Countries

PNC   Power Reactor and Nuclear Fuel Development Corporation

RBMK   High-power Channel-type Reactor (Russian: Reaktor Bolshoy Moshchnosti Kanalnyy)

SCAP   Supreme Commander for the Allied Powers

STA   Science and Technology Agency (1956–2001)

TEPCO   Tokyo Electric Power Company

USIS   United States Information Service

UTH   University of Tokyo Hospital

Technologies and Terminologies

ABWR   Advanced Boiling Water Reactor

AGR   Advanced Gas-Cooled Reactor

ARS   Acute Radiation Syndrome

ASTRID   Advanced Sodium Technological Reactor for Industrial Demonstration

ATR   Advanced Thermal Reactor

AVB   Anti-Vibration Bar

BWR   Boiling Water Reactor

DDFP   Diesel-Driven Fire Pump

ECCS   Emergency Core Cooling System
           (encompasses multiple safety systems)

ERC   Emergency Response Centre

FAC   Flow-Accelerated Corrosion

FBR   Fast Breeder Reactor

HPCI   High-Pressure Coolant Injection

HWR   Heavy Water Reactor

IC   Isolation Condenser

IVTM   In-Vessel Transfer Machine

JPDR   Japan Power Demonstration Reactor

JRR-1   Japan Research Reactor 1

LNG   Liquid Natural Gas

LOCA   Loss-of-Coolant Accident

LWR   Light Water Reactor, encompassing BWRs, PWRs
           and others

MOX   Mixed-Oxide Fuel

MSIV   Main Steam Isolation Valve

OFC   Off-Site Centre

PORV   Pilot-Operated Relief Valve

PWR   Pressurised Water Reactor

RCIC   Reactor Core Isolation Cooling

RETF   Recycle Equipment Test Facility

RHR   Residual Heat Removal System

SIB   Seismic-Isolation Building

SPEEDI   System for Prediction of Environmental Emergency
           Dose Information

SRV   Safety Relief Valve

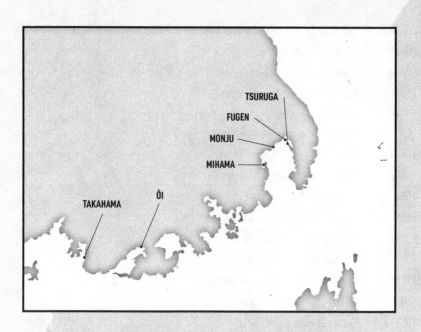

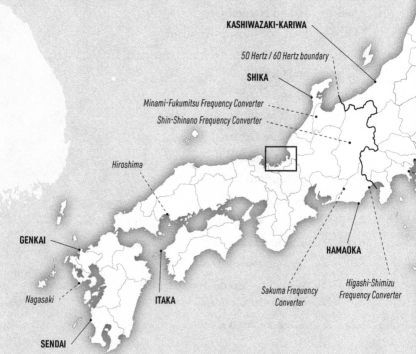

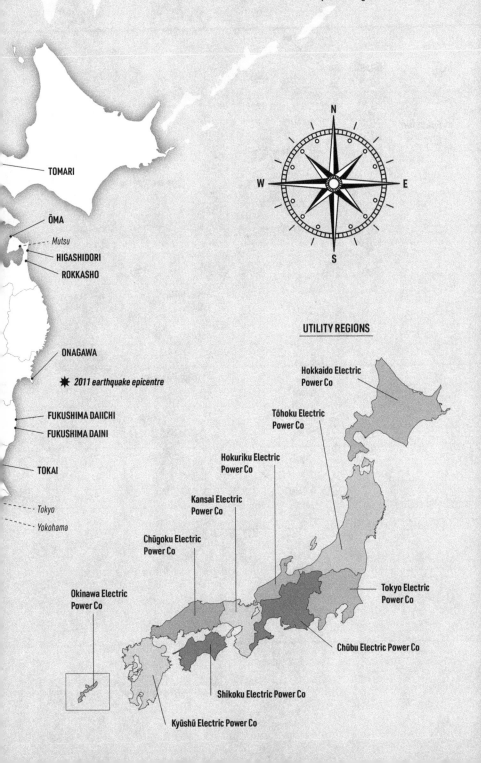

TOMARI

ŌMA

Mutsu

HIGASHIDORI

ROKKASHO

ONAGAWA

✴ *2011 earthquake epicentre*

FUKUSHIMA DAIICHI

FUKUSHIMA DAINI

TOKAI

Tokyo

Yokohama

UTILITY REGIONS

Hokkaido Electric
Power Co

Tōhoku Electric
Power Co

Hokuriku Electric
Power Co

Kansai Electric
Power Co

Chūgoku Electric
Power Co

Okinawa Electric
Power Co

Tokyo Electric
Power Co

Chūbu Electric Power Co

Shikoku Electric Power Co

Kyūshū Electric Power Co

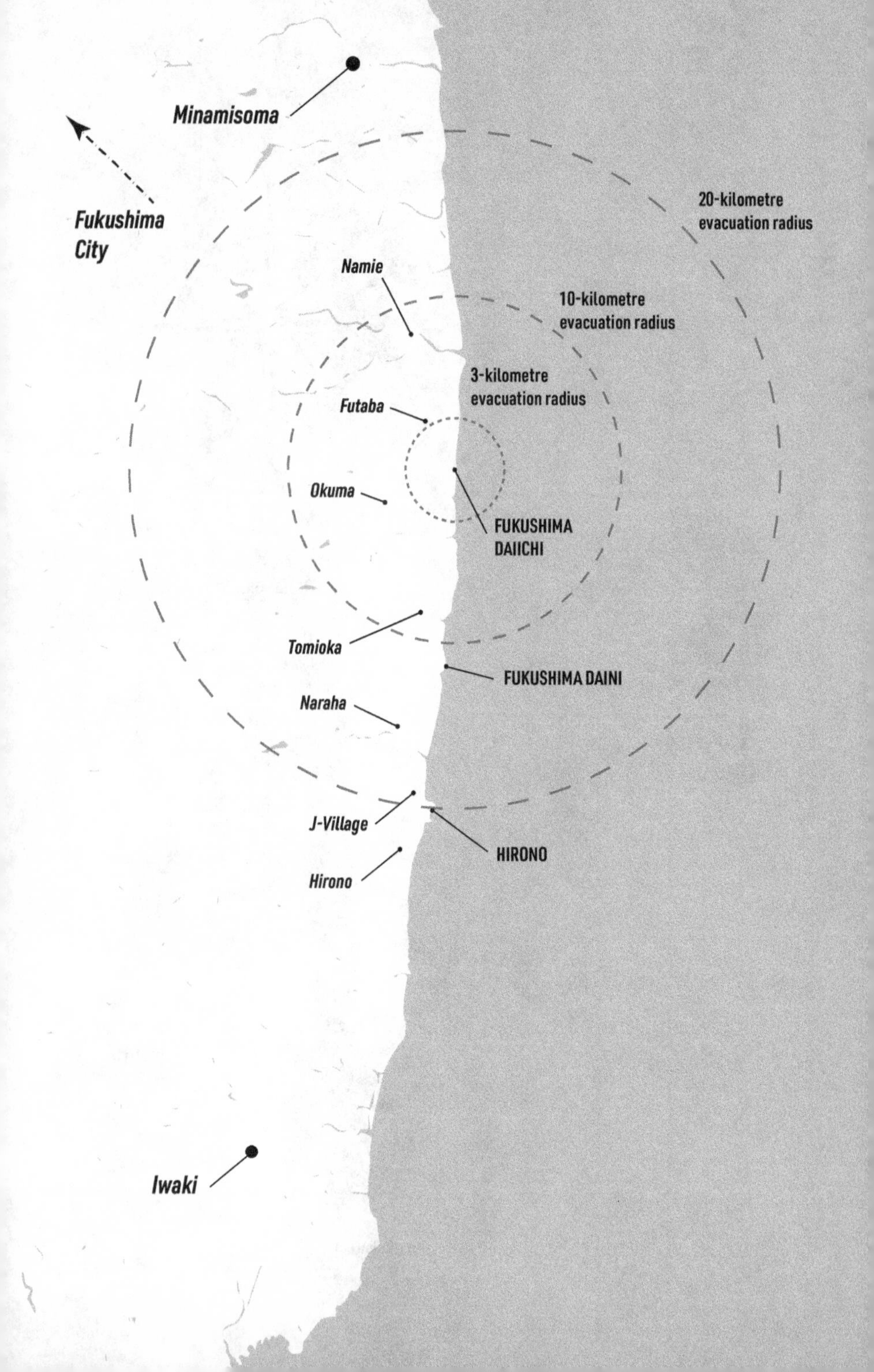

Minamisoma

Fukushima
City

20-kilometre
evacuation radius

Namie

10-kilometre
evacuation radius

Futaba

3-kilometre
evacuation radius

Okuma

FUKUSHIMA
DAIICHI

Tomioka

FUKUSHIMA DAINI

Naraha

J-Village

HIRONO

Hirono

Iwaki

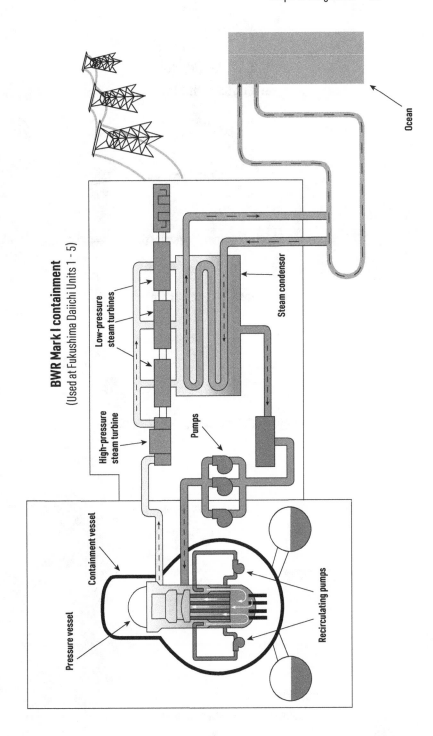

BWR Mark I containment
(Used at Fukushima Daiichi Units 1 - 5)

Low-pressure
steam turbines

High-pressure
steam turbine

Pumps

Steam condensor

Containment vessel

Pressure vessel

Recirculating pumps

Ocean

Boiling Water Reactor
Mark I Containment

Refueling floor

Steel containment vessel dome

Pressure vessel (reactor core)

Concrete containment vessel / drywell

Concrete reactor building
(secondary containment)

Wetwell / torus
Pressure suppression chamber

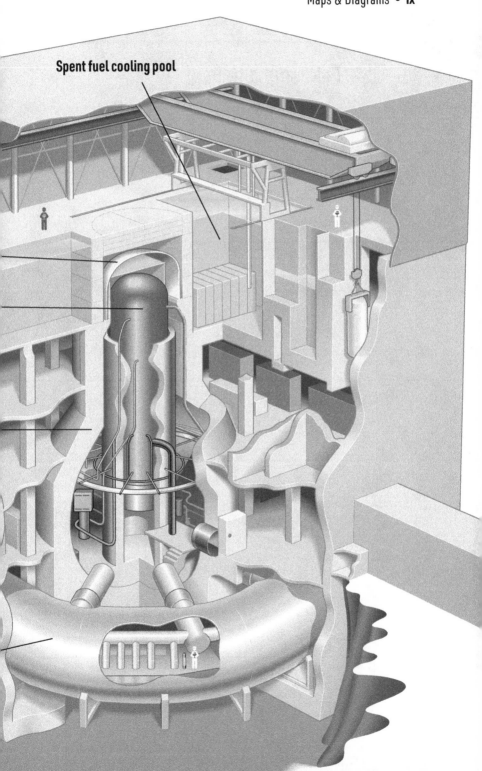

Spent fuel cooling pool

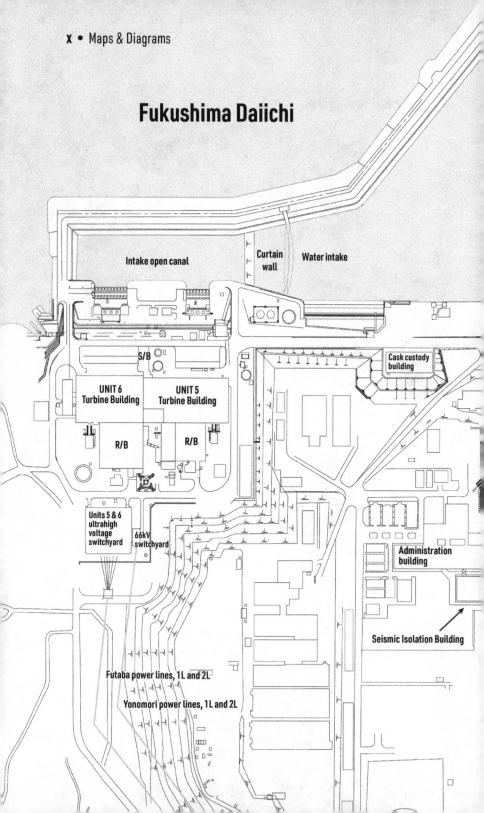

Fukushima Daiichi

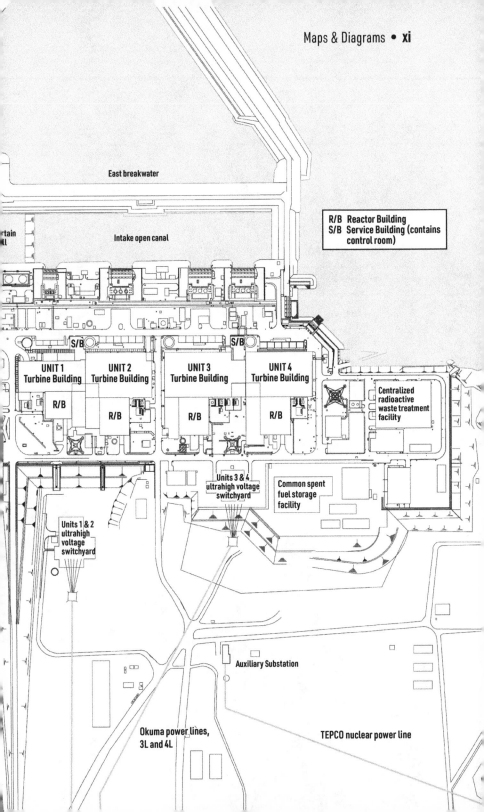

East breakwater

Intake open canal

| R/B | Reactor Building |
| S/B | Service Building (contains control room) |

S/B

S/B

UNIT 1
Turbine Building

R/B

UNIT 2
Turbine Building

R/B

UNIT 3
Turbine Building

R/B

UNIT 4
Turbine Building

R/B

Centralized radioactive waste treatment facility

Units 3 & 4 ultrahigh voltage switchyard

Common spent fuel storage facility

Units 1 & 2 ultrahigh voltage switchyard

Auxiliary Substation

Okuma power lines, 3L and 4L

TEPCO nuclear power line

PREFACE

On the afternoon of 11th March 2011, an earthquake shook the Pacific Ocean floor with a force so violent it redistributed Earth's mass, altering its figure axis and shortening our day/night cycle by 1.8 microseconds. From the epicentre, 75 kilometres east of Japan, rose a towering, unstoppable wall of water. This tsunami breached Japan's coastline within 40 minutes, crashing against the Onagawa Nuclear Power Station – the closest nuclear facility to ground zero – with waves as high as 14 metres (46 feet). Tremors rocked the hardened facility far beyond its maximum tolerances, but Onagawa survived the dual onslaught relatively unscathed and even became a refuge shelter for those made homeless in the chaos. Just over half an hour later, the tsunami reached a second power station: Fukushima Daiichi. The Tokyo Electric Power Company (TEPCO), Daiichi's owner and one of the world's largest and most influential utilities, claimed they took every conceivable measure to prepare for such an event. It wasn't enough – a deluge of water overwhelmed Fukushima's inadequate coastal defences with ease, crippling the site's ability to cool its six reactors, plunging the government into a catatonic crisis and instigating the worst nuclear disaster in 25 years.

2011's Great East-Japan Earthquake may have been a natural trigger, but Fukushima Daiichi's downfall was very much a preventable, man-made event: one Japan should have been well prepared for. Minor earthquakes hit the country every single day, after all, and major ones with accompanying tsunami come every few years.[1] "Of all the places in all the world where no one in their right mind would build scores of nuclear power plants," began one *Japan Times* article almost seven years earlier, "Japan would be pretty near the top of the list."[a] Indeed, the Japanese rank among the least likely people to espouse atomic energy and were its most passionate detractors until a post-World War II, CIA-backed propaganda campaign turned public perception around. From the very beginning, however, the

---

1 Take a look at the Japan Meteorological Agency's earthquake monitor: https://www.jma.go.jp/en/quake/quake_local_index.html. At the time of writing, 30 have occurred in the last week alone.

Japanese energy industry failed to afford due consideration to the serious threat posed by tsunami while being plagued by negative foreign influence, safety violations, technological setbacks and cultural peculiarities that stifled genuine independent oversight.

Accidents are inevitable in any industry, but rather than form substantial accident-mitigating countermeasures, power utilities and the government focused on accident prevention almost exclusively. It didn't work, leading Kyushu and Tōhoku University researchers to conclude in 2018 that "Japan has had more nuclear accidents of greater severity than [all] other countries."[2, b] Insiders proclaimed their technology infallible and, when significant accidents did occur, they were often concealed or explained away with a convenient scapegoat. Meanwhile, a powerful clique of politicians, business leaders, media figures and academics prevented any widespread critical discourse. Along the way, what began as an ambitious and cutting-edge dream fell apart as technological difficulties, inadequate regulation, a box-ticking approach to safety inspections, cover-ups and mounting public opposition hampered the industry's ability to fulfil its goals.

The United States and most other countries stopped building nuclear plants and all but abandoned civilian nuclear power after the 1986 Chernobyl disaster, yet Japan ploughed ahead. Why? Simple: the island nation is burdened with a crippling lack of natural resources for energy, forcing it to import 96% of its fuel in 2011. Other than wood and copper, coal is the only resource Japan has ever possessed in abundance, resulting in more than one national security crises. The Miike coal mine (Japan's oldest and, eventually, largest) operated in some form since as early as the Kyōhō era of 1716–35 until it closed in 1997, taking the local economy with it.[3, c] Addressing this problem became a relentless obsession to find a self-sustainable alternative, and nuclear power offered the solution.

While one kilogram of coal can generate 12 kilowatt-hours of electricity, the same amount of uranium generates 24 *million* kilowatt-hours, meaning you could power the whole world for a year with just 7,000 tons of the latter in a perfect system.[d] With a domestic nuclear fuel enrichment and recycling program, Japan

---

2   Of course, this comes with the caveat that less responsible countries may not report their nuclear accidents.

3   Miike was the location of a major disaster on 9th November 1963, when an accidental explosion killed 458 people by the explosion or by carbon monoxide poisoning caused by the explosion; 839 others were injured and suffered permanent brain damage from the gas.

envisaged a completely self-sufficient fleet of power stations without having to rely on imported fuel. The Fukushima disaster derailed popular opinion of the technology, which had remained favourable among Japanese people since the 1950s. Yet, despite widespread calls to abandon nuclear development, the government continues to push forward. So how did it reach this point? To fully comprehend what went wrong and how the disaster was a product of Japan – just as the Chernobyl disaster was a product of the Soviet Union – a detailed history of the energy industry and nuclear power's introduction and growth in the Land of the Rising Sun is invaluable.

I will never forget that day in March 2011, sitting at home watching videos of the tsunami swallowing everything before I scoured the internet for updates on Fukushima Daiichi. It prompted me to write about Chernobyl, setting me on the path to where I am now. When I began looking for a topic for my sophomore book, something that could retain my interest over what would take years of work, Fukushima was not the obvious choice. I find industrial disasters fascinating, so I first researched non-nuclear incidents such as the devastating Bhopal chemical leak in India and China's Banqiao Dam failure from 1975, which ultimately claimed over 200,000 lives. I read Fukushima accident reports out of curiosity, with no intention of writing about it, but in doing so I noticed tantalising snippets about Japan's electrical and nuclear past. I knew virtually nothing about Japanese history and had no specific goal when I started, other than to learn, and I was often surprised by what I found.

So here it is, the result of my search. While I continue to be a firm supporter of nuclear energy as a clean, scalable power source, the topic is rarely free of controversy, so I want to emphasise, as I did with my book *Chernobyl 01:23:40*, that what follows is intended to be neither anti- nor pro-nuclear, and I hope nobody will take it as such. Any technology, industry, government or country placed under a microscope will reveal innumerable problems; I have simply collated the most interesting information I could find in the most objective manner possible. I had intended to write another accompanying travelogue, but it soon became clear that there just wouldn't be space for it. One of my main goals with *01:23:40* was to write a book that carefully explained every concept and allowed a complete lay-person to *really* understand what had caused a nuclear accident, without being tossed technical terminology out of context. Countless people wrote to me expressing their gratitude for this approach, so I have attempted to do that again here.

I've included a multitude of footnotes for people who appreciate supplemental information, and English-language references have been prioritised as much as possible for those who are particularly keen to learn more. (Footnotes are in arabic numerals; endnotes, where the sources are listed, are in roman numerals.) Condensing 170 years of history into one book has proven to be even more difficult than I imagined and has inevitably meant heavy simplification, but I have done my best to avoid creating falsehoods. I hope you find the material as engrossing as I do.

All Japanese names are written in the Western order, with given names first and family names second.

CHAPTER 1

Forced into a New Age

i. Opening Japan

It is perhaps ironic that foreign demands for a share of Japan's scant energy resources helped usher in its industrial revolution. The islands enjoyed over two centuries of economic self-sufficiency and peaceful isolation from the outside world, beginning in the 1630s. During this time, foreign relations with other nations (non-Asian, in particular) were kept to a strict minimum, and few were permitted to enter or leave the country. Indeed, among Western powers, only the Dutch held trade rights to transport goods such as spices, porcelain, silk and books back to Europe. Dutch ships docked at a single location – a purpose-built artificial island called Dejima, at the city of Nagasaki – and never set foot on the mainland, bar special occasions. Similarly, save for a few hand-picked workers, ordinary Japanese citizens were not permitted onto the island.

This insular period, known as Sakoku ("closed country"), came to an abrupt end in 1853. Two centuries of seclusion engendered military and technological stagnation, and while the islands fought off would-be conquerors and traders aboard wooden galleons, the Western powers gradually mastered metal and steam. The new generation of ships reduced travel time between California and Japan from almost two years to a mere eighteen days but, though Dutch visitors regaled Japan's elite with fanciful tales of steamships and railroads, none had ever laid eyes upon them.[1]

On 24th November 1852, Commodore Matthew Perry of the US Navy departed from Norfolk, Virginia, bound for East Asia. Millard Fillmore, America's 13th president, had entrusted Perry with a vital

---

1 This is a simplification and not strictly true. John Manjiro (and possibly others) had seen such things, but these examples were extraordinarily rare.

mission: compel the Japanese to acquiesce to a trade agreement, using gunboat diplomacy if necessary. Secondary benefits of such a treaty would include Japanese coaling and supply stations for the US's burgeoning fleet of steamships and ensuring the rescue and safe passage of any shipwrecked sailors.

Matthew Calbraith Perry was born on 10th April 1794 in South Kingston, Rhode Island, to a proud naval family. His father and older brother both commanded vessels in the navy and he soon followed in their footsteps, enlisting when he was scarcely 15 years old. He participated in the War of 1812 against the British, the Second Barbary War (1815) off the coast of North Africa, and the Mexican-American War (1846–48), earning the respect of his superiors. By the time Fillmore decided to force the isolationist Japanese to open their borders, the 59-year-old Perry had ascended to the navy's highest rank of commodore and possessed extensive diplomatic experience.[2]

An expeditionary force of four ships – the steam paddle frigates USS *Susquehanna* and USS *Mississippi* with sails furled, and two sailing sloop-of-war middleweights, the USS *Plymouth* and USS *Saratoga* – approached Japan on 8th July 1853 on a hot and hazy[a] morning, then followed the densely forested coastline towards Edo Bay. The city of Edo (meaning "estuary," a reference to the region's topography) held around one million people, ranking it among the world's largest at the time. Japan's rulers knew of the Americans' impending arrival, having been forewarned by the Dutch a year prior, but this did nothing to assuage the fear and awe felt by all who caught sight of their great black ships. The *Susquehanna*, Perry's sleek and imposing flagship, dwarfed her Japanese counterparts, which resembled 16th-century galleons more than modern war machines.[3] The country's leaders had banned the construction of any vessel weighing over 100 tons and with more than a single mast, while even Perry's smallest ship exceeded this weight limit by a factor of eight.[b] At 78 metres (256 feet) in length and almost 14 metres (45 feet) wide, with masts towering over 30 metres (100 feet) above the water, the *Susquehanna* was a technological marvel triple the size and, at 2,450 tons, well over 20 times the weight of Japan's largest ships. A

---

2   In fact, Perry was the second choice for the job. The mission's original commander was Commodore John H. Aulick, but he was relieved of command before it began.

3   Perry set sail from the United States with the *Mississippi* as his flagship but changed the flag to the *Susquehanna* in May before arriving in Japan.

meagre assembly of small craft tried to intercept the American fleet steaming along the coast, but could only watch as Perry maintained speed, unhindered by the headwind.

After several days of stand-off, Perry led a diplomatic mission ashore to the town that became Yokohama, accompanied by a retinue of 300 officers, sailors and marines.[c] The Japanese, in turn, mustered an imposing force of at least 5,000 samurai – the legendary warrior class – to greet the foreigners, followed by a brief ceremonial exchange of letters, with the formalities exchanged in Dutch.

Perry departed Japan after several more uneventful days, promising to return in a year for an official response to President Fillmore's letter, though Fillmore himself had left office by now. Both Fillmore's and Perry's letters to the Japanese leadership, though polite and respectful in tone, were primarily a threat: open your borders and trade with us, or else.[4] Perry went so far as to include a white flag with his letter, with the assurance that if hostilities did take place and Japan wished to surrender, "you should put up the white flag that we have recently presented to you, and we would accordingly stop firing and conclude peace with you."[d]

Japan was ruled by the Tokugawa Shogunate, a feudal military dictatorship with a shōgun as head of state. Successive shōguns from the Tokugawa family ruled over roughly 200 *daimyō* ("lords," most of whom, by this late in history, were little more than aristocratic figureheads) and their retained samurai. The shogunate was already in decline and, with 60-year-old shōgun Ieyoshi Tokugawa lying on his deathbed from an ailing heart throughout Perry's visit (and dead within days of his departure), the country's top officials were rendered paralysed with indecision. His unpopular and sickly fourth son, Iesada Tokugawa, assumed the role of shōgun, but *de facto* rule fell to the Council of Elders, whose senior councillor was 33-year-old Masahiro Abe. Abe was the fifth son of a daimyō from modern-day Hiroshima Prefecture, who became his home domain's daimyō in 1836 at just 17. It was Abe who decided, behind the scenes, to accept Perry's letters in Ieyoshi's stead, if only to buy some time.

"This was Japan's first desperate crisis from abroad since the Mongol invasions in the late thirteenth century," wrote historian Professor D. Y. Miyauchi in 1970, "a crisis which was to set in

---

4   MIT's Visualizing Cultures page describes Perry's approach: he "resolved to assume the most forbidding demeanor possible within the bounds of proper decorum."

motion the final disintegration of the once mighty Tokugawa power in Japan."ᵉ Leaderless, overwhelmingly inferior and threatened with total destruction, a rift tore in Japanese society, forcing Abe to take the unprecedented step of soliciting opinions not only from the nation's daimyō and their retainers but also the 22-year-old figurehead emperor Kōmei and his imperial court. Thus, for the first time in centuries, an emperor gained a foothold in national decision-making.

Should they open their borders to the West or defend the customs of their isolationist society and fight off the aggressors? Some open-minded pragmatists knew that, prior to sakoku, Japan had prospered, and thought reintroducing foreign trade would bring unprecedented wealth and knowledge. Others rejected this assertion, believing that self-sufficiency would continue to serve them well, though few expected victory in open warfare. Everyone knew that China – their fearsome neighbour – had fared poorly in the First Opium War (1839–42) and were adamant to avoid the similar fate of unfavourable, exploitative trade treaties, which brought an end to over 2,000 years of dynastic rule. A few, such as Daimyō Naosuke Ii, urged a more cunning approach: open Nagasaki to trade, gain the knowledge, abilities and technology to build a steam navy of their own, then retake control. "When one is besieged in a castle," he argued, "to raise the drawbridge is to imprison oneself and make it impossible to hold out indefinitely."ᶠ Emperor Kōmei refused to consent to a treaty, preferring seclusion. Indeed, few sought any outcome, satisfied (as most were) with the status quo, and a swift capitulation to Perry's demands could insight an uprising against the shogunate.

As it happened, the return of American ships to Edo Bay on the morning of 12th February 1854 cut their deliberations short, far sooner than promised.⁵ The US force came ashore late in the morning of 8th March, after almost a month of stonewalling by both sides. Officials ushered each party into a reception hall built for the occasion, to a thunderous salute of ships' guns from both nations. Beyond the hall, in a more intimate chamber reserved for just a handful of men, negotiations commenced.

From the outset, according to the official US record, the Japanese acquiesced to "the proposals of [the US] government concerning

---

5 The Americans had heard of a Russian visit to Japan, where they had also attempted to force a treaty, prompting the early return.

coal, wood, water, provisions, and the saving of ships and their crews in distress," which had been some of President Fillmore's primary goals.[g] The Americans, in return, expressed their gratitude and a desire for a treaty similar to that which they established with China. The following week passed with low-level discussions (including the purchase of supplies from the Japanese, the currency exchange to be decided by equivalent weight of coin) and the bestowing of gifts for Japanese officials and Shōgun Iesada Tokugawa.

Perry hand-picked the expansive collection "to demonstrate the technological superiority and advanced civilization of the United States," most famously including a working quarter-scale classic American 4-4-0 steam train replica, together with passenger carriage and tracks.[h] Along with Perry's imposing steamships, train engines were new to Japan's shores, leading few to disguise their curiosity. Onlookers watched as joyous volunteers rode the train, perched atop the passenger car's roof, their robes blowing in the breeze.

Amid further strenuous negotiations, and with Perry refusing to concede on any points, the Japanese relented and agreed to allow American ships docking rights at three ports around the country, with near-immediate effect. Their resistance then collapsed, and the negotiators agreed to almost all demands. On Friday, 31st March 1854, both parties signed the Treaty of Kanagawa and set Japan's industrial revolution in motion.[6]

* * *

The British, Russians and French signed similar "unequal treaties" with Japan within three years. Humiliation and culture shock spread throughout the land like a virus, causing an ever-greater malady to its host with each new painful concession. With no apparent end to the setbacks wrought by such profound changes, the shogunate's once-irrefutable and unbreakable hold over Japan gradually weakened through a combination of constant and often unrealistic foreign demands; massive economic instability, rising inflation and unemployment from the sudden influx of foreign currency; a succession of mighty earthquakes and tsunami in the

---

6 The Treaty of Kanagawa did not include specific details for trading rights. These were signed into force in 1858 with the Treaty of Amity and Commerce, known as the "Harris Treaty" after US diplomat Townsend Harris.

Nankaidō[7] and Tōkai regions, killing close to 100,000 people in the mid- to late-1850s; a wave of cholera from outsiders that brought sickness and death on a massive scale; rearmament by the country's once-cooperative daimyō seeking to reinforce control over their domains; and many other complex social factors. Intense disaffection manifested among the people in a social movement known as Sonnō Jōi: "Revere the Emperor, Expel the Barbarians."

Iesada Tokugawa left no direct heirs after his death in August 1858. His adopted son, Iemochi Tokugawa, continued the line of succession, but the 12-year-old's appointment as shōgun was controversial and his influence weak. As the shogunate bloodline diluted[8] and its capitulation to foreign powers caused their political strength to wane, Emperor Kōmei, who had publicly opposed a treaty with the United States, found his symbolic position morph into real power. This culminated with his "order to expel barbarians" on 11th March 1863. The government ignored his order, prompting sporadic assassinations and other attacks by Kōmei's allies – many of whom were young, ideological samurai – against the shogunate, foreigners and their ships. A series of naval engagements (coupled with obvious military inferiority) then blunted the Sonnō Jōi movement for a time, but it did not disappear. Like his adopted father before him, Iemochi approached death childless at the young age of 20, having almost crippled his government thanks to incompetent advisors and political infighting. He adopted a relative, ten years his senior, who, in August 1866, became Shōgun Yoshinobu Tokugawa: the last shōgun.[9]

Emperor Kōmei succumbed to smallpox on 30th January 1867, leaving his 14-year-old son Prince Mutsuhito, who adopted the era name Meiji ("Bright Government"), as Japan's new emperor. (Emperors are posthumously called by their era name, so this is how I shall refer to each. They are simply named "the emperor" during their lifetime.) Yoshinobu Tokugawa, unlike his recent predecessors, was an intelligent, educated, strong-willed and shrewd shōgun, but fate conspired against him. By year end and little more than a year

---

7   Nankaidō encompasses Shikoku, the smallest of Japan's four main islands, plus modern-day Wakayama Prefecture and part of Mie Prefecture on Honshu.

8   "Diluted" is perhaps the wrong word. In fact, the shogunate family clan was extensive but had suffered from centuries of inbreeding.

9   Yoshinobu had been a candidate for adoption back when Iesada chose a successor, but he lost out to Ii Naosuke's preferred choice of Iemochi. Naosuke was made prime minister at the time but was assassinated two years later.

after assuming power, with his government de-legitimised by succumbing to Western imperialism and losing control of the daimyō, he bowed to public pressure and resigned as the country's leader, ending over 260 years of shogunate rule. He spent his twilight years as an amateur photographer. This power shift prompted an 18-month civil war, known as the Boshin War, in which imperial allies defeated the shogunate army and restored supreme control to the emperor. The Meiji Restoration, the most transformative period in Japanese history, had begun.

Emperor Meiji moved the imperial court – and, in doing so, changed the nation's capital city – from Kyoto, where it had resided for over a millennium, to Edo, which he bestowed with the new name Tokyo, meaning "Eastern Capital." The fledgling government's new policies demonstrated, within a few short years, a complete about-face among what had been ardent opponents of Western technology and style of governance. This apparent betrayal of honour-bound principles – the *Jōi* element of Sonnō Jōi – was not as unexpected as it may appear, however, as Japan was primed for modernisation and industrialisation, as most forward-thinkers knew. The nation boasted widespread educational institutions, government promotion of industrial and agricultural growth, established communications and transportation networks, and, excluding perhaps the years since Perry's arrival, peace and political stability.[i] Just as Emperor Meiji took office, the most powerful daimyō sent a letter to the imperial court, arguing that Japan should accept external influence and avoid "the bad example of the Chinese, who fancy themselves alone to be great and worthy of respect, and despising the foreigners as little better than beasts, have come to suffer defeat at their hands."[i] There was also a sense that many global advancements aligned with the natural order of things and that, rather than being intrinsically 'Western' ideas, they were universal.

The government abolished its old feudal domain system and combined the regions into 47 prefectures in 1871.[10] Later that year, around 100 government officials, military leaders, prominent scholars, statesmen and students – under the umbrella title of the Iwakura Mission, after the group's leader, Tomomi Iwakura – travelled around Europe and the United States until 1873 to re-negotiate treaties and learn about the world and the latest tech-

---

10 It actually took a while to settle on this number, but the main process happened in 1871.

nologies. They concluded that economic self-sufficiency through educational reform, industrialisation and military strength was not only achievable but was vital to securing Japan's long-term future. To help promote this radical new belief, Emperor Meiji's government adopted the ancient Confucian slogan *Fukoku kyōhei* ("Rich Country, Strong Army") as its principal goal and driving force. Military reform introduced a requisite four-year national conscription for all males and the abolition of the legendary samurai warrior class.

Japan found itself in the advantageous position of being able to pick and choose the best elements from the world's industrialised nations and then adapt them for its own use. A new army mimicked the Prussian army, for example, while the police force copied the French approach. The penal code was first based on imperial China's version of the Confucian legal principles but was changed in 1907 to a German model. The postal service, too, underwent reform, modelled on the government-operated British example of a nationwide flat fee for deliveries.[k] By the late 1870s, the Meiji government had erected a spider's web of telegraph lines between all major cities, nationalised the country's most valuable natural resources (including its 11 largest coal mines) and built a fleet of cutting-edge factories and shipyards using public funds and the latest technologies imported from Europe.[11,1] Such initiatives worked to an extent, with coal production soaring from 600,000 tons of coal in 1875 to 13 million by 1905, but the enormous expenditure and typical bureaucratic inefficiencies of government rendered them unprofitable. Most sites were handed back to the private sector for a pittance a few decades later to reduce the national deficit and inflation. These efforts had not been in vain: the economy began to thrive.

---

11 In December 1868, the Japanese government contracted British engineers to erect the nation's first telegraph network. Sharp minds viewed this revolutionary technology as critical to national defence and to maintaining peace across the still-factional and unstable lands. One major problem faced by local Japanese officials during Perry's visits was the interminable delay while they waited, sometimes for days, to receive instructions from the central shogunate decision-makers. With telegraphy, long-distance communication occurred almost instantaneously. A year later, excited workers sent the first communications between Tokyo and Yokohama's courthouse, 32 kilometres away. This success brought a great push to connect urban areas, and by the mid-1870s telegraph lines snaked across almost 7,000 kilometres from one end of the country to the other and were transmitting over 600,000 telegrams a year. The postal and telegraph services merged a decade later by making telegraph systems available in many local post offices. Fukuzawa Yukichi, a leading intellectual at the time, remarked, "When we think about the function of the telegraph, it is the nerve system of the country. The Central Office is like its head and the branch offices are its nerve endings."

ii. Pre-Atomic Electricity

On 25th March 1878, attendees at a banquet commemorating the grand opening of the Central Telegraph Office at Tokyo's Institute of Technology watched in awe as electricity provided lighting for the first time in Japanese history. It arrived via a Duboscq arc lamp,[12] a French technology introduced to Japan by English professor William E. Ayrton and his student, Ichisuke Fujioka.[m] An arc lamp uses a principle similar to lightning in the atmosphere, whereby electricity jumps – arcs – between two carbon electrodes, producing a brilliant light up to 200 times brighter than a conventional bulb. Arc lamps saw widespread use as street lighting in cities like Washington and Paris but were impractical for home use due to their extreme brightness, short life and fire risk. American inventor Thomas Edison began production of his incandescent light bulb at the Edison Electric Light Company in 1880, the culmination of 80 years of tireless innovation by many great minds.[13] This world-changing technology was more practical than the arc lamp and used electricity to heat a filament (of carbonised bamboo in those days) in a vacuum bulb until it glowed to create light.

The Tokyo Electric Light Company (known today as Tokyo Electric Power Co, or TEPCO – the company that operated Fukushima Daiichi) became Japan's first electric power company in 1883. Its chief engineer was none other than Ichisuke Fujioka. Born in 1857, the first son of a samurai, Fujioka studied telegraphy and electrical engineering at the Imperial College of Engineering (later the Faculty of Engineering at Tokyo University) under Professor Ayrton.[14] After an inspirational visit to the Edison Electric Light Company in 1884, Fujioka began experimenting with his own prototype inventions. He designed the first domestic electric generator, the first electric elevator (in Tokyo's 12-story Ryōunkaku building, or "Cloud-Surpassing Tower" in English, Japan's first Western-style skyscraper), and

---

12 While this first lighting is usually claimed to have been with an arc lamp, respected scholar Edward Seidensticker's *History of Tokyo 1867–1989: From Edo to Showa* describes this original light show as using "an electric bulb, which burned out in fifteen minutes, leaving the assembly in darkness. In 1882, an arc light was successfully installed before the Ginza offices of the Okura enterprises ..." Seidensticker knows far more about Japanese history than I do, but given the weight of other sources claiming it was also an arc lamp in 1878, I suspect he was incorrect on this occasion.

13 The company was founded two years earlier.

14 The college was a Meiji-era institution under the Ministry of Public Works, initially staffed largely by foreign engineers and physicists. The engineering building itself was the first substantial Western-style building in Japan.

electrified and exhibited the country's first electric train, emblazoned with Tokyo Electric Light Company, at the 1890 National Expo.[n,15] Five years later, he oversaw construction of the first electric railway in Kyoto, followed by a second line in Tokyo.

Electric railways and electric power companies were often closely intertwined in those first decades, with the railway companies supplying consumer electricity along their tracks. Fujioka joined forces with Masakazu Miyoshi, one of the country's most prominent electric appliance manufacturers, and in 1899 the pair established Hakunetsu-sha Co ("Incandescent Co"), Japan's first incandescent bulb manufacturer. The company renamed itself the Tokyo Electric Co (unrelated to TEPCO) in 1899 and, decades later, in 1939, merged with Shibaura Engineering Works Co to form Tokyo Shibaura Electric Co. People called this new entity Toshiba for short, and in 1978, the company officially changed its name to match.[o]

Fujioka, who came to be known as "Japan's Edison," helped design and build some of Tokyo Electric Light's earliest power stations, the first of which came online in Tokyo in November 1887.[p] Less than six years after the world's first consumer power station commenced operation in Victorian London, public electric lighting dazzled homes in a country that had been isolated and primitive a mere four decades earlier. The new power station used a small 125 Hertz, single-phase, 25-kilowatt (kW) coal-fired thermal power plant. To put that in perspective, the shower in an average modern-day home has a rating of around nine kilowatts, while Fukushima Daiichi's six reactors boasted a combined capacity of over 4,500 megawatts (MW), or 4.5 million kilowatts. Kobe Electric Light and Nagoya Electric Light were founded next, both in September 1888, but were beaten to providing actual supply by Osaka Electric Light Co, founded in 1889 and delivering its first public electricity the same year. Most major cities had electric utilities by 1890, with a total capacity of around 1,500 kW.[q, r]

High rates made early adoption exclusive to industry and the rich, with more widespread use hindered by limited long-range distribution and poor supply for rural areas, where the traditional oil lamp remained dominant for years to come. This problem was solved, in part, by the emergence of small-scale hydroelectric plants to power remote settlements. Harnessing the energy of running water dates back to the fourth century BC, but in 1870 the English industrialist

---

15 The train itself was an American import, but Fujioka did the electrification.

William Armstrong developed the world's first hydroelectricity by damming a lake on his Northumberland estate, then harnessed the power generated by a Siemens dynamo to operate a hydraulic lift and laundry machinery. In 1878, he used his hydroelectricity to light an arc lamp and subsequently lit his entire home – the world's first with proper electric lighting. Japan's own first hydroelectricity came ten years later, at a Miyagi Prefecture textile spinning factory with its own power supply. It was common practice for factories and other businesses to run their own small, private power stations in those days, and public service electricity still only accounted for 39% of electricity generated even 20 years later, in 1907.[s]

Kyoto's population declined by almost one-third after losing its status as capital city. In 1881, Kyoto Governor Kunimichi Kitagaki embarked upon an ambitious and astonishingly expensive civil engineering project (valued over $9 billion in 2020 money) to reverse the city's ailing fortunes.[16] The project would harness water from nearby Lake Biwa (Japan's largest freshwater lake) to provide drinking water, irrigation and inland water transport. A young genius named Sakuro Tanabe, one of Fujioka's peers from the Imperial College of Engineering, travelled to the lake to perform land surveys in 1881 and '82, then modelled the concept for his thesis. His work became the basis for the entire project, and he was hired as its chief engineer at the age of 23.

Tanabe – together with another engineer named Bunpei Takagi, who would go on to introduce fire-extinguishing technology to Japan – visited a new American hydroelectric plant at a silver mine in Colorado several years later.[t] The pair were so impressed they decided to incorporate the technology into the Lake Biwa project midway through construction. The Keage Power Station was thus born.[u] Keage gained fame not only as the nation's first public service hydroelectric power station but also as part of the first large-scale civil engineering project to be overseen by a Japanese chief engineer (Tanabe), as most major projects were supervised by foreigners at the time. Though people harboured suspicions about how electricity could be created from water and worried that it would somehow deplete the water's nutrients and harm their rice crops, Keage commenced operation in June 1897 with a capacity of 1,760 kW and remains in use to this day.[v, w]

---

16 The project cost ¥1,250,000 at the time, which is over ¥1 trillion ($9.2 billion) in 2020 terms.

* * *

Thomas Edison's Edison General Electric Company merged with the Thomson-Houston Electric Company in 1892, to become the global engineering icon the General Electric Company. The number of individual Japanese electric light users topped 10,000 by 1894, then jumped ten-fold over the next dozen years.[x] By the end of the century, Tokyo Electric Light, as the leading force among its peers, installed power generators for each electricity provider across eight of Japan's nine regions – all but Okinawa. Six of these generators were built by German company Allgemeine Elektrizitäts Gesellschaft (known as AEG, or "General Electricity Company" in English).[17] AEG's equipment used a European-standard 50 Hertz system, adopted in 1891 once the foibles of high-frequency electricity became clear.[y] Tokyo Electric Light outfitted the city of Tokyo and, gradually, all of eastern Japan with this 50 Hertz system.

After an electrical fire on 20th January, 1891 destroyed the new Imperial Diet (the Japanese parliament building) less than two months after its completion, the government stepped in and assigned the Ministry of Communications a supervisory role over the fledgling power industry the following year. A further five years passed before the ministry introduced its first regulations in May 1896, however, because that industry was entirely private and little understood by people outside it.[18, z, aa] This early legislation standardised voltages and focused on safety precautions for generators and electric railways but otherwise took a hands-off approach. Instead, local city governments took over supervision of the utilities operating in their areas but left the new industry to more or less run itself.[ab] That same year, General Electric sold Osaka Electric Light Co (OELCO) Japan's first electric generator to operate at 60 Hertz, the frequency used across North and South America. OELCO used its own influence to push the 60 Hertz standard, which became the dominant frequency across western Japan, just as 50 Hertz did in the east.[ac] The two competing systems pushed their way outwards until they met on opposite banks of the Fuji River in central Japan, then split north following the Itoigawa Shizuoka Tectonic Line – roughly

---

17 Although AEG was a General Electric affiliate and descended from a company co-founded by Edison and German industrialist Emil Rathenau, it is otherwise a separate company.

18 This seems to have been assigned to the Ministry of Communications because the telegraph made use of electricity and already fell under their remit; therefore, the electricity industry did too.

the boundary of Shizuoka and Nagano Prefectures' eastern edge. The eastern 50 Hertz side today covers 43% of the country, while the western 60 Hertz side covers the other 57%.[ad]

By the early 1900s, a total of 76 electricity companies – only one of which, in Kyoto, was government-owned – operated over 2,000 electric motors. This situation stood in stark contrast to the wholly government-controlled telegraph industry, which remained under state or semi-public control from its inception until 1985, when the Nippon Telegraph & Telephone Public Corporation was privatised. As impressive as the electric power industry's explosive growth was for the emerging nation, however, the difficulty of transmitting high-voltage power over long distances hindered its expansion.

The Americans solved this problem in the 1890s with the invention of a disc insulator, which allowed transmission line capacity to exceed 40 kilovolts (kV, 40,000 volts). These lines were installed in 1895 at the new hydroelectric Adams Power Plant Transformer House in Niagara Falls, New York: the world's first large-scale, alternating-current electric power station. Adams revolutionised the industry, and its technology allowed Japan to take full advantage of its own remote mountain river systems. The first of this new generation of hydroelectric plants was the 55 kV Komabashi Power Plant in Yamanashi Prefecture (built by Tokyo Electric Light in 1907 using imported AEG generators), which transmitted power 76 kilometres eastwards to Tokyo.[19] Energy companies could now generate electricity in the most efficient locations and then transmit it across great distances, disrupting local suppliers' monopolies and allowing Tokyo Electric Light to cut light and power rates by 12% and 22% respectively the following year.[ae]

Cheap foreign imports of coal and other resources damaged the homegrown market by the late 19th century, forcing many Japanese mines to close or downscale. The outbreak of the Russo-Japanese War of 1904–05 then strained demand against a limited supply, and the price of native coal skyrocketed. Energy issues notwithstanding, Japan defeated Russia with relative ease, making Russia the first European power in modern times to lose to an Asian nation. This sent a shocking message to the West that the dormant Japan of 50 years prior had grown into a force to be reckoned with.

---

19 I've seen several spellings for the Komabashi Power Plant, including Komohashi and Komahashi. I'm fairly confident the actual spelling is Komabashi, not least because TEPCO themselves call it that on their list of hydroelectric plants, but the other spellings are just as common.

The invention of cheap, low-power tungsten filament bulbs encouraged public adoption of electric lighting after 1906, but prices remained out of reach for many. In 1909, 34-year-old entrepreneur Yasuzaemon Matsunaga founded the Fukuoka Electric Railway Co in Fukuoka Prefecture's capital city, together with his friend, business partner and fellow future industry-shaping visionary Momosuke Fukuzawa.[af, 20] With Matsunaga acting as executive director and Fukuzawa as president, the venture arose from an agreement, years in the making, between city authorities and several private companies to bring such a railway with overhead power lines to the area. Using the railway's own small coal power station, the first trams began to run in March 1910.

The eldest son of a merchant, Yasuzaemon Matsunaga III was born on 1st December, 1875 on the tiny island of Iki, near Nagasaki.[ag] At 18, he quit Tokyo's Keio University when his father died and returned home to manage the family estate. The young Matsunaga had a brilliant mind for business, perhaps inspired by a childhood reverence for his grandfather's trading company and tireless work ethic, but by 1887 he had grown restless and returned to Keio, leaving everything to his younger brother. He met the wealthy Fukuzawa through an associate and the pair became firm friends, but when Matsunaga confided that he had lost his appetite for education, Fukuzawa urged him to quit his studies for a second time and return to the world of business. He did just that, first working at the Bank of Japan for a year – via a door-opening recommendation from Fukuzawa – before leaving to enter business with Fukuzawa himself as head of the Kobe branch of a lumber business.[ah] Matsunaga only worked there for a year before establishing his own coal trading company, which he ran for eight years prior to his first foray into electricity at Fukuoka Electric.

Meanwhile, Osaka Electric Light Co began construction of the Ajigawa West Power Plant in April 1909, using five 4,500-horsepower, 60-Hertz steam turbine generators supplied by General Electric rival Westinghouse, cementing a partnership that endures in the region to this day.[21] Two years later, the government combined

---

20 The Fukuoka Electric Railway Co was one of the predecessors of the West-Japan Railway Company (Nishi-Nihon Ryokaku Tetsudō Kabushiki-gaisha). Momosuke Fukuzawa was born Momosuke Iwasake but changed his name when he married into the rich Fukuzawa family. He had befriended Yukichi Fukuzawa, the founder of Keio University, and eventually married his daughter.

21 Ajigawa West became the first of several large-scale plants. The plant began operation in 1910, followed by the nearby East power plant in 1914. They were eventually merged into one, which operated until 1964.

various scattered pieces of local and national legislation to create the Electric Power Industry Law. Though the new law required all power stations to hold a license and companies to share their transmission lines, Tokyo Electric Light retained its dominant position as the country's largest utility. Japanese hydroelectricity, with its round-the-clock natural energy generation, surpassed coal power output for the first time in 1912, producing 233,000 kW vs 229,000 kW, and within two years Tokyo's electricity came from Fukushima Prefecture's Inawashiro hydro plant at a distance of 228 kilometres: the third-longest transmission in the world.[ai, aj]

Fukuoka Electric Railway lasted just two short years during this tumultuous time before it merged with Kyushu's third-oldest and third-largest power company, Hakata Electric Light, in November 1911.[22] Hakata's president happened to be another founding executive of Fukuoka Electric, so Matsunaga retained his position in charge within the new company. The pair then merged with Kyushu Electric Light the following June to form an even larger business named Kyushu Electric Light Railway Co. Fukuzawa, flush from stock market investments, became its largest shareholder but retired into an advisory role because he had also taken over Nagoya Electric Light Co, so Matsunaga again became an executive director. The former envisioned a nationalised electric industry and successfully argued that hydro power was the only way forward as the railway side of the business became less of a focus.

In 1915, at a time when only 5.5% of Berlin's households were electrified, Kyoto became the first Japanese city (and among the first in the world) to illuminate *all* of its homes with electric lighting.[ak] Prices dropped, demand rose, and the subsequent two decades saw massive infrastructure and economic growth, propelled by the electrification and mechanisation of Japanese industry – predominantly cotton spinning and silk reeling – and the rapid expansion of electric railways.

The end of World War I, in which Japan fought alongside the Allies in a minor role, brought another major industry shift. The country's hydro plants had been built along major rivers and were designed to maximise output during the cold, dry winter months, when water levels dropped and demand peaked, but they were "run-of-the-river" type plants, in that they lacked dams for water storage.

---

22 Hakata is the area of Fukuoka City from where the company originated, hence the name.

This meant that the flow rate could not be controlled, resulting in a huge over-supply of electricity during the rest of the year, when water output was far higher.[al] This situation, plus an influx of new energy suppliers covering ever-greater distances and stagnating demand during the post-war depression, forced the roughly 700 smallest and weakest businesses to merge into larger companies. One such company was Nagoya Electric Light, which, in April that year, and with Momosuke Fukuzawa still as its president, was absorbed into Kansai Electric Co. In December, Kansai merged again, this time with Yasuzaemon Matsunaga's Kyushu Electric Railway Co, before changing its name to Tōhō Electric Power Co in June 1922.[23] Fukuzawa resigned to pursue other things and the 46-year-old Matsunaga became vice president of what was now a massive company.

This period of power company mergers ushered in the demise of several originals, including Osaka Electric Light Co in 1923. It also saw Tokyo Electric Light wither, with poor management and corruption eventually forcing the massive Mitsui Bank, which financed the business, to step in and install a director of its own so the company could repay its debts.[24, am, an] Soon, just five main entities jostled to supply Japanese commercial electricity, together accounting for well over half of national electricity demand. These five comprised the three largest traditional utilities – Tokyo Electric Light Co, Tōhō Electric Power Co and Ujigawa Electricity Co, a predecessor of Japan's present second-largest utility Kansai Electric Power Co – plus two wholesale electricity companies: Daido Electric Power Co (which acquired OELCO's assets), and Nippon Electric Power Co, which supplied power to the utilities from their fleet of hydro plants in the Chubu mountains.[25, ao] All fought for dominance in the "Electric Power War" as their capacity exceeded demand.

Matsunaga, by this point, had earned a reputation as a clear-headed, no-nonsense, consumer-first manager who worked tirelessly to provide the best possible service to his customers, no matter

---

23 The companies were a relatively great distance from one another, but the idea was to ease the raising of large sums and to reduce overlapping expenses.

24 TEPCO president Shohachi Wakao siphoned considerable company funds for his own personal political ambitions. At the time he served double duty as the Director of General Affairs of the Rikken Seiyūkai, one of the main political parties of pre-WWII Japan.

25 The City of Osaka local government actually bought the business side and some assets of Osaka Electric Light Co, for ¥64.65 million as part of a pre-existing agreement. Daido acquired part of OELCO's business and most of its assets.

the expense. He was the first major power company executive to address, as early as 1923, the inefficiency of the system popularised by his friend Fukuzawa of almost exclusively building hydro plants. Matsunaga's radical alternative envisioned a vast, interconnected national electricity grid, with backup coal plants making up any shortfall at a time when thermal power was thought to be in decline. When combined with his innovative approach to securing funding and managing costs, Tōhō Electric incurred fewer costs than their competitors, including Tokyo Electric Light: savings which were passed on to consumers.[ap] This diligence paid off, and Tōhō Electric soon dominated the southern Japanese island of Kyushu. In May 1928, Matsunaga (now Tōhō Electric's president) published a book of his thoughts on how the industry should be structured, arguing that to further reduce costs for suppliers and their customers while ensuring an appropriate supply surplus, nine companies should each hold a monopoly over one of the nine regions of Japan. Although his argument had little impact at the time and, in fact, isolated him from his peers, he showed incredible foresight of events to come.[26]

By 1932, with the value of Japan's currency crumbling during a depression and the relentless undercutting of the Electric Power War damaging the energy companies themselves and risking intervention from their creditors, the industry opted to end its infighting and to self-regulate. A cartel called the Electric Power League was established by mutual agreement of the government and five main power companies to formalise these changes, backed up by the revised and expanded Electric Power Industry Law of 1932, which prevented the government from seizing control of the industry.[27, aq] The new law also changed the industry's "legal character" into a public utility rather than a purely private supply-and-demand venture, henceforth requiring companies to seek government permission to change their rates rather than dictating them to customers.[ar] Geographical boundaries were drawn to minimise overlap and wasted invest-

---

26 Matsunaga commented on the proposal in a newspaper article decades later: "What I proposed then was virtually what we see today as the result of the post-war reorganization ... I talked about the division of the nation into nine electricity districts with one company operating in each; mergers of small companies or, where that was not possible, pooling capacities; monopoly on supply allowed in each district; improvements nationwide in load factor (ratio of electricity consumption to peak demand) and in bulk rates; privatizing state- and other public-operated thermal power facilities, many of which were owned by the Ministry of Railways; official approval required in setting electric rate[s]; and establishment of a public works committee to act as a supervisory body, among others."

27 The Electric Power League is also known as the Japan Electric Society and the League of Electric Power Companies, depending on the translation.

ment, establishing regional monopolies across the country, while the Electric Power League helped minimise costs by, for example, making joint purchases of coal.

The two decades prior to this time saw one other fundamental change. With the weak Japanese yen, electric power became the country's only industry to be majority-funded by banks and foreign bonds (from 10% of capital in 1903 up to 70% by 1929), partly because the government waived the limit on foreign borrowing in 1927.[as] This dependence made the banks nervous, and the electrical engineers who founded and ran the most influential power companies were increasingly forced out by businessmen from the banks who excelled at financial management but were ignorant of the technologies they now controlled.[at] Japan's electricity industry grew into the third-largest in the world by the late 1920s, but the incredible pace matched against the new management meant growth took priority over all other considerations. Japan's leaders, meanwhile, became more aggressive on the world stage, quitting the League of Nations – a United Nations precursor established at the end of World War I – in March 1932.

Self-regulation of Japanese electricity companies didn't last either, as the government bought out every utility's foreign debts in 1934, granting bureaucrats huge influence over the industry. Two years later, Japan became the world's third-largest coal producer, at 40 million tons in one year.[au] In anticipation of further international conflict, word spread among utility executives that the government planned a forced takeover of the power industry using a proposal of public control but private ownership because of insufficient money to actually buy the companies outright. These companies' shares plummeted at the news, but they were determined not to relinquish control. As this went back and forth, Matsunaga made national headlines in January 1937 after calling the government's advocates for state control the "scum of the earth" in the Nagasaki Chamber of Commerce.[av] The government proposal collapsed. Following the outbreak of hostilities with China in the Second Sino-Japanese War in mid-1937, however, mismanagement of national energy resources – including a lack of suitable extra labour and mining equipment – meant mines struggled to keep pace with a sudden increase in demand, and the coal supply was soon drying up.[28, 29]

---

28 This war continued until it merged with World War II when Japan attacked Pearl Harbor.

29 The bulk of Japan's coal mines were clustered together in a few small, remote areas. As such, they relied

This problem filtered down to certain industries, which began facing shortages in the summer of 1938, by which time the Electric Power League had dissolved.

To right the ship in a time of war, the politicians regrouped and pushed through a revised control bill. A new privately owned but state-controlled electricity generation and transmission mega-company called Nippon Hassoden Co[30] was formed under the Ministry of Communication in March 1938 with the forced consolidation of the 33 largest power companies, including Tokyo Electric Light and Tōhō Electric Power.[31] The new company's creation came one month after the death of Momosuke Fukuzawa, but closely matched his vision of one nationalised power company: Nippon Hassoden controlled just over half the total installed capacity of Japan.[aw] The government proposed taking over operations in April 1939, providing just over a year to arrange and coordinate the transfer. Compensation to private individuals for losing ownership of their facilities came in the form of shares in the new company, which was capitalised at around ¥780 million (around $3.2 billion in 2020). The change came "over determined opposition" by the power industry but enjoyed public support, thanks to the promise of lower rates.[ax] Total national electrical output now reached 7,560,000 kW (7.56 MW), of which hydro power accounted for 56%.

Nippon Hassoden was classed as *tokushu gaisha*, a "special-purpose company," and given a sweeping monopoly throughout much of Japan with the aim, as defined in the law, "to manage the generation and transmission [but not distribution] of electric power, in order to lower the cost of electricity, ensure an adequate supply of power and promote a wider range of its use."[ay] The company's main purpose was meeting the electricity demands of the military-industrial complex and other national requirements. Increasing the number of hydro plants to conserve the national coal reserve became a long-term goal because coal was required to produce petroleum. While these changes did help scale demand and eliminated in-fighting between suppliers, eminent historian Takeo Kikkawa argues

on rural labour, which was not in inexhaustible supply. Large wage rises to attract more people from far and wide did not entice sufficient numbers.

30 Or, to use the company's full name, Nippon Hassoden Kabushiki Kaisha Co, the "Japan Electric Power Generation and Transmission Co." This full name was rarely ever used, however, and it is almost always simply called Nippon Hassoden in both Japanese and English.

31 All power plants over 10,000 kW, under-construction hydro plants over 5,000 kW, 7,000 kilometres of distribution networks and 94 transformer stations.

that this system was not "economically rational" for the following reasons: "(1) It had the effect of suppressing originality and ingenuity in management and sapped the vitality of private electric enterprises. (2) It followed a one-dimensional system of integrated generation and transmission heavily dependent on hydroelectric power, which made it difficult to maintain a stable power supply and to realise cost-effectiveness. (3) It completely separated generation and transmission on the one hand, and distribution on the other."[az]

While Nippon Hassoden struggled to get the coal power sector back on track, a great drought – said to be the worst for 170 years – caused a downturn of hydroelectricity beginning in December 1938, and the energy shortage spiralled. The utilities' former managers aggravated the situation by maliciously using up their stocks of coal in the intervening year before losing control without replenishing any of it. There were even reports of mining operations "impeded by the lack of electric power" because hydro power was cheaper than a mine running its own coal plant. "During 1939," as a February 1940 issue of *Far Eastern Survey* noted, "the shortage of electric power spread from a few large consumers of water power, especially the electrochemical and electrometallurgical concerns, to virtually all Japanese industries. In the last months of 1939 – with the exception of some slight relief in December – things became so acute that the government stepped in and rationed power supplies from the viewpoint of national defense."[ba] The mines kept busy and coal production reached an all-time high in 1940 with 57 million tons, but demand still outstripped supply, forcing the government to acquire coal from Canada, India and the black market (causing a scandal by itself).

In July 1941, US president Franklin Roosevelt enacted the latest in a series of punishing economic warfare tactics which, according to historian George Morgenstern, "froze Japanese assets in the United States, thus bringing commercial relations between the nations to an effective end. One week later Roosevelt embargoed the export of such grades of oil as still were in commercial flow to Japan."[bb] Japan only produced around 7% of its own oil at the time, an amount equivalent to a single day of production in the US, which provided 80% of Japan's supply.[bc] The embargo strangled the Eastern nation, backing it into a corner.[32]

---

32 Japan had anticipated this move and attempted to side-step it in 1937 with the Synthetic Oil Industry Law, which increased funding for research on turning coal into fuel, but progress was slow

Nippon Hassoden struggled to coordinate a stable national electricity supply during the drought. With Roosevelt's sanctions wreaking havoc, the Japanese government felt it had no choice but to take complete control over national electricity generation, transmission *and* distribution. It forcibly integrated all remaining significant facilities into the company, beginning in September 1941 – just two months before Pearl Harbor.[bd] Prior to this time, historians believe Japan peaked at 818 electric power companies, over half of which dealt with power distribution.[be] All the remaining smaller businesses were, again according to Kikkawa, "forced into dissolution, with very few exceptions."[bf] The government formed nine power distribution companies from the behemoth Nippon Hassoden to distribute electricity across all regions of the country: Hokkaidō, Tōhoku, Kantō (Tokyo), Chūbu, Hokuriku, Kansai, Chūgoku, Shikoku and Kyūshū. Yasuzaemon Matsunaga could not tolerate the loss of control over his life's work and retired into seclusion.

iii. Atoms for War

In 1921, a 31-year-old Japanese physicist named Yoshio Nishina travelled to Copenhagen to conduct research under Nobel Prize–winning titan of nuclear physics, Niels Bohr. Upon his return home in 1929, Nishina went back to work at Tokyo's National Institute of Physical and Chemical Research, from whence he set up his own research laboratory. After the fall of Nazi Germany in April and May of 1945, the Allied powers turned their full attention and combined might east, and blockading the supply of imported fossil fuels – particularly oil – was a key element of their strategy. Japan, starved of vital resources, adopted desperate war strategies to compensate: a lack of aircraft fuel, for example, prompted the use of kamikaze as a battle tactic. In addition to military research, the laboratory founded five years earlier by Nishina, himself now considered the founding father of Japanese physics research, was tasked with discovering what it would take to build a nuclear weapon.

Nishina's research built upon the work of American physicist James Chadwick who, back in 1932, had discovered the final missing piece of the atomic puzzle: neutrons. Chadwick's peers soon learned that atoms – the infinitesimal building blocks of everything in the

and didn't amount to much.

universe – are comprised of a nucleus (a grouping of protons and neutrons) surrounded by electrons. The physical mass of all things is determined by the composition of the atoms they're built from. This discovery instigated a flurry of other revelations that fell into place like dominoes. Neutrons can be fired at other atomic nuclei, causing those struck atoms to split apart and release enormous amounts of energy and extra neutrons, in a self-sustaining reaction that requires outside influence to suppress, rather than increase. This process of releasing atomic energy is called "nuclear fission." Then came the further discovery that the likelihood of fission occurring can be increased by the use of a "moderator," which slows the pace of speeding neutrons in the moderator's vicinity and thus improves the chances of it hitting a nearby nucleus. Uranium, the natural metal mined and used as fuel for most nuclear reactors, is not sensitive enough to its own neutrons to initiate a fission chain reaction by itself and requires a moderator for the process to work.

* * *

To say Japan suffered for attacking the United States is an obvious understatement; bombs rained down on cities around the clock, causing utter devastation. The most severe instance occurred during the night of 9th March, when approximately 100,000 people – almost all civilians – died in a punishing napalm bombing raid on Tokyo.[33] And yet, even with the relentless loss of life and Nishina's weapons program making little headway, the Japanese would not yield. Their military and political leaders knew for months that defeat was inevitable, but could not accept the US's terms of surrender, which would see the Emperor likely executed as a war criminal. Across the ocean, Roosevelt died in April 1945 from a brain haemorrhage after decades of declining health from an autoimmune disease, but his replacement, President Harry Truman, feared a catastrophic loss of life if forced to invade Japan by land.[34] He made one of the most controversial decisions in history: to drop the most cataclysmic weapon ever devised onto a city full of people.

---

33 Tokyo was among 64 Japanese cities to be hit with napalm towards the end of the war.

34 It's impossible not to put a footnote here. Truman knew that the bomb was not necessary to end the war and that the Japanese had been trying to negotiate for better peace terms for some time. He did fear the loss of American lives if he didn't use it, but why he didn't simply alter the peace terms is still debated.

A small army of Allied scientists and engineers, equipped with the knowledge of atomic structures, toiled throughout the war on classified research for the US government's "Manhattan Project," with the singular aim of constructing a nuclear bomb. The weapon harnessed fission's potential for a staggering, instantaneous release of energy in the form of heat by using explosives to smash two pieces of uranium together. Early in the morning on 11th August 1945, a lone B-29 Superfortress crossed into Japanese airspace unchallenged and released the weapon. Residents of the western city of Hiroshima died in sickening numbers as the explosion and subsequent fires flattened and incinerated everything within five square miles, save for a smattering of crumpled concrete ruins. Sixty percent of the population were killed or badly injured.

Truman issued a public statement, warning the Japanese that if they did not surrender, "they may expect a rain of ruin from the air, the like of which has never been seen on this earth. Behind this air attack will follow sea and land forces in such numbers and power as they have not yet seen, and with the fighting skill of which they are already well aware." He also acknowledged that, though fundamentally destructive when used as a weapon, the ability to manipulate matter at an atomic level was "the greatest achievement of organized science in history.... The fact that we can release atomic energy ushers in a new era of man's understanding of nature's forces." He was somewhat deceptive in describing Hiroshima as a military base rather than a city, though it *was* the headquarters of the Second General and Chūgoku Regional Armies, but most important was the threat: give up, or this won't stop. Days earlier, the Soviet Union had informed Japan of their intention to abolish a four-year-old neutrality pact and immediately invade. The Eastern nation clearly had no chance, but still refused to surrender. Another atomic bomb fell; another 70,000 souls perished. It was over.

* * *

The United States took effective control of Japan, placing five-star general Douglas MacArthur in charge of an Allied occupying force under the title of Supreme Commander for the Allied Powers

(SCAP).[35] The Japanese government ran day-to-day operations, but important decisions required approval from MacArthur's office. For now, Japan would pursue an energy policy of "coal first – oil second" to help get the iron and steel industries back on their feet.[bg] Media reporting on the nuclear bombings and the effects of radiation on citizens was banned almost from the outset, and foreign journalists' access was limited. SCAP even forbade public discussion and the dissemination of photographs of the bombings to reduce the chances of an uprising, leaving its victims the only people on earth unable to truly grasp the nature of what had happened to them.[bh]

The *hibakusha* (literally "atomic bomb-affected people") from the two ruined cities would not find sympathy from their fellow man. Instead, injured, dying and misunderstood, they were shunned by society. Ordinary people grew paranoid that the ailing victims were somehow harmful or contagious, a belief not helped by a Japanese government report describing acute radiation syndrome as an "evil spirit." Japanese doctors knew virtually nothing about radiation at the time, and their American counterparts knew little more. Even members of the Manhattan Project team had extremely limited experience of its effects on living tissue. "The chief effort at Los Alamos [a Manhattan Project research facility] was devoted to the design and fabrication of a successful atomic bomb," said Dr Stafford Warren, the project's chief medical officer and one of the first Allied officers to visit Hiroshima.[bi] "Scientists and engineers engaged in this effort were, understandably, so immersed in their own problems that it was difficult to persuade any of them even to speculate on what the after-effects of the detonation might be." After visiting Hiroshima and Nagasaki, Dr Warren concluded that radioactive fallout (i.e. radiation other than that beamed out by the initial explosive reaction) was not a major health hazard.

Two years later, following his experiences in charge of safety for the Operation Crossroads nuclear testing in the Pacific, Dr Warren had changed his mind. He published his thoughts and conclusions in the 11th August 1947 edition of *Life* magazine, writing that the "atomic bombs caused damage by a number of massive effects. The first is an instantaneous shower of penetrating radiation. The second is a shock wave and blast of air which rush out from the point of

---

35 The SCAP acronym in Japan also referred to the general occupying force.

explosion. The third is the release of great quantities of radioactive fission products."[bj] He described how soldiers sent to retrieve equipment would enter areas around Bikini Atoll in the Pacific Ocean after atomic bombs had been detonated. They underwent decontamination upon their return, but, he noted,

> occasionally when a man had taken off his protective gloves in a "hot" area, the safety section had to dissolve the outer layer of skin from his hands with acid. Clothes worn in the target area were often measured with Geiger counters and found too contaminated to clean. Hundreds of pairs of shoes and gloves and tons of other clothing were sunk a mile in the ocean because there was no other way to keep them permanently away from human beings.[bk]

In observing the effects of continual exposure to lingering but less intense radiation on humans and animals, unexpected problems materialised. He continued (emphasis in original):

> The area of *slight* contamination was spreading outside the target area. The algae in the water, moreover, had absorbed radioactive particles and passed them on to little fish. These died at the end of the second week and were eaten by larger fish. These died in the third week, their decaying bodies passing radioactivity back to the algae. When algae collected on the hulls of the expedition's supporting ships, the radioactivity was sometimes strong enough to be detected through the steel hulls. In some instances, it became necessary to move bunks back from the hulls to protect the men who slept in them.[bl]

Dr Warren felt reassured by the knowledge that "the contamination of Bikini could have no serious consequences because the atoll is isolated from human habitation" and that "the contaminated fish were not the migrating kind." He worried, however, about "what might have happened if Bikini had been a populous harbor with a wind blowing in from the sea. Fission products equivalent to tons of radium would have been spread over the city. Most of its people would have inevitably died.... The only defense against atomic bombs still lies outside the scope of science. It is the prevention of atomic war."

* * *

The Allied occupation force considered Japan's monopolistic *zaibatsu* – the all-powerful industrial and business conglomerates, literally "financial clique" – to have applied a significant negative influence in nudging the country towards militarism, so ordered their dissolution and restructuring. SCAP hoped this would not only rid the country of the anti-competitive elements that drove it to war but would also strike a delicate balance between recovering Japan's industry by encouraging competition while not interfering too much in that recovery – all while (they hoped) retaining quality, efficiency and production quotas. Parliament ratified the "Policy on Excessive Concentration of Economic Power in Japan" in December 1947.[bm] As part of this new law, the shares of *zaibatsu* holding companies owned by their controlling families were bought out with government bonds and sold by auction to the general public.[bn] Even banks were forced to sell their *zaibatsu* shares.[bo] In particular, the policy targeted the "Big Four," referring to Japan's most powerful *zaibatsu* with the strongest military and political ties – Mitsui (Japan's largest pre-war company), Mitsubishi, Sumitomo and Yasuda – which dominated everything from shipbuilding, mining, steel fabrication and banking to the engineering and chemicals industries. Over 300 businesses made the list, including most large banks. When the Japanese government objected to this latter point, SCAP agreed to exempt banking institutes and most of the smaller businesses. In the end, dozens of gigantic companies were dissolved, including Nippon Hassoden Co and the nine power distribution companies formed only a few years earlier.

Power generation dipped in 1944 and '45 after Allied bombing destroyed 44% of Japan's coal plants during the war. A year later, dry summer months and shortfalls in mining forced the government to funnel most hydro and coal electrical output to the industrial sector as Japan strained to rebuild.[bp] By 1948, however, the country's fortunes had changed, and electricity supply exceeded the highest levels seen during World War II. Nevertheless, the utilities struggled to meet demand and economic growth stalled, forcing the government to petition first the United States and then the World Bank for loans to buy industrial electricity machinery.[bq, br]

This money then passed to the utilities, which dealt directly with US power companies like Westinghouse and General Electric. Power industry administration fell under the jurisdiction of the Ministry of Communications before the war but was transferred over to the new Ministry of Munitions in November 1943 and then the Ministry of International Trade and Industry (MITI) in May 1949, which made the reorganisation and reconstruction of the power industry a primary goal, at the Americans' suggestion.[36] In a slight twist on what came before, according to historian Mikio Sumiya, they "recommended the creation of unified management of power generation, transmission and distribution in each region" of Japan.[bs]

SCAP brought 73-year-old Yasuzaemon Matsunaga out of retirement in November 1949 after removing many of his former colleagues from the electricity industry for their support of the nationalist wartime government. The bureaucrats, political parties, industrialists and utilities had all fought over what should become of Japan's energy infrastructure for several years by this point, and Matsunaga was something of a last-ditch effort to resolve the matter. He was made chairman of the Electric Power Industry Reorganization Council and began planning for another paradigm shift. Nevertheless, the council out-voted Matsunaga and submitted a proposal in February 1950, mainly written by Nippon Steel president Takashi Miki, who feared price rises under private ownership. Matsunaga's views were included as a sign of respect. SCAP, however, vetoed the report, preferring Matsunaga's plan.

Japan's parliament, known as the Diet, fought against this decision, and progress stalled until General MacArthur himself wrote a letter to Prime Minister Shigeru Yoshida ordering action. Yoshida bypassed the Diet, and the Electric Utilities Reorganization Ordinance and the Public Utilities Industry Ordinance were both issued in November 1950, thus shuffling the administration and oversight bodies once again and forming a new system that remains in place today. The format was essentially what Matsunaga had proposed back in 1928. The following month, he took on yet another crucial position, becoming vice chairman of the Public Utilities Commission, the group overseeing the reconstruction and distribution of the facilities and assets of all Japanese public utilities.[bt]

As of May 1951, unified power generation, transmission and

---

36 Known as the Ministry of Commerce and Industry prior to May 1949.

distribution came under the exclusive control of nine individual private company monopolies divided across nine regions. Electricity rates required government approval, thus preventing the typical monopolistic price excesses, though they did immediately go up by 30% due to continuing shortages. This approach stood in stark contrast to that used in many post-war European countries, including France and Britain, which nationalised their electricity industries in 1946 and 1948, respectively. The new Tokyo Electric Power Co (TEPCO, Tōkyō Denryoku Hôrudingusu Kabushiki-gaisha) took control of Nippon Hassoden Co's assets in the Kanto region encompassing the capital city, thus continuing the legacy of Tokyo Electric Light. TEPCO relied on long-distance hydro power to supply around 80% of Tokyo's electricity during the early 1950s but, much like its peers, lacked funds for new power plants, let alone research and development.

The government – backed by industrialists such as Miki but against the wishes of Matsunaga and the utilities – aimed to correct this situation in September 1952 with the formation of the Electric Power Development Co (EPDC, or Dengen Kaihatsu Kabushiki-gaisha in Japanese, known today as J-Power) to shoulder some of the burden during the broader economic recovery. The various parties compromised with 67% of EPDC's money provided by the Ministry of Finance and the rest split between the nine power companies. The EPDC, based in Nippon Hassoden's old headquarters, began by building the Sakuma Dam in the remote forested mountains of western Shizuoka Prefecture. Many others followed, but the company was only permitted to sell electricity to the utilities, not directly to consumers.

By November 1952, just seven months after the Allied occupation of Japan had ended, the nine utilities established the Federation of Electric Power Companies (FEPC) to help organise and "promote smooth operations within the industry."[bu] The FEPC immediately began analysing the question of whether Japan should strive for self-sufficiency and negate any vulnerability to external influences by sticking with domestic coal and hydro power, or cast their net wider. External influences, once again, decided for them.

iv. Atoms for Peace

Peace prevailed for a few brief years while humanity licked its wounds, but a new kind of conflict arose from the mud and ashes, one without direct engagements between adversaries, where the omnipresent threat of nuclear annihilation kept a paranoid and delicate balance: the Cold War. Everyone soon recognised the incredible potential of nuclear fission as not just a weapon but as a source of near-infinite energy that could change everything. Even in Japan, a mere five days after the destruction of Hiroshima and one day after surrendering, the country's largest newspaper reported that "US heavy industry expects nuclear power to change the means of production drastically" and that people already anticipated that "it will replace coal, oil, and water" as the singular power source.[bv] The US's first generation of nuclear-powered electricity would make front-page news six years later.

Despite the optimism, Allied governments feared the Soviet Union's own nuclear weapons build-up and banned further nuclear research in Japan, worried the Japanese might still aspire to develop such weapons of their own. An arms race was underway, the victor apparently being whoever had the biggest and best nuclear arsenal. The new president of the United States, former army general Dwight D. Eisenhower, had, at one time, criticised the use of doomsday weapons but now found himself normalising them. In March 1955, for instance, he told reporters at a press conference, "Where these things are used on strictly military targets and for strictly military purposes, I see no reason why they shouldn't be used just as exactly as you would use a bullet or anything."[bw] He ignored the inescapable fact that a typical bullet has a diameter of 9–12 millimetres, whereas the United States was manufacturing bombs with a destructive diameter of well over ten kilometres. Eisenhower's casual embrace of nukes proved to be an unpopular approach to foreign policy that received global condemnation. He and his senior advisors decided on a propaganda campaign to shift popular opinion after a Defense Department psychological consultant advised him that "the atomic bomb will be accepted far more readily if, at the same time, atomic energy is being used for constructive ends."[bx] The resulting Operation Candor both promoted the positive uses of nuclear technology and rallied public support for nuclear armament.

One key element – an ambitious technology-sharing campaign dubbed "Atoms for Peace" – aimed to win over the national and international public. Eisenhower announced the program during a televised speech to world leaders and almost 2,000 journalists at the UN General Assembly in Geneva on 6th December 1953. He began by warning the Soviet Union that the path both nations walked would not end well for anyone. "Today," he declared, "the United States' stockpile of atomic weapons, which, of course, increases daily, exceeds by many times the total equivalent of the total of all bombs and all shells that came from every plane and every gun in every theatre of war in all the years of the Second World War." But, despite the terrible ruin these weapons posed, Eisenhower reassured attendees that the United States did not "wish merely to present strength, but also the desire and the hope for peace." Americans wanted to be "constructive, not destructive," and Eisenhower believed that "it is not enough to take this weapon out of the hands of the soldiers. It must be put into the hands of those who will know how to strip its military casing and adapt it to the arts of peace." He proposed forming a neutral International Atomic Energy Agency (IAEA) which would disseminate nuclear technologies "to the needs of agriculture, medicine and other peaceful activities. A special purpose would be to provide abundant electrical energy in the power-starved areas of the world." While tacitly justifying continued nuclear weapons development, Atoms for Peace would, in theory, bring great technological benefit to the world by demonstrating the peaceful, practical alternative uses of uranium and plutonium. Such advances were expected to include nuclear-powered ships, planes, trains and even nuclear cars.

Physicists and engineers, now revered by some as heroic figures for their almost God-like scientific abilities, spent the years after the war designing and testing experimental nuclear "reactors" that could contain and harness the release of atomic energy. These reactors applied the same water-boiling principles as coal power stations but used uranium nuclear fuel rather than coal to create steam, which was directed through rotating turbines to generate electricity. Though the basic idea was similar to fossil fuels, nuclear power plants were (and still are) unfathomably complex and beyond all but the most advanced nations. The United States pledged to supply the technology, fuel and know-how to countries that agreed to use them exclusively for peaceful purposes. Officials recognised that this avenue of sharing was fraught with potential unforeseen dangers, so only countries deemed trustworthy were eligible. The passage of time altered global alliances: India, Iran, Iraq and Pakistan all received nuclear assistance at this time, and all, in the subsequent decades, either developed their own nuclear weapons or used the technology for international political manipulation. Irrespective of future actions, however, the program proved to be popular, with 37 nations signing contracts and a further 14 expressing interest.[by]

Many of Japan's intellectual elite and highest-ranking government bureaucrats saw huge potential in Atoms for Peace. The general sense was that the United States had defeated Japan in the war in part because of the former's mastery of superior technology, most decisively its manipulation of the atom.[bz] This outlook notwithstanding, for two years after SCAP's April 1952 departure and the restoration of Japanese autonomy, prominent scientific minds on the government-formed Japan Science Council and broader scientific community were conflicted on what course to take. Some, such as physicist Taketani Mituo, believed that "the Japanese, being the casualties of atomic warfare, are entitled to have the strongest say in the development of atomic power ... [and] possess the greatest moral right to carry out research. Other nations are obliged to help Japan's effort."[ca] Even those opposed to the idea, such as physicist and Hiroshima bomb survivor Mimura Yoshitaka, did not object on principle but instead argued that "Japan should not embark on nuclear research until the tension between the US and the USSR is eased.... If that [means] there is a delay to Japanese civilization, so be it."[cb] Some idealists argued they should develop their own nuclear technologies rather than depend on imports from the outset, but the pragmatists – not to mention the industrialists, whose motivation was invariably money – recognised that the gulf in capabilities would be too great if Japan hoped to keep pace with the West. Importing technology, therefore, was the sensible choice.

The citizenry felt less conflicted. Though many had seen the numerous anticipatory newspaper articles on nuclear power developments during the preceding years, most still harboured a profound resentment of atomic energy after the war; few were persuaded by Eisenhower's promises. Their buried rage and animosity towards the Americans would explode a few months after his speech.

* * *

From the moment the tuna fishing boat *Daigo Fukuryū Maru* (*Lucky Dragon 5*) set sail on 22nd January 1954 from Yaizu in Shizuoka Prefecture, what should have been a routine trip went badly. The boat's engineer forgot a spare crank, a crucial backup component for the antiquated and unreliable engine. Rather than head back, which was considered bad luck, they set course for a different nearby port to acquire a replacement. Then, with the crank aboard, the *Lucky Dragon 5* ran aground on a shallow sandbar while departing for the second time, and attempts by a passing mackerel vessel to tow her back into the water failed. That evening's high tide refloated the small wooden boat, but the 23-man crew had lost an entire day. Next, the ship's master announced they were not sailing to the calm fishing grounds of the Banda Sea off the coast of Indonesia as each man had thought. Instead, they set course south-east towards Midway Island, with its pounding, relentless ocean swells.

They hit harsh weather on day two. "The seas were rough, and our old ship tossed like a leaf tossed about on the huge waves," remembered fisherman Ōishi Matashichi decades later in his book on the incident.[cc] "I can't forget how frightening it was. For three days and three nights, huge seventy-foot waves surged toward us, and the ship disappeared up to its masts in the troughs. At night, lit up by the searchlight, we could see the giant waves pressing in on us. Far from toughening us, it made us think we were standing at the gates of hell." The 4,000-kilometre journey took two weeks.

On their second day of actual fishing, they lost 170 of the boat's 330 fishing lines (each around 300 metres long), which may have snagged on coral reefs. In a panic, the men squandered four days combing the treacherous water for any sign of their lost lines but only recovered 22. Faced with yet another setback, the crew rallied, determined not to waste the trip. The *Lucky Dragon* headed south in search of better fishing grounds – towards a region in the Pacific Ocean used by the United States for nuclear weapons testing. The little boat's officers knew of the Americans' activities, but examined their navigational charts and agreed on safe fishing grounds outside the US's 145-kilometre exclusion boundary around Enewetak Atoll, established in August 1952. They were unaware that, five months earlier, the American government had notified Japan's Maritime Safety Agency of an eastwards expansion of that boundary to incorporate the area around Bikini Atoll.^{cd} Patrolling US aircraft combed the area for stray vessels but did not spot the *Lucky Dragon*. In fact, the vessel was far closer to the testing area than her crew realised: 140 kilometres north of an unremarkable reef off Namu Island, in the Bikini Atoll.[37] Sitting upon this reef, inside a utilitarian two-storey structure – little more than a shed – rested an unremarkable metal cylinder lying on its side, approximately five metres long by 1.5 metres wide and weighing just over ten tons. It was a new generation of atomic weapon: a thermonuclear bomb.[38]

At 6:45 a.m. on 1st March 1954, as depicted in the title of Ōishi Matashichi's book, *The Sun Rose in the West*: a blinding orange fireball lit the dark morning sky, visible for over 400 kilometres (250 miles). Within a single second, the explosion engulfed an area seven kilometres wide. Fierce heat and an earth-shattering shockwave hurtled away from the epicentre at hundreds of miles per hour, vaporising everything it touched. The physicists and engineers who designed and built the weapon quickly realised something had gone wrong: the explosion far exceeded their expectations. Typical US nuclear tests in the early 1950s ranged in size from 1 to 500 kilotons, though most were below 100 kilotons. For comparison, the Hiroshima bomb was 16 kilotons, while the lone anomaly among these weapons was the experimental 10 megaton "Mike" test, part of Operation Ivy in 1952. It was the original full-scale thermonuclear device and was equivalent to 10 million tons of TNT. This new test, Castle Bravo, was projected to have a more modest (by gigantic-explosion standards) 6-megaton yield but actually detonated with a prodigious 15 megatons – by far the most powerful explosion ever at the time, and still the largest nuclear bomb ever tested by the United States. It dwarfed the Hiroshima bomb and was calculated as being 93,650% more powerful.

As the *Lucky Dragon*'s crew stared in awe, floating far outside the prohibited zone, the waves' rhythmic sloshing collapsed under a deafening roar. The captain leapt into action and shouted at his men to retrieve the trailing fishing lines in case the explosion caused a tsunami. It didn't, but 90 minutes later, fine white ash, carried by

---

37 Other respectable sources put the distance at 115 kilometres / 72 miles – still a long distance.

38 A new generation, that is, relative to the two atomic bombs dropped on Japan. The United States had first tested hydrogen bombs in 1952, but they had gone mostly unnoticed by the general public.

the prevailing wind, began silently falling from the sky.[ce] As the crew hauled in the long ropes by hand, they contemplated the strange substance coating the deck. "We had no sense that it was dangerous," said Matashichi. He continued:

> It wasn't hot; it had no odor. I took a lick; it was gritty but had no taste. We had turned into the wind to pull in the lines, so a lot got down our necks into our underwear and into our eyes, and it prickled and stung; rubbing our inflamed eyes, we kept at our tough task.[cf]

The ash was fallout from the explosion: dangerous radioactive particles created by the fission reaction, mixed with earth kicked up by the blast. That evening the crew "complained of serious pain," recalled Matashichi:

> Headache, nausea, dizziness, diarrhea. Our eyes turned red and were itchy with mucus. Beginning about the third day, our faces turned unusually dark, and many small blisters appeared on the parts of our bodies – wrists, ankles, waists at belt level – where ash had settled as we worked. It was strange: they looked like burns, but they weren't very painful. I think it was a week before our hair began to fall out. Pull, and hair would fall out in bunches, roots and all; but there was no pain. All this made us shudder.[cg]

Their condition had further deteriorated by the time they returned to Japan on 14th March at 5:30 a.m., and when the ship's owner met them at the harbour, he ordered them straight to hospital. The lone doctor on duty that Sunday morning suspected radiation sickness, according to Matashichi, but after applying some ointment to their skin, he discharged the men anyway. Several of the crew returned to hospital the next day, but it took until the 28th for them all to be hospitalised.[ch] The next morning's newspapers reported widespread contamination of tuna fish from the Castle Bravo explosion, but fish caught by the *Lucky Dragon 5* and other Japanese vessels that showed radioactive contamination had already been distributed across the country. Frantic teams armed with Geiger counters scoured Tokyo's fish markets, but some of the missing tuna was never recovered, causing its market price and sales to plummet for weeks over health fears. The Japanese Ministry of Health and Welfare ordered radiation tests on fish caught within a 2,500-kilometre radius of the explosion, soon discovering that almost 900 fishing boats with a total of 20,000 crew members had been exposed to radiation from Castle Bravo.[ci] Of those, 77 boats full of fish – 135 tons in all – were declared unsafe for consumption.

The story spread. People knew, of course, that nuclear bombs produce radiation upon detonation, but few grasped the full effects, believing it to be localised and short-lived. This was something else entirely: lethal poison raining from the skies across vast distances. They called it death ash. Ordinary men and women across the globe suddenly faced the terrifying reality that nuclear war was a real possibility and that they, too, could be victims.

The American government squandered the opportunity to address public paranoia and salvage its reputation among Japanese people with its apathetic responses. Lewis Strauss, head of the US Atomic Energy Commission (AEC), even denied that the *Lucky Dragon*'s crew were afflicted by radiation and instead claimed in a press conference that their skin lesions were "thought to be due to the chemical activity of the converted material in the coral rather than to radioactivity." Instead, he suggested that the little fishing boat may have been a disguised Russian spying vessel – an absurd position in the face of Japanese doctors' experience with radiation after Hiroshima and Nagasaki. John Dulles, Eisenhower's secretary of state, warned Strauss in private to consider "the tremendous repercussions these things have.... The general impression around the world is we are appropriating a vast area of the ocean for our use, depriving other people of its use ... it would be a good thing if something could be said to moderate the wave of hysteria. It is driving our Allies away from us." Members of the Japanese and American governments worried the incident could derail public relations between them, and the Atoms for Peace initiative in particular. "All the effort we painstakingly put into it seemed to get lost ... [as] the *Lucky Dragon 5* incident turned the Japanese against the program," said the then-head of the US Information Agency in Tokyo in a 1994 documentary by the Japan Broadcasting Corporation (Nippon Hōsō Kyōkai, better known as the NHK).[cj]

Soviet scientists achieved a world-first when they connected a nuclear power station to the public electricity grid just six months after Eisenhower's "Atoms for Peace" speech, on 27th June 1954, at the Institute of Physics and Power Engineering in Obninsk, 90 kilometres from Moscow. The reactor was an experimental precursor to the RBMK (or "high-power channel-type," from the Russian name) Chernobyl-type reactors that were eventually installed in all Soviet nuclear power stations, but with an electrical output of only around five megawatts, it was small and underpowered. It also required more energy to run than it could generate, though this was a closely guarded secret. What mattered was that Soviet science appeared to have beaten American efforts to commercialise nuclear power by four years, making it a potent propaganda victory.

While Soviet leaders could build nuclear facilities wherever they wanted without question, the American government had a big PR problem, even among its closest allies. The whole world knew how atomic energy had erased two entire cities from existence and could not ignore the reality that nuclear weapons could end all life on earth. Even worse, a clear demonstration of the dangers of radioactive fallout was now forefront in the public consciousness. The United States regarded Japan as pivotal to its diplomatic influence in the East. As extreme as it sounds, at a mere 40 kilometres from Russia at the closest point, Japan itself could have been at risk of rejecting democracy and sliding down a slippery slope to communism if its people rejected peaceful uses of this new technology.

Indeed, America grew seriously concerned that Russia and the Japan Communist Party (JCP) would gain a propaganda victory from the *Lucky Dragon* accident. On 22nd March 1954, US Assistant to the Secretary of Defense G. B. Erskine submitted a memo to the National Security Council Operations Coordinating Board: "A vigorous offensive on the non-war uses of atomic energy would appear to be a timely and effective way of countering the expected Russian effort and minimizing the harm already done in Japan."[ck] Thomas Murray, commissioner of the AEC, pointed to Japan in September, saying "now, while the memory of Hiroshima and Nagasaki remain so vivid, construction of such a power plant in a country like Japan would be a dramatic and Christian gesture which could lift all of us far above the recollection of the carnage of those cities."[cl] Reactions were mixed – some found the idea exploitative and distasteful, while others considered it a fitting symbolic olive branch to the Japanese. Hiroshima mayor Shinzō Hamai numbered among the latter, saying, "the fact that Hiroshima will become the 'first nuclear power city' will comfort the souls of the dead. The citizens themselves, I think, would like to see death replaced by life."[cm] It mattered not: Eisenhower's government opposed siting a plant in Hiroshima, and that specific proposal went nowhere.

Despite Japan's three main newspapers (the *Asahi*, *Yomiuri* and *Mainichi*) printing a consistent message in support of peaceful nuclear power throughout the occupation, anti-nuclear sentiment across Japan rose.[cn] Hiroshima hosted the first World Conference against Atomic and Hydrogen Bombs a year after being bombed, attended by 30,000 people, and the conference still convenes annually. Six months after Castle Bravo, on 23rd September 1954, the *Lucky Dragon 5*'s radio operator, Aikichi Kuboyama, died of liver failure after his immune system disintegrated from radiation damage. The Japanese people flew into a frenzy. Anti-nuclear protesters poured onto city streets across the country and housewives from a book club in Tokyo's Suginami ward began circulating petitions calling for a ban on nuclear weapons, gaining 270,000 signatures after two months. Their efforts inspired a remarkable 32 million people – a full third of the population – to sign similar petitions within a year, including over one million from Hiroshima Prefecture (half the prefecture's population at the time). The event even inspired the world's longest-running fictional movie character – Godzilla – who, born from a Pacific Ocean hydrogen bomb test, came to Japan and annihilated Tokyo. "The theme of the film, from the beginning, was the terror of the bomb," said the film's producer, Tomoyuki Tanaka. "Mankind had created the bomb, and now nature was going to take revenge on mankind."[co] The film concluded with a character expressing the opinion that further nuclear weapons tests would cause another Godzilla to rise. Countering this national disdain by winning over the hearts and minds of Japanese citizens would be a vital propaganda coup for the United States.

CHAPTER 2
Baby Steps: 1950s–1960s

i. Propaganda

John Jay Hopkins, president and CEO of American military hardware and pioneering nuclear power systems manufacturer General Dynamics, gave a speech before the US National Association of Manufacturers on 1st December 1954 in which he proposed a "Plan for the Development of International Atomic Energy under the Leadership of American Industry."[a] General Dynamics had launched the world's first atomic submarine, the USS *Nautilus*, earlier that year, and the company was keen to expand its operations. Hopkins wanted American industry and the US government to join forces to supply nuclear reactors to "have-not" nations to increase their standard of living, with a focus on Asia and Africa. This approach was similar to one of Eisenhower's pledges in his "Atoms for Peace" speech, but hearing it from the leader of one of the Western world's biggest industrial companies, rather than a politician, had a "considerable impact among leading Japanese industrialists."[b] According to a US Central Intelligence Agency (CIA) report, the State Department was "not persuaded, however, of the wisdom of official endorsement" of Hopkins's plan.[c] The report noted that the "proposal differs in important respects from our established national policy regarding cooperation with other countries in the atomic energy field, which does not envisage a broad program of grant aid. We would run the risk of encouraging false hopes as to the prospects of United States financing of atomic power plants, and also as to the imminence of atomic power."[d]

A year later, in December 1955, 37-year-old nationalist politician and former Imperial Navy lieutenant Yasuhiro Nakasone addressed a special committee in Japan's House of Representatives on the promotion of science and technology. "Nuclear power used to be a violent animal, but has now become a farm animal," he declared in a speech proposing an atomic energy bill. "Japan should increase its national strength through the promotion of nuclear power in an effort to gain a rightful place in the international community."[e] Nakasone had campaigned to persuade the US to share its nuclear technology since before the Allied occupation officially ended in 1952, believing that the minuscule fuel requirements compared to coal power made nuclear ideal for Japan.[f] He had witnessed the unprecedented power of nuclear energy – what he called "the largest discovery of the 20th century" – first-hand at the bombing of Hiroshima in 1945 and wrote "I still remember the image of the white cloud…. That moment motivated me to think and act toward advancing the peaceful use of nuclear power."[g]

Nakasone knew that Atoms for Peace presented an unmissable opportunity. Without the fluctuating whims of foreign exchange rates or the heavy cost of importing millions of tons of coal and oil to power its factories and homes, Japan could swiftly recover and prosper. "When I learned that Eisenhower had switched to a policy of using nuclear energy for peaceful purposes," he recalled later in life, "I thought to myself, 'Japan can't fall behind. Nuclear energy is going to define the next era.'"[h] In March 1954, around the time the *Lucky Dragon 5* returned with its irradiated crew, Nakasone convinced a receptive Parliament – many of whom shared his concern about and keen interest in embracing a new long-term energy supply as the solution to Japan's lack of natural resources – to grant a ¥250 million budget (around $14.6 million in 2020) for science. Of this, ¥235 million went towards research and construction of an experimental nuclear reactor. Yasuhiro picked the number after the nuclear fuel – uranium-235 – but it ballooned to ¥5.1 billion the following year and increased year-on-year thereafter.

TEPCO saw a change on the horizon and established its own Nuclear Power Generation Department in 1955. The historic US-Japan Agreement for Cooperation Concerning the Civil Use of Atomic Energy was signed by both parties in November and passed on 16th December 1955, whereby the United States would supply two experimental (i.e. non-commercial) nuclear reactors to Japan for their own research, provided any used enriched uranium fuel be returned to the US.[i] Natural uranium mined from the earth is not "potent" enough to sustain a fission reaction in ordinary reactors, so it requires refinement – "enrichment" – before use. Three days later, the Diet approved the Atomic Energy Basic Act, to "secure energy resources in the future, achieve scientific and technological progress, and promote industry by encouraging the research, development and utilisation of nuclear energy, thereby contributing to the improvement of the welfare of human society and of the national living standard."[j]

The act stipulated that Japan's nuclear research was for peaceful purposes only, to be performed independently of whichever party controlled the government. The following is a simplified list of government bodies created during this time.[1] To oversee and administer the fledgling industry, the Diet approved the creation of two main government bodies: an independent, bipartisan Japan Atomic Energy Commission (JAEC) to dictate policy and form long-term nuclear plans for the nation, and a Science and Technology Agency (STA) to promote nuclear power research and development and to create the infrastructure for a complete national nuclear power industry.[2] The JAEC had its own Nuclear Safety Commission (NSC) to "plan, deliberate on and determine matters related to the research, development and utilisation of nuclear energy" and safety obligations to all.[k] Basically, the NSC created guidelines used to assess and write safety regulations, but absolute responsibility for safety and upholding safety regulations did not rest with any one organisation. Two new quasi-governmental organisations were created to aid the STA: the Japan Atomic Energy Research Institute (JAERI), to provide the necessary technology and training, and the Atomic Fuel Corporation (AFC) to conduct research on nuclear fuels, enrichment and waste. While (at least in the early years) the STA heavily influenced the general direction chosen by the JAEC, rather than the other way around, overall regulation of the nuclear industry and authority over plant construction permits fell under the jurisdiction of the Ministry of International Trade and Industry (MITI). The prime minister held the legal authority to approve licences for nuclear businesses, though in reality such approval fell

---

1   Numerous related laws and organisations have been omitted to keep the list from becoming too confusing. These include the Act for the Regulation of Nuclear Source Material, Nuclear Fuel Material and Reactors (often simply referred to as the "Nuclear Reactors Regulation Act"); the Act for Prevention of Radiation Hazards; the Electricity Business Act (which applied to the entire power industry, not just nuclear); the Act on Special Measures Concerning Nuclear Emergency; the Act for Technical Standards of Radiation Hazards Prevention; the Radiation Review Council; and many others.

2   The JAEC's bipartisanship meant that members had to be approved by both houses of the Diet.

to the STA. The multitude of overlapping governing and oversight groups created a problem that would trickle down and eventually enable TEPCO to ignore recommendations that could have prevented the Fukushima disaster.[1] Much like the original electric industry regulations from 1896, the JAEC and NSC were advisory rather than regulatory, despite their *de facto* law-making nature. It was generally considered unnecessary to make laws because companies would adhere to all suggestions. Thus began the rocky first phase of Japanese nuclear regulation.

* * *

One man was given the rank of cabinet minister and took control of both key organisations, as both chairman of the Atomic Energy Commission[3] and director of the Science and Technology Agency: the 70-year-old and highly influential media mogul Matsutarō Shōriki. Born on 11th April 1885, Shōriki entered Tokyo University – Japan's most prestigious place of learning, traditionally attended by its future political elite – to study law in 1907, when he was 22.[m] While there, he met and befriended many people who would propel him forward later in life. Shōriki graduated in July 1911 but failed the civil service entrance exam, having dedicated more time to judo than his studies. An influential friend from his hometown, who happened to be chief cabinet secretary for Prime Minister Prince Saionji Kinmochi, wrote Shōriki a letter of recommendation that bestowed him a job in the Statistics Bureau. This position initially bored the young man, but he soon became fascinated by numbers and crime statistics. Following this interest, he joined the Tokyo Metropolitan Police as an inspector in 1913, where he gained a reputation as an intelligent, determined and inventive leader who nonetheless possessed an ardent hatred of communists and socialists. He rose swiftly through the ranks until a distant formerly high-ranking relative recommended Shōriki as director of the Secretariat of the Metropolitan Police Board (equivalent to the chief of staff) in 1921. In this role, he became acquainted with many prominent bureaucrats, sowing the seeds of his own future political aspirations.

---

3   Shōriki served in this position from January 1956 until December 1956, then again from July 1957 to June 1958.

The magnitude 7.9 Great Kantō Earthquake struck two years later, on 1st September 1923, killing over 100,000 people in Tokyo and destroying around 400,000 buildings, 90% of them from fire. In the chaos, a rumour spread – probably true on a small scale, though it was quickly exaggerated – that communist Koreans living in Tokyo and Yokohama were taking advantage of the chaos by killing, looting and poisoning wells, prompting the murder of up to 10,000 Koreans by angry mobs. Shōriki was later accused of encouraging the slaughter. His biography from the early 1950s, by contrast, paints his role as that of a noble hero, noting that "he investigated the rumours [of Korean misdeeds], found they were false, and when batches of Koreans were arrested by his hot-headed associates, he ordered them freed."[n] Whether there is any truth to this seems to have never been definitively answered. Later research implicates him in the rounding up of those same people, but the police *did* switch from warning the populace about Koreans to warning of the gangs out to kill the Koreans midway through the massacre, then gathered them up for their own protection, so it's difficult to judge.[o]

An almost-successful assassination attempt three months later on the young prince (and future emperor) Hirohito by communist anarchist Daisuke Nanba derailed Shōriki's career.[4] Nanba lost his life, the prime minister and his entire political cabinet resigned, and Shōriki was dismissed from the police.[5] He was saved from a stickier fate by extensive political connections, most notably home minister and former Tokyo mayor Gotō Shinpei.[6] Shōriki sought a new avenue of influence, so, with a loan from Shinpei, bought the struggling 50-year-old *Yomiuri Shimbun* ("Yomiuri Newspaper") in February the following year. In little more than a decade, his changes – such as switching from objective left-wing to sensationalist right-wing reporting and adding sports pages and games – increased annual sales from 56,000 copies in 1924 to over 1 million in 1938, placing it among Tokyo's largest newspapers.[p]

His newspaper became an unofficial state mouthpiece, which strongly encouraged the nation's drift towards militarism in the

---

4 Nanba was also partially motivated to avenge the death of Shūsui Kōtoku, who had been executed for the attempted murder of Emperor Meiji.

5 Shōriki is occasionally (and incorrectly) described as having resigned.

6 The home minister was a senior government cabinet position in the Meiji period. The job was focused on internal security at the time, in particular on Bolshevik-inspired labour unrest and anarchists such as Nanba. Shinpei served in this cabinet position twice under two separate prime ministers: from 1916–18 and again from 1923–24.

1930s, while the man himself was said to be an ardent promoter and enabler of Japan's fascist military goals during WWII.[7] His propaganda skills were even put to official use before and during the war, when he was appointed to a number of government organisations.[q] At war's end, Shōriki joined the second wave of officials accused of war crimes by the Allies (and his own staff). Curiously, however, his was the sole name "included on the basis of evidence compiled [in Japan]," rather than being part of the War Crimes Office's list made with assistance from the US State Department.[r] In any event, he spent almost two years in Tokyo's Sugamo Prison. His friends claimed that "Shōriki's policy was to leave the selection of news and editorials to his editorial staff and he seldom did the selection himself," though this contradicts reports by the same staff and was written in support of having him released.[s] After two years and the CIA's "exhaustive search" of the evidence, all accusations against him were branded as having been "ideological and political [in] nature."[t,] [u] Freed from captivity, he set about making up for lost time: he returned to the *Yomiuri Shimbun*, established the *Nippon TV* network with covert help from the US government (television was seen as good for pro-US and anti-communist propaganda) and, at some point no later than 1952, himself became a secret CIA informant.[v]

Ever since the *Lucky Dragon 5* incident had incited anti-US and anti-nuclear weapons protests across Japan, Shōriki feared an upswell of public support for communism and the Soviet Union. The USSR had mirrored Atoms for Peace with its own campaign to distribute peaceful nuclear power among satellite states and potential allies and was now eyeing Japan. He was not alone. As a declassified CIA document notes, "the bitterness people feel toward [the] United States for its use of the bomb is deep-seated, and taking advantage of this existing sentiment, the Communists clearly are succeeding in their anti-US and peace propaganda offensives. Given the intense pressure of the Soviet peace offensive, both Japan and the United States will suffer seriously if this emotional trend of public sentiment is left to follow its own course."[w] To seize control of the narrative, Shōriki directed his media outlets to bombard the public with articles and discussions promoting a magical future of infinite clean, cheap electricity from imported British and American nuclear power.

---

7 Japan was not really fascist, but this is a handy and mostly correct label. Japan's particular brand of fascism was known as Shōwa statism, an amalgamation of various Japanese right-wing political ideologies.

He turned to the United States for assistance. A further CIA document from 31st December 1954 (a year before Shōriki was appointed to the JAEC) reported that the *Yomiuri* newspaper and *Nippon TV* "desire [to] engage in [a] full extended campaign promoting President Eisenhower's 'Atoms for Peace' proposal, and are soliciting our guidance and assistance."[x] Whether he was exploiting the Americans or vice versa, or if both were mutually beneficial to each other, remains unclear.[8] What is clear is what drove him. As one CIA officer noted, referring to Hidetoshi Shibata, Shōriki's business partner at *Nippon TV* and others, and arguably the brains behind the man, "Shibata has admitted quite frankly that his and Shōriki's motives are, in large part, political." The officer goes on to note, in the classic clipped style of such reports, that the ambitious Shōriki "desires to be power in Democratic Party, dreams of Cabinet post.... Both believe they will gain in prestige through association with important US figure with positive program. Both convinced that atom issue key question in coming elections; if conservative candidates cannot point to constructive peaceful program relating to atomic energy, they will suffer severely at polls." Shōriki joined the aforementioned Democratic Party in 1955, and by the end of the year was seriously considering running for the lofty position of prime minister.[9] He was helped by what the CIA now considered to be "perhaps, in terms of stirring the mass mind, the most influential" media empire in the country.[y] In his ambition, Shōriki went so far as to offer a ¥400 million ($23 million in 2020) bribe to Ichirō Kōno, the cabinet minister of Agriculture and Forestry, if Kōno could "secure" the top government position for Shōriki. Kōno's joking response was that Shōriki "should give him the money first, and then he would consider the matter."[z]

Shōriki read the *Yomiuri* every day and exerted enormous control over its content, maintaining contact "not only with desk chiefs, but also with individual reporters," allowing him to sway public opinion to his will whenever he felt it necessary, in a manner similar to Rupert Murdock.[aa] "When an article appears which is at variance with his own opinions, he demands the name of the writer and personally upbraids him for not following the paper's editorial line,"

8   The CIA certainly considered that Shōriki "regarded [us], in the main, as a channel for exploitation." CIA document volume 2_0043, p. 2.

9   The Japan Democratic Party merged with the Liberal Party (both right-wing conservative parties) in November 1955, forming the Liberal Democratic Party that has held almost uninterrupted power ever since.

according to one CIA report from June 1956.[ab] That said, another report from a few weeks later implies a hands-off approach to the newspaper's day-to-day operation, stating that "at the present time, however, it does appear … that the editorial staff and column writers enjoy a greater amount of freedom in their selection and treatment of newsworthy topics [than during WWII]."[ac] He was also not above attacking the United States via the *Yomiuri* when he wasn't getting his own way.

The latest iteration of Shōriki's long-running propaganda campaign kicked off on New Year's Day 1954 with a *Yomiuri* newspaper article entitled "Finally, the Sun Has Been Captured."[ad] His stated intention was to "calm the storms of violent controversy and combat the activities of the leftists and rightists."[ae] In another article, the *Yomiuri* tried to reframe Japan's position as a victim of nuclear energy, arguing that the Japanese, of all people, should be the ones to promote peaceful uses of the technology that did them so much harm – a belief shared by many people in Japan.

Meanwhile, Shōriki's connections paid dividends. In February 1955, he was elected to join the Lower House in the Diet, as had been promised to him at least a month earlier. Next, Shōriki and his lieutenant, Shibata Hidetoshi, together with the CIA's backing, invited John Hopkins of General Dynamics, along with Nobel-winning physicist Ernest Lawrence and former AEC director of reactor development Lawrence Hafstad, to Japan as "Atomic Peace Emissaries, in order to revive the hopes of the deflation-oppressed Japanese in reconstructing their economy."[10, af] Hopkins accepted the offer, prompting the *Yomiuri* newspaper to announce his visit on its front page. On 9th May 1955 the trio travelled to Tokyo, where they met politicians and industry leaders and delivered packed lectures on the peaceful uses of atomic power, all of which was reported with zeal in the *Yomiuri*.[11] Two weeks later, the Organising Committee of Japan Atoms-for-Peace Council, with Shōriki as its principal representative, sent a letter of thanks to Hopkins, writing that "within a week after your arrival, the sudden surge of ardour on the part of our leaders and of the public made it possible for the Hatoyama cabinet

---

10 Shōriki acting as head of a "committee of prominent politicians, businessmen and academicians" (CIA document vol. 2_0038, p. 1). Shibata Hidetoshi, as a trusted partner, was often the one directly involved in communications with US representatives and in some respects played a more prominent role than Shōriki himself.

11 So packed that thousands more reportedly watched Hopkins's speech from televisions set up outside the hall.

formally to accept enriched uranium – previously considered a most difficult decision – and to initiate moves leading to the signing of an agreement with your country."[ag]

Shōriki's Japan Atomic Energy Commission released its very first long-term plan (the Long-Term Plan for the Research, Development and Use of Nuclear Power) in 1956, going on to release a new long-term plan roughly every five years. This first plan stipulated an industry ideology of *kokusaku minei*, or "national policy, private management." Government would lay the path, while private business would walk it. Among the JAEC's principal concerns was the world's perceived limited supply of fossil fuels and uranium, writing "the basic policy dictates the reprocessing of spent fuels be conducted in Japan as far as possible…. Japan's effort to develop nuclear power shall aim to develop the fast breeder reactor (FBR) which is deemed to be the most suitable atomic reactor for Japan from the viewpoint of effective use of nuclear fuel resources."[ah]

Fast breeders are "closed-loop" reactors, meaning they create more plutonium fuel via reprocessing than they consume and can extract 100 times more of the energy in uranium compared to other reactors.[ai] This was a tantalising and almost infinite source of energy, one first proposed back in World War II by Manhattan Project scientists. Geologists discovered Japan's first uranium deposit at Ningyo-toge in a remote corner of Okayama Prefecture in November 1955, but the total indigenous supply was expected to be small, so government bureaucrats considered fast-breeder and reprocessing technology to be essential. While Japan took baby steps towards this goal, its more immediate intention was to import light water reactor (LWR) technology such as that already operational in the United States. Some, such as Japan's UN ambassador Renzō Sawada in January 1955, tried to steer Japan towards the broader United Nations and developing domestic abilities rather than importing mature technology from the United States specifically, but the business community favoured the latter approach.[aj]

The Hopkins event had been a promising success and so, despite continuing protests across the country, the US government went ahead with the proposed "Peaceful Uses of Atomic Energy" travelling exhibition, with the *Yomiuri* newspaper as a co-sponsor.[12] The exhibition followed two similar exhibitions in 1950 – the

---

12 The *Yomiuri* may have been the sole sponsor for the Tokyo exhibition, though various organisations did sponsor different locations. Different sources provide different statements.

America Fair near Osaka and the Japan Trade and Industry Fair in Kobe – both of which, in part, extolled the virtues of the atomic age. After weeks of daily promotion in local and national newspapers and TV stations, the exhibit opened on a beautiful sunny day at Tokyo's Hibiya Park on 1st November 1955, to enormous crowds.[13] To commemorate the occasion, US Ambassador to Japan John Allison read a message from President Eisenhower that "the exhibit stands as a symbol of our countries' mutual determination that the great power of the atom shall be dedicated to the arts of peace."[ak] Prime Minister Ichirō Hatoyama also conveyed a message via Chief Cabinet Secretary Ryutaro Nemoto, and the prime minister formally endorsed the event. "Praise the greatness of atomic science in the present century, and welcome the exhibit as an educational movement." As chairman of the JAEC, Shōriki spoke too. Wearing his trademark round-rimmed glasses, he declared the exhibit "an historic curtain-raiser on the atomic age in Japan."[al]

The park's great hall, the largest auditorium in Tokyo, was stuffed with information displays and animated scale models of nuclear power plants, reactors, an atomic train layout (drawing comparisons to Perry's model steam train), atomic structures, and radiation-related medical treatments and devices, all explained by local guides in strictly favourable terms. Pro-nuclear media and the various elements of the exhibit itself repeatedly highlighted that nuclear power promised a wonderous utopian future where domestic energy would be in uninterruptible, limitless supply. The message was loud and clear: Japan need no longer fret over its lack of natural energy resources. Nuclear power could provide cheap and reliable electricity, with few maintenance outages. At no time did anyone mention what would happen if something went wrong at a nuclear power station, nor that anything ever *could* go wrong.

Six weeks later, having played host to as many as three-quarters of a million people, the "overflowing" exhibition wound down in Tokyo as a complete success.[am] The combined efforts succeeded to some extent in separating the word "nuclear," referencing nuclear weapons, from "atomic," which was framed as being for the betterment of humanity and fundamentally peaceful in nature. Even some anti-nuclear weapons organisations proved to be in favour of peaceful atomic energy.

---

13 As many as 3,500 stories on the peaceful atom appeared in the media, almost ten per day. Source: *Japan Viewed from Interdisciplinary Perspectives: History and Prospects*, Yoneyuki Sugita (ed.), p. 270.

For the next few months, the exhibit travelled from the island of Kyushu in the far south to Hokkaido in the far north, stopping for weeks at a time in the cities of Fukuoka, Hiroshima, Kyoto, Nagoya, Osaka, Sapporo and Sendai. Two prominent *Yomiuri* rivals, the *Asahi* (Japan's largest newspaper) and *Chugoku*, contributed their own sponsorships and positive coverage. The Hiroshima exhibit opened on 27th May 1956 at the city's most iconic building, the Peace Memorial Museum, and was co-sponsored by numerous local private, public and political bodies, including the Hiroshima City Council, the Prefectural Government and Hiroshima University.[an] The choice of venue offended many of Hiroshima's citizens, prompting loud objections and public debate. Residents were particularly upset that Hiroshima museum artefacts were removed to make way for exhibition displays and that vast amounts of money were being spent on the exhibition and its promotion while the local government wouldn't provide funds for atom bomb victims. Still, even in Hiroshima, where one would expect righteous resistance, people overall seemed to accept and even embrace the idea of using atomic energy for peaceful and progressive endeavours, as the exhibition attracted over 100,000 visitors. They believed that perhaps Japan, of all nations, should be the one to pioneer this technology. Prominent local officials came out in force to praise the entire concept of atomic energy in Hiroshima. "We are entering a splendid era," said the leader of the Prefectural Chamber of Commerce in the *Chugoku* newspaper. "It is good that I achieved old age [to see it]. [This era] is full of wonder and [we are laying] the infrastructure to make it happen."[ao] Upon the exhibition's end, the city elected to continue showing atomic energy items at the museum because of the keen interest it attracted, though this move also drew criticism from some, who labelled it a museum of horrors.

In total, the "Peaceful Uses of Atomic Energy" exhibition drew over 2.5 million people nationwide during its run. The United States Information Service (USIS) reported that "the change in opinion on atomic energy from 1954 to 1955 was spectacular. Through an intensive USIS campaign, atom hysteria was almost eliminated and by the beginning of 1956, Japanese opinion was brought to popular acceptance of the peaceful uses of atomic energy.... Substantial progress has been made in improving Japanese opinion towards the US."[ap] Shōriki himself, though clearly biased and unafraid of hyperbole, later said that "the bloodshot eyes of the Japanese, sparkling with hatred of atomic energy, changed overnight to serene eyes adoring the goddess of peace!"[aq]

During all this time, Japan's industrial powerhouses formed five main groups for developing nuclear power. Each were former *zaibatsu* organisations that had survived the post-war efforts to weaken them and emerged almost as powerful as ever. They endured because the system that replaced the *zaibatsu* setup, while different in a few ways, was essentially the same in many respects, and the "banking system," as it came to be known, was (and still is) where Japan's largest banks replaced the old family leaderships. These five groups moved to strengthen existing cooperation with their American and British nuclear manufacturing counterparts. These groupings were Mitsubishi Atomic Power Industries and Westinghouse; Hitachi, Nissan and General Electric; Sumitomo Atomic Energy Industries and United Nuclear; Mitsui, Toshiba and General Electric; and Daiichi (this was the name of a company and had nothing to do with Fukushima) and Nuclear Power Group (UK).[14, ar]

The door opened for Japan to embrace nuclear power, and momentum drove it forward, but anti-nuclear unions and grass-roots protestors, fearful of the technology whose weaponised form many had witnessed first-hand, scrambled to slam it shut. Japan's oldest newspaper, the *Mainichi Shimbun*, denounced the campaign, writing: "First, baptism with radioactive rain, then a surge of shrewd commercialism in the guise of 'Atoms for Peace' from abroad."[as] Huge numbers of Japanese people reflected this sentiment across the country. Also, while some hibakushu saw promise in turning something bad into good, the feeling was not at all universal, with the lack of understanding about the effects of radiation and its conspicuous omission from any of the exhibitions being a principal concern. Nevertheless, the non-stop propaganda efforts had the desired effect: a 1959 US government report on USIS-Japan influence campaigns stated that the percentage of Japanese citizens who equated "atom" with "harmful" dropped from 70% in 1956 to 30% in 1958.[at] In effect, the Japanese psyche underwent an important shift: nuclear weapons were still to be opposed, but peaceful uses of atomic technology were henceforth often embraced. The country was set on the path towards nuclear energy, though for the time being that path did not lead to the United States.

---

14 First Atomic Power Industry Group (FAPIG-Daiichi). Merged with Kangyo in 1971 to form Daiichi Kangyo.

* * *

The Americans weren't the only ones courted by Shōriki and the *Yomiuri*; his newspaper also invited prominent British nuclear physicist Sir John Cockroft and engineer Sir Christopher Hinton to Japan, the latter of whom visited in May 1956. During his two-week stay, Hinton met the prime minister and Emperor Hirohito, among many others, then gave travelling lectures on Britain's atomic prowess, two of which were broadcast on Shōriki's *Nippon TV*. In these lectures, Hinton made a point of mentioning that, during their early efforts, British scientists had "very much less information about the chemistry and properties of uranium than we have now."[au] He argued that Japan would greatly benefit from British expertise employed in and gained from their Calder Hall power station, currently being built under Hinton's supervision. He also sang the praises of the various reactor types being developed in Britain, including the Calder Hall gas-cooled Magnox-type[15] ("heavy and cumbersome but safe, simple and almost conventional in design – the slow speed reciprocating engine of the reactor world"), the fast-breeder reactor ("highly rated, smaller and lighter but using materials and techniques which are as yet not fully established") and the sodium/graphite reactor ("capable of higher ratings than the gas-cooled systems [and] producing heat at a high enough temperature to use the steam pressures associated with the most up-to-date coal-fired stations").[av] Commenting on the 'light water reactors' adopted by the US, Hinton acknowledged their benefits but criticised their large size and the "serious corrosion difficulties for which we have no immediate solution." The Americans, in contrast, loudly proclaimed the Magnox to be a poor man's reactor that used graphite purely because Britain lacked the capabilities to produce the enriched uranium fuel required in more advanced light-water designs.

Before he left, Hinton proposed that a Japanese delegation visit the United Kingdom to view nuclear technology and plans for a possible Japanese reactor, only for the British government to delay the visit by several months because those plans were unfinished. Nevertheless, he was enthused by the overall response to his visit,

---

15 Magnox reactors are so named because of the "magnesium non-oxidising" alloy used to clad the natural uranium metal fuel elements.

and the Japanese began deliberations on a purchase of Britain's Magnox reactor by early July 1956.[aw] Britain's ambassador to Japan, Sir Esler Dening, was less excited, remarking that the Japanese would "try to derive the maximum advantage from our knowledge and experience and to give us as little as possible in return."[ax] This view was shared by much of the UK's nuclear industry after Japanese diplomats came and went from Britain without any firm agreements, but, given that it was a brand-new technology which few knew anything about, such Japanese caution was hardly a surprise.

The Japanese government, Shōriki and the nine utilities hoped to use American light water reactors from the outset because of Japan's reliance on the US military for national defence and the existing good relations with General Electric and Westinghouse, but negotiations stumbled. US export embargoes of its enrichment technologies and uncertainty surrounding the fate of spent uranium fuel (which could be used to fabricate plutonium for use in nuclear weapons) became the main sticking point in choosing British over American reactor technology. This roadblock, despite an existing US-Japan nuclear cooperation agreement, arose because of the American government's concerns about Japan and its private civilian companies gaining access to the enriched fuel. The know-how for enriching uranium was a highly classified secret at the time and the US planned to keep it that way. British Magnox reactors, by contrast, used natural, unenriched uranium that did not require the same sophisticated steps to use, though this point was somewhat academic, as the Magnox design had the ability to "produce sufficient plutonium for an estimated 30 [nuclear] weapons per year," according to a June 1965 US government report.[ay] Nevertheless, the ability to use natural uranium from anywhere in the world made it an attractive choice for the Japanese.

The realisation in June 1956 that Japan would not get the American technology it wanted, again according to CIA documents, "came as a considerable shock to Shōriki and prompted his overtures for the procurement of a reactor from the British."[16, az] By September, the US government deemed it "highly probable" that the Soviet Union would exploit the disagreements by encouraging the adoption of their own reactors in Japan, given the USSR's existing offers of assistance to Egypt, India, Indonesia, Iran and others. The US lifted

---

16 A team of scientists headed by Dr Marvin Fox of the Brookhaven National Laboratories in New York's Long Island visited Japan in June. Fox told Shōriki that the Americans were five years behind.

their restrictions later that same month and agreed on a contract to deliver 20 American reactors the following year, but their construction would not occur for some time.

ii. Nuclear Power Comes to Japan

"Atomic scientists, by a series of brilliant discoveries, have brought us to the threshold of a new age," remarked Queen Elizabeth II, glancing at her notes, as she opened the world's first industrial-scale nuclear power station on a cold and windy but bright October morning in 1956. Calder Hall held two (soon four) 180-megawatt reactors at the colossal Windscale nuclear site in the north-west corner of England, the pride of British engineering.[17] Nuclear power was no longer a fantasy, it was "real" and hailed across the world as a magical new technology that would usher in an age of almost limitless cheap and clean energy. Indeed, the British government expected to power half the country with nuclear plants within a decade. Attendant representatives of Japan and dozens of other nations came away suitably impressed; Britain was not only the global leader in the field but was five years ahead of the United States – a position it intended to exploit. "The British maintain that they will be producing electrical power at a feasible cost at Calder Hall," Shōriki said in a newspaper statement at the time. "American experts have at least six similar projects in the works but they could not tell me when they would be able to match Calder Hall."[ba]

Britain took a dim view of Japanese economic expansion after WWII, even using its influence with other Commonwealth countries such as Australia and New Zealand to resist US pressure for most-favoured-nation trading agreements.[bb] They did, however, find mutual interest in Japanese requests for British military, civilian and trainer aircraft from manufacturers Rolls Royce, Hawker and de Havilland, builders of the world's first commercial jet airliner: the Comet. Ironically, many potential sales did not come to fruition, in part because British companies feared damaging relations with their American customers by appropriating Japanese business.[bc] Military procurement also fell through because the US controlled Japan's military spending.

---

17 Unit 2 was complete at the time of this opening ceremony but did not achieve its first "criticality" until December 1956. Units 3 and 4 followed in January and December 1958, respectively.

Nuclear power was different. Not only did this marvellous new technology hold clear game-changing economic potential, but Britain had established itself as the world-leader and was keen to maintain this position. The Japanese government, meanwhile, came to believe non-American technology held a poignant political advantage; exploiting the US-UK nuclear rivalry came as a bonus. And a rivalry it was. As written in a priority letter from the American embassy in Tokyo dated 10th July 1956, "the Japanese have tended to view the American and British nuclear energy programs as essentially competitive. As a result, the most controversial issue today in the nuclear energy field is this choice between US and British power reactors."[bd] Although many were thus keen to see Britain's reactor technology in Japanese power stations, Japan's industrial leaders felt otherwise. "The electrical power industry has worked very closely with its counterparts from the US," the embassy letter continued, "and it is inclined, if at all possible, to look for assistance and finance from the US in the development of atomic power in contrast to Shōriki's current 'love affair' with the British power reactor."[be]

* * *

In July 1956, Egypt's coup-installed president, Gamal Nasser, ordered his troops to seize control of the Suez Canal from the Universal Maritime Suez Canal Company, which built the canal nearly a century earlier and now ran it independently of the host nation. Though the canal runs through Egypt, the company was owned by Britain and France, which viewed it both as a critical source of Middle Eastern oil and as the main route to Britain's Commonwealth empire.[18] In addition to nationalising the canal, Nasser closed the conduit to Israeli shipping and blockaded the Straits of Tiran – Israel's only outlet to the Red Sea and Indian Ocean.[19] Enraged, Israel invaded eastern Egypt in a dramatic multi-pronged land and aerial attack on 29th October, defeating Egypt's military but not before it blocked the canal.[20] Britain and France,

---

18 Legally the canal belonged to Egypt, but it had been granted 99 years' concession and was not due to revert to Egyptian government control until November 1968.

19 Nasser's stated reason for nationalising the Suez Canal was to help fund the Aswan Dam across the River Nile. The Soviet Union offered an unconditional loan for the project.

20 Israel's navy helped fight off an Egyptian counterattack on the Israeli city of Haifa, but was not really involved in the initial invasion, while Israeli P-51 Mustang fighter aircraft went so far as to use their

under the prearranged false pretence of breaking up the fight, then dropped paratroopers along the canal to reinforce Israel on 5th November.[21] The invaders were forced to withdraw under humiliating political pressure, while the Egyptians hailed Nasser a hero. The canal itself remained closed for six months, causing a global increase in the cost of oil. Japan was not directly affected, as it was on the right (i.e. southern) end of the canal to receive oil supplies. Yet, faced once again with an energy resource problem, national discourse returned to the topic of bolstering Japanese self-sufficiency and limiting dependence on a resource it didn't control. Nuclear power was the obvious solution.

The International Atomic Energy Agency, the world's central intergovernmental forum for scientific and technical co-operation in the nuclear field, accepted its first 56 members on 29th July 1957, including Japan, the United States, the Soviet Union, Germany, Afghanistan, India, South Korea, and the United Kingdom. Weeks later, on 27th August, the Japan Atomic Energy Research Institute (JAERI) commenced operation of Japan's first nuclear reactor: the experimental Japan Research Reactor 1 (JRR-1). It was a small, liquid-fuel, 50-kW Boiling Water Reactor (BWR) design imported from the United States, housed in a nondescript single-storey building with a windowless two-storey cuboid lodged in the middle. Today the reactor is a museum piece. Back then, JRR-1's main uses were the study of reactor physics and the training of 680 operators, engineers and researchers.[bf] These people would go on to play important roles in the development of Japan's nuclear industry, which would mimic the existing power industry structure of nine private companies with regional monopolies – in stark contrast to Shōriki's desire to have nuclear technology fall under government control.

All parties compromised on 1st November 1957 with the formation of the Japan Atomic Power Company (JAPC; Nihon Genshiryoku Hatsuden in Japanese) to kick-start the domestic nuclear industry and run its first power stations, in a manner reminiscent of the Electric Power Development Co's creation five years earlier.[22]

---

wings and propellers to sever Egyptian telephone communication lines hanging from pylons: brave, stupid and impressive.

21 The United States even sided with the Soviet Union against Britain and France in a UN vote on the matter – an unusual occurrence.

22 The company's name is sometimes abbreviated to JAPCO but appears as JAPC on their website. Why "Company" in this case is shortened simply to C and not CO, as with other Japanese utilities, I do not know.

From left to right:
JRR-1, JRR-2, JRR-3

Also similar to EPDC was the 20% public / 80% private split on the initial investment, with half the private side covered by the utilities and the rest by various heavy electrical machinery manufacturers.[bg] This way, no single utility shouldered the risky expense of building a nuclear plant. Future government funding would concentrate on research and development for more advanced nuclear technologies, with businesses financing commercial applications. Today, seven utilities own over 86% of JAPC, with TEPCO holding by far the largest share. At this time in Japan (around 1957), on the eve of the nuclear age, the balance of energy generation from the three main sources was coal 39%, hydro 37% and oil/gas 19%, with the final 5% coming from other sources.[bh]

On 7th October 1957, Britain suffered the world's first major nuclear accident, at Windscale (the site where Calder Hall and various other facilities were located), when uranium fuel in one of its two plutonium-production reactors caught fire and burned for several days. (These were separate from the Calder Hall reactors, which were more advanced.) After multiple failed attempts to extinguish the blaze by various means, Deputy Manager Tom Tuohy shut off the flow of CO_2 reactor coolant, thus starving the fire of oxygen and saving the facility. Despite some misgivings about the event and the new Magnox reactor design's apparent teething problems and potential safety issues (it did not feature the secondary containment vessel that is standard on most designs, for instance), the new reactor was the only proven option available, so a delegation of Japanese government representatives met with their British counterparts eight months later in London on 16th June 1958 to sign the Agreement for Co-operation in the Peaceful Uses of Atomic Energy. The following April, it was agreed by tender that Japan's first commercial nuclear power plant would be designed by General Electric Company (GEC, a different company from Edison's General Electric) in collaboration with Simon Carves Ltd, two 19th-century British engineering firms.[23] Together, the two companies partnered with 14 Japanese businesses in a consortium known as the First Atomic Power Industrial Group to build the facility under the supervision of their British counterparts.

As part of the early preparations for a national nuclear industry,

---

23 GEC was one of three UK companies to tender for the job. The two others were Associated Electrical Industries (AEI, acquired by GEC in 1967) and English Electric. Unlike GEC, Simon Carves still exist today and is now a Mitsui subsidiary.

the Science and Technology Agency, now headed by Yasuhiro Nakasone, in 1960 tasked the Japan Atomic Industrial Forum (a collective of business and lobbyist groups who speak for the atomic energy industry) with preparing a report on probable liability costs in the event of a nuclear accident.[24] Their worst-case scenario calculation was more than double Japan's entire national budget at the time and would likely have forced the permanent evacuation of the surrounding area. Though the report was quickly buried without being made public (it resurfaced in 1999), it influenced the decision to site nuclear power stations in rural areas, away from urban population centres.[bi] Japan lacked accessible inland freshwater lakes and rivers in depopulated areas and of sufficient properties to support the massive volume required to cool a nuclear reactor, making remote coastal regions the only option.

From 1960 onwards, the Ministry of International Trade and Industry (MITI) conducted surveys along the coastline using location standards borrowed from the US, and compared various potential sites against a range of factors.[bj] These included topography, characteristics of the local geology and any historic earthquake activity, available space for existing road and shipping access, reasonable proximity to major electricity demand, the insurability of a given location, and the possible environmental impact from discharged cooling water, which would be above the local ambient temperatures. The extent and adequacy of any geological surveys were limited by the available technology, and the standard of fault tests was questioned as early as the 1980s. A nuclear site topographer, speaking to UPI Asia on condition of anonymity, claimed a site he had examined in 1988 was not analysed using industry-standard practices and "they deliberately neglected active faults."[25, bk]

In the few areas that met these criteria, the deciding factor tended to be local population density and the attitudes and malleability of local residents and authorities. Or, as professor of political science and director of the Security and Resilience Studies Program at Northeastern University, Daniel P. Aldrich, puts it, the "weakness of local civil society."[bl] Mountains cover 73% of Japan, but the govern-

---

24 Nakasone was also head of the Japan Atomic Energy Commission during this time, from June 1959 to July 1960.

25 Anonymous comments should always be taken with a grain of salt, but this person's criticisms are not isolated. They are the only comments from someone who claimed to have actually carried out these inspections rather than being from a university geologist or something similar; hence the comment is worth mentioning.

ment still sought out areas with fewer than one-third of the national average population density to build power stations. Over the last 50 years, scholars have determined that small, rural Japanese communities lack the power to resist massive undesirable construction projects for a number of reasons, including peer pressure to abide by authority figures, a belief that their opinion doesn't matter or cannot influence an expected outcome, and hurdles preventing the formal recognition of non-profit organisations that might otherwise stand in the way of such projects. Japan's law of eminent domain (property acquisition by government) also does not require market-value compensation for the land, which means communities are pressured to sell at the initial higher offer price.[bm] This forced sale has been used multiple times, for example with Narita International Airport, dams and thermal power plants, but never with a nuclear plant. Instead, the government sought "weak" towns and villages which lacked a strong community spirit, then convinced them to accept a nuclear plant. Communities around the small coastal village of Tōkai-mura (meaning East Sea Village) in Ibaraki Prefecture accepted nuclear technology in the beginning, but utilities and the national government would encounter problems down the line at other locations. Regardless, of the locations deemed suitable, Tōkai-mura was the obvious preference because of the existing JAERI campus nearby. Thus, Japan's first nuclear power station, run by JAPC, would be built at and named Tōkai.

* * *

Nuclear was not the only fuel type on the rise. The demand for cheap foreign crude oil for Japanese power stations rose at a rate of over 25% per year from almost nothing during the 1950s. From 1953 to 1955, MITI attempted (unsuccessfully, by and large) to limit adoption of heavy oil to save the domestic coal industry and prevent coal prices from inflating. Collieries across the nation struggled to remain competitive, with some going so far as selling raw and "low-calorie" inferior coal, which had previously been discarded, to utilities for burning in coal-fired power stations.[bn] The iron, steel and electric power industries, in particular, had found various benefits in switching from coal to oil, primarily for increased efficiency through precise temperature control, and each thus resisted the move. Coal power held over 90% of the thermal power market, with oil in single digits as late as 1955, but a decade later coal was down to 44.9% with oil rocketing ahead to 52.1%.[bo] Government economic analyses shifted by 1956 and '57, with the acknowledgement that Japan would struggle to match the exponentially rising energy demand using coal alone, that electric utilities' use of oil power stations would make electricity cheaper, and hence that oil was vital for sustaining the country's stable economic growth. This plan made sense, but the price and availability of oil – unlike domestic coal – soon proved to be less than stable.

Kyoto University became the first Japanese educational institution to establish a nuclear engineering department in 1958, with the University of Tokyo following suit in 1960.[bp] A second 10-MW reactor named JRR-2 commenced operation in October of that same year, this time a Heavy Water Reactor (HWR) based on the CP-5 design from the US Argonne National Laboratory in Illinois. JRR-2's main use was for neutron beam and medical irradiation experiments. By September 1962, JRR-3 – another 10-MW design – reached criticality (a self-sustaining fission chain reaction) and was proudly announced as the first nuclear reactor built exclusively using Japanese scientists and engineers with domestic technology by businesses such as Hitachi, Toshiba, Mitsubishi Heavy Industries and Fuji. The reactor was powered by Canadian uranium supplied by the IAEA at a price of $35.50 per kilogram (nearly $300 in 2020) – the first time the newly founded agency sold nuclear fuel to a member state – and was later upgraded to JRR-3M for an additional 10 megawatts in 1990.[bq] JRR-3 received the American Nuclear Society's Historic Landmark Award in 2007 for its outstanding contribution to nuclear technology development and neutron science. All three reactors were built next to each other in a small pine forest on the same road overlooking the ocean at JAERI's campus near the coastal village of Tōkai-mura in Ibaraki Prefecture, 100 kilometres north-east of Tokyo and about 110 kilometres south of where Fukushima Daiichi is now. The JAERI campus was a popular destination for schools and tourists by the early 1960s, offering guided tours to thousands of visitors.[br]

The United States launched the world's first nuclear-powered surface ship in 1959, and people in Japan hoped to develop their own. The JAEC, the Ministry of Transport and the shipbuilding industry argued that the domestic technology did not exist, so a team of government and maritime industry experts flew to the US, Britain, France, Norway and the IAEA in September and October 1959 to consult and gather resources for an ambitious nuclear shipping project. The JAEC greenlit a fourth research reactor at Tōkai-mura upon their return: a 2.5-MW "swimming pool-type," cunningly named JRR-4, for research into a future nuclear ship.[26] JRR-4 began operation on 28th January 1965 but, unlike its larger brethren, it was designed to be started up, operated at various power levels and shut down on a regular daily basis.[bs] Its power output was upgraded in 1976 after the nuclear ship launched, taking it to 3.5 MW and a new life in medical research.

* * *

As the nuclear industry took its first significant baby steps, opposition sentiment evolved into a more organised national movement, propelled by widespread outrage over several unrelated health crises caused by industrial pollution, namely "Minamata disease" and "Yokkaichi asthma." The former first appeared in April 1956 at the southern city of Minamata, a 70-kilometre stone's throw from the former shōgunate foreign-trade city of Nagasaki. Its symptoms included numbness of the hands and feet, impaired hearing, sight and speech, insanity, paralysis, coma and even death. A research team from Kumamoto University concluded in November of that year that ingestion of fish contaminated with heavy metals from industrial waste dumped into Minamata Bay was the cause. Another team from the Ministry of Health and Welfare confirmed mercury as the source three years later, but the finding was ignored. The nearby Chisso Corporation had been dumping industrial waste into the bay since before World War I, and methylmercury specifically in concentrations viable for mining since as early as 1932. The notorious practice had already resulted in two separate compensation agreements over toxic fish in 1926 and 1943, but as one of the country's largest and most advanced chemical factories, it had enormous clout with the government.[bt] Knowing

---

26 The "swimming pool-type" means that the reactor's core is submerged in an open pool of water and can usually be looked into from above. This is safe to do because the volume of water acts as a neutron shield as well as a moderator, but these designs are always used for research or training, not for commercial electricity.

all this, government advisors found scientists who attested to other causes, such as rotten fish, thus providing them with an excuse to side with Chisso. Nothing changed until the company altered its production method in 1968, after which the mercury dumping stopped.[27] Well over 2,000 people – not to mention countless fish, birds and household pets – died after ingesting fish poisoned by the mercury. Just as Chisso and others stopped dumping mercury, the government suddenly admitted in September 1968 that the metal had caused the disease all along.

The other health crisis, Yokkaichi asthma, came as a direct result of Japan's shift from coal-fired to oil power stations. By 1959, the city of Yokkaichi held one of Japan's largest petrochemical complexes, occupied by dozens of companies and built on the site of a former Imperial Navy fuel depot bordering residential districts.[bu] Oil and natural gas refineries still abut houses there to this day. The waters off Yokkaichi were popular fishing grounds at the time, but fishermen found that the fish they caught stank of chemicals and, by the early 1960s, doctors noticed a sharp local increase in the number of severe respiratory afflictions. An investigation found that Middle East oil contained over 3% sulphur which, when burned in unprecedented quantities, had formed a poisonous cloud over Yokkaichi far beyond the legal limit for such pollutants.[bv] Domestic coal also had a high sulphur content (albeit far less so than oil, at around 1%), which further hastened its decline and replacement. Much worse was the aforementioned inferior quality raw coal, which held a sulphur content of 4.78%, but its relative scarcity in comparison to oil at the time meant this fact was largely overlooked.[bw] Residents petitioned the local city and prefecture governments, which pioneered aggressive anti-pollution measures that were eventually adopted by the wider Japanese government in the late sixties and seventies, but these national regulations were undermined for years due to a focus on economic expansion.[28, bx] Those who stood against the industries that supported this expansion were often seen as bothersome communists interfering with Japan's remarkable growth.[by] A general Japanese cultural unwillingness to *really* stand up in outrage has also been proposed as a reason for this movement failing to gain serious traction.[bz]

---

27 A second outbreak occurred at Niigata in 1965 from dumping by a second chemicals company and was named Niigata Minimata disease.

28 The 1967 Basic Law for Environmental Pollution, for example, was a symbolic solution that did little to solve the actual problems by being ambiguous and unenforced.

These and similar experiences during Japan's "high-growth era" from the late fifties to the seventies – with particularly high growth from 1960 to 1973, when the economy grew by 10% each year – created a heightened sense of opposition to, and fear of, all possible sources of industrial environmental health crises. Nuclear power had not contributed to any of these problems and was less of a focus than chemical pollution at the time, but it nevertheless proved to be no exception. This was a somewhat contradictory position to take since nuclear power stations produce no pollution, but it was bundled in as a general fear of radiation. The government – the Ministry of International Trade and Industry (MITI) specifically – and electric utilities saw opposition to Japan's vital power infrastructure as a serious issue and took action to improve public relations.

* * *

Oil prices declined by the early 1960s as memories of the Suez Canal incident faded, making the enormous investment needed to build a commercial nuclear plant less appealing to utilities. At the same time, domestic coal production reached its post-war peak and declined from this point. Even so, it was clear that the demand for electricity would only increase, as Tokyo had just become the world's first city to exceed 10 million residents and nuclear power was a safe bet for the future. Its running costs would not fluctuate in the way the price of oil did, nor did nuclear power pose the sorts of geological, environmental and human geographic challenges (such as the necessity for flooding vast areas) that hydro power did. The country also neared the limit of suitable dam sites, and the price of coal was steadily increasing.

Construction of JAPC's Tōkai Nuclear Power Plant began on 16th January 1960, a short distance from the JAERI campus.[ca] The Magnox reactors had the unusual characteristic of being cooled by carbon dioxide (CO_2) gas propelled by powerful fans rather than pumped water, as in almost all other reactor types. Britain's engineers reasoned that CO_2 allowed for higher reactor temperatures than water cooling, making the steam-driven turbines more efficient. In practice, however, the fuel cladding's magnesium alloy could not operate above 650°C (1,200°F) and was initially limited to an average 414°C to retain strength and longevity. Subsequent operation revealed high oxidation rates of some steel components (when electrons move from metal to oxygen molecules in the coolant, causing metal corrosion), and the reactors were then further limited to around 360°C.[cb]

Tōkai Unit 1 was a 157-megawatt design based on GEC's Hunterston "A" Power Station under construction on Scotland's west coast, but – as was the case with each Magnox site, because no single construction consortium had the resources to design and build two plants at once – its GEC-Simon Carves design was somewhat unique and customised to suit the needs of the customer. In this case, the designers made extensive modifications to provide extra protection against earthquakes (sometimes late in the process, after Japanese academics and the public expressed concern about building a nuclear plant in such a quake-prone part of the world).[cc] These protections included stacking the graphite moderator blocks in a hexagonal lattice instead of square, providing heavy reinforcement of the concrete foundations and superstructure, and implementing special supports and "earthquake restraints" for the steam generators and reactor that fit "between the biological shield and a cylindrical lip welded to the top of the pressure vessel" to prevent the core and its 870 tons of graphite and 187 tons of fuel from shaking apart.[cd, ce, cf] They also lowered the entire assembly's centre of gravity to reduce the effect of horizontal acceleration in an earthquake by flipping aspects of the base Hunterston design upside-down to allow fuel insertion from above the core instead of below.[29, cg] A network of accelerometers sat around the building to record any ground tremors and provide as much warning as possible.[ch] If an earthquake of sufficient force tripped these sensors, hoppers above the core would drop thousands of boron-steel balls down chutes into dedicated channels in the reactor, independent of those for ordinary control rods. This action could terminate the fission reaction even if the earth beneath the plant opened up and the huge 18.4 metre-wide (60-foot) reactor somehow came to be tilted at a 45° angle.[ci] Moreover, the reactor building's concrete central structure ensured that, if indeed there was movement, the steam boilers, gas-cooling circulators and other components vital for safety would move together as one, even if an earthquake wrenched the turbine hall away.[cj]

---

29 Note, Hunterston itself was a departure from the conventional Magnox design, which allowed fueling and defueling from above. Hunterston's core was higher up than the turbines but the Tōkai core sat at the same low level as its turbines.

Actually building the ambitious facility proved more challenging than anticipated (doesn't it always?) and delay after delay dogged Tōkai, partly because of the untried earthquake modifications. British steel plates 80 and 94 millimetres thick for the reactor's spherical pressure vessel – the central container for the fuel and coolant, which must be exceptionally strong – were found to be defective on arrival and had to be replaced with Japanese steel.[ck] The innovative central fuel channels, hollowed out to facilitate cooling, did not function as intended, to the frustration of JAPC, which complained that Tōkai was being used as a testbed for British research. Failures continued in other areas when some construction subcontractors struggled to attain the ultra-high standards required at a nuclear facility.[cl] Things didn't fare much better after completion, as the fast pace of developing cutting-edge nuclear technologies, the dilution of personnel with key skills through the use of multiple companies, and those companies creating non-interchangeable designs, all threw up unforeseen problems. Defective components had to be replaced after failing on start-up, problems occurred with the cooling pools for spent fuel cartridges, and the innovative on-line refuelling system (to allow the exchange of individual fuel cartridges while the reactor was at power) proved to be unreliable.[cm, cn]

The British embassy had to step in as mediator when JAPC asked GEC to cover the cost of all additional work as the project budget skyrocketed. The embassy found GEC to be at fault and, mindful of Britain's reputation in the nuclear engineering field, told JAPC they would not have to pay. The plant's new operators, though excellent and experienced scientists and engineers, also had to contend with many new concepts. "Since there was no precedent in the management of radiation control in Japan from the standpoint of [a] commercial power station," wrote a group of JAPC employees in 1969, "we were obliged to grope in the dark in establishing a [radiation control] policy at our station at first. Not being expert by nature it was necessary for us to deal with problems open-heartedly as unique matters which would crop up one after another in the course of practical business at [the] site."[co]

In 1961, as the new Tōkai power station took shape, members of Parliament passed the Act on Compensation for Nuclear Damage, prompted by the Windscale fire. It accepted that "where nuclear damage is caused as a result of reactor operation … the nuclear operator [i.e. the utility] who is engaged in the reactor operation

… shall be liable for the damage," and "no person other than the nuclear operator … shall be [held] liable."[cp] In effect, this meant the likes of General Electric, Westinghouse, and (later) Toshiba, Hitachi and Mitsubishi would be off the hook for damages arising during a plant's life. With prescient foresight, however, the act included the standard insurance get-out-of-jail-free line: "except in the case where the damage is caused by a grave natural disaster of an exceptional character, or by an insurrection."[cq] In other words, no pay-out would be forthcoming if an earthquake or tsunami ever destroyed a plant. The act reflected the public/private-ownership nature of the only nuclear utility at the time – the Japan Atomic Power Company (JAPC) – and conceded that the government "may" contribute to any insurance pay-outs but that they were not legally required to. This approach was good in the sense that utilities should pay for their own mistakes and oversights but is and was clearly unworkable in reality, given the potential damages involved in a major accident, which is one reason critical analysis of potential disasters was generally frowned upon.[cr] As such, the act limited utility liability to $14 million (over $120 million in 2020), and any sum beyond would require a Diet debate.

iii. Turnkey

The issue of the country's two halves – the 50-Hertz and 60-Hertz divide discussed earlier – came up again and again from the 1920s through the early 1960s. All agreed it should be resolved, with utilities in one half or the other replacing all their equipment, or everyone ditching everything and starting fresh with a new, forward-looking standard, but stubborn pride and the astronomical cost prevented any capitulation from either side. The compromise, decided in February 1962, was to construct a 300-MW converter station to bridge the gap, located 65 kilometres from the Fuji River frequency boundary, beside EPDC's hydroelectric Sakuma Dam power station. It was the world's first high-voltage DC converter for synchronising separate AC power grids and commenced operation three and a half years later, enabling true national energy sharing for the first time.

Three more converter stations followed, in line with Japan's ever-expanding electricity needs. First, in 1977, was the Shin-Shinano

Tōkai Nuclear Power Plant, Unit 1

Substation (built by Hitachi and Toshiba but owned by TEPCO) in the landlocked Nagano Prefecture. Another converter was added to the site in 1992, doubling its capacity from 300 to 600 megawatts. Second was the 300-MW Minami-Fukumitsu converter in Toyama Prefecture, which came online in March 1999 and connects the networks operated by Chūbu Electric Power Co and Hokuriku Electric. The country's fourth and final converter is the Higashi-Shimizu Frequency Converter Station, also operated by Chūbu Electric and built just 14 kilometres from the Fuji River, which finally commenced temporary operation in 2006 after more than two decades of troubled construction. It did not begin its full-capacity operation of 300 MW until 2013.

JAERI authorised construction of Asia's very first electricity-producing nuclear reactor, the experimental Japan Power Demonstration Reactor (JPDR), after the institute's early successes with other research reactors. Built by General Electric and Hitachi, and designed by GE subsidiary Ebasco (originally the Electric Bond and Share Company) with early earthquake protection measures, the JPDR was a critical research platform for a full nuclear power station and Japan's first proper light water reactor with early earthquake defences: a modest 12.5-megawatt BWR that went critical for the first time on 22nd August 1963 after almost three years of construction.[cs] With these facilities, each built at one centralised location, JAERI drove "the development of fundamental nuclear technology and provided Japanese companies opportunities to accumulate nuclear technologies."[ct] Japan possesses many other small research reactors today, but this and JRR-1, 2, 3 and 4 were the most historically significant.

The US government conceded to Shōriki's demands for American light water reactors several months after Japan signed its cooperation agreement with Britain back in the 1950s. The Japanese had displayed a willingness to embrace America's technological competitors with the Magnox deal, and encroachment by the Soviet Union would be a complete disaster for Western interests. Indeed, the USSR subsequently impressed all attendees of the April 1958 Japan International Trade Fair in Osaka with a range of scientific displays, including their reactor technology and a model of Sputnik 1 – the world's first satellite, which had launched six months earlier. The United States agreed to supply the requisite enriched uranium fuel, removing the original roadblock. Once the fuel was used up, or "spent," Japan would ship it back to America for reprocessing. For the island nation, depending on American goodwill for fuel was undesirable in the long term but acceptable until domestic technology progressed to the point of self-sufficiency.

Then the market changed. As the sales of household electrical appliances like televisions, washing machines, vacuum cleaners and refrigerators took off, and the typical peak-and-trough of winter-to-summer energy use flattened with the invention of air-conditioning, engineering advances meant coal-powered thermal plants became more efficient. ^{cu} This development made the bloated and ever-increasing construction costs of nuclear plants, already far greater than their operating costs (the opposite of coal), a core issue that threatened to derail the technology before it took off. That all changed in December 1963 with the US introduction of the Oyster Creek "turnkey" plant in New Jersey: a radical payment concept introduced by General Electric (GE) whereby utilities paid a fixed fee, decided through bidding, that was competitive with fossil fuels. The vendor company would then take on the burden of all materials procurement, construction, commissioning and staff training. With everything complete and operational, they would simply "turn the key" over to the utilities, which gained a working power plant.[30] Oyster Creek was a $600 million, GE-designed 650-megawatt Boiling Water Reactor (BWR): purportedly far cheaper and more powerful than the reactor still under laboured construction at Tōkai. Westinghouse quickly adopted the turnkey concept, and together the two companies altered the market almost overnight by building plants at a loss to stimulate the nuclear market.[31]

Meanwhile, the Tōkai nuclear power plant (NPP), according to Sir Christopher Hinton, "was disastrously managed and … the job did untold harm to the prestige of British heavy engineering in the Far East."^{cv} In fact, the plant did untold harm to British nuclear engineering prestige across the globe and contributed, along with far fewer than expected foreign orders for Magnox plants, towards dashing British hopes of retaining nuclear power engineering dominance. The Tōkai NPP eventually commenced commercial operation two years late, on 25th July 1966, just a few weeks after TEPCO had

---

30 As in the key to the door, although I've also seen this described as turning the key to start producing power.

31 General Electric and Westinghouse actually lost a lot of money through the turnkey system, but they gained in establishing their reactor designs as the de facto choices around the globe.

settled on a US-designed turnkey BWR for the planned Fukushima Daiichi power station.[32] Though Tōkai Unit 1 contributed over 28 terawatt-hours to Japan's electricity supply over the course of its life, it was the second and last Magnox reactor built beyond Britain's shores. (The first was Italy's Latina Nuclear Power Station, located 140 kilometres north-west of Naples, which operated from 1963 to 1987: one of many plants decommissioned early in the wake of the Chernobyl disaster.) The world's last Magnox reactor shut down in 2015.[33] British plans to sell the Magnox successor, the Advanced Gas-cooled Reactor (AGR), to the Japanese or any other foreign government failed to materialise because of US dominance, confining Britain's future role in Asia to the provision and reprocessing of Tōkai's new and spent fuel.

This all mattered little to the Japanese. Gone were utilities' fears over construction costs at nuclear sites, where the only certainty was uncertainty, and within a decade US utilities alone ordered well over 200 reactors.[cw] By the end of 1966, the Diet and the electric utilities agreed on an output target of 30–40,000 megawatts of light water (i.e. not gas-cooled) nuclear power by 1985. They didn't quite manage this goal, only reaching 28,000 MW by mid-1987, but nuclear power *did* succeed in surpassing oil for the first time in 1985.[cx] Within months of Tōkai Unit 1's commissioning, GE and Westinghouse each began work on two separate nuclear power plants close to the remote coastal city of Tsuruga, on either side of a picturesque mountainous peninsula. For the foreseeable future, GE would supply the designs for all of Japan's BWRs, while Westinghouse would supply all the Pressurised Water Reactor (PWR) designs, based on the US Navy's submarine reactor from the USS *Nautilus*. Each built their respective plants in cooperation with companies such as Hitachi and Toshiba, which worked towards the goal of designing and building their own plants.

On 24th November 1966, at the end of an idyllic bay on the peninsula's north-eastern tip, workers began construction on a 357-megawatt BWR at Japan Atomic Power Co's (JAPC's) Tsuruga Nuclear Power Plant, embedded in the side of a hill. Sixty-nine days later and seven kilometres away, on the western side, a fleet

---

32 Generation for Fukushima Daiichi was planned to begin in 1963–64 but would not happen until July 1966.

33 Magnox reactors were never intended to last very long, and the replacement Advanced Gas-cooled Reactor (AGR) design was being planned as early as 1964.

of orange construction vehicles rolled onto a forested cape and began to dig the foundations of Kansai Electric Power Company's (KEPCO's) 320-megawatt PWR at the new Mihama Nuclear Power Plant, named after a nearby fishing village. Japan's gross national product ranked second in the world by year's end, trailing only that of the United States.

One analyst observed that this was "a time of enthusiasm.... The government and industry were making strides, local communities were benefiting from infrastructure development, and little was known of the difficulties that lay ahead."[cy] JAPC chairman Dr Tamaki Ipponmatsu, the much-respected person most responsible for the contracting, construction and operation of the Tōkai NPP, some years later made a more nuanced observation: "The public wanted to regard nuclear power as the herald of a brave new age of modern science and technology; yet it was afraid that this radioactive mammoth might reproduce Hiroshima and Nagasaki in some unpredictable way.... The technology of nuclear power generation throughout the world was very young, and not many people could give definitive explanations to answer many of the problems raised."[cz]

The JAEC released a new long-term plan in 1967 in which the commission reaffirmed its ultimate goal of creating a closed-loop, self-sufficient nuclear energy cycle. The first step, after years of building conceptual mock-ups, was to build an experimental fast breeder reactor in the 1970s named Jōyō ("eternal sunshine"), then use Jōyō to develop a full-scale working prototype named Monju during the 1980s. A new type of reactor called an Advanced Thermal Reactor (ATR) would be developed in the interim to gradually replace the light water reactors (like the BWR and PWR), with the first demonstration prototype to be named Fugen. Eventually, once fast breeder technology reached commercial application, FBRs would take over as the standard reactor type as older sites were decommissioned. Jōyō was the old name for the area around its planned construction site in Ibaraki Prefecture. For the two prototype reactors, Japan chose the names of noted enlightened figures in Buddhism. According to that faith, the Bodhisattva Fugen (representing "practice") is closely associated with Monju, who represents "wisdom and enlightenment."

Both pages:
Mihama Nuclear Power Plant

Tsuruga Nuclear Power Plant,
Unit 1

Tsuruga Nuclear Power Plant, Unit 1 (Mark I BWR)

CHAPTER 3
Growth: Fukushima and the 1970s

i. Number One

On 1st April 1967, after nine months of land reclamation and a modest ground-breaking ceremony, the years-long work of erecting Japan's fourth and TEPCO's first nuclear power station commenced in the eastern prefecture of Fukushima. The total cost would reach ¥14.8 billion, equivalent to $561 million in 2020, 32% of which was foreign capital.[a] For a few years, the plant was known just as Fukushima, but it became Fukushima Daiichi or "Fukushima Number One" (sometimes abbreviated to 1F) after planning work began on Fukushima Daini (Number Two / 2F). Daiichi Unit 1 became the first of six reactors at the site, with a thermal output of 1,220 megawatts, translating to an electrical output of 400 MW. TEPCO continued their long-standing relationship with General Electric, opting for a second-generation[1] BWR Series 3 reactor in a turnkey plant designed by GE subsidiary Ebasco.[2] Unit 1 was a copy of an existing Ebasco/GE design for the proposed Santa María de Garoña power plant in Spain on the belief that any problems would be resolved by the time work began in Japan. Unfortunately for TEPCO, lengthy delays meant

---

1   Second-generation reactors essentially encompass all reactors built after the early prototype reactors (of which the Magnox design was one) until the year 2000. These include PWR, CANDU (Canada's own reactor design), BWR, AGR, RBMK (the Chernobyl reactor), and VVER (the Russian design now used in place of RBMKs). Of the BWRs, Dresden Nuclear Power Plant's first unit was a BWR-1, which began operation in 1960 in Illinois and was the first privately financed nuclear power reactor to be built in the United States. Oyster Creek Nuclear Power Station, completed in 1969, was a second iteration BWR-2. They're now up to BWR-9.

2   Ebasco was created by GE in 1905 and was responsible for many nuclear power stations but was sold to Foster Wheeler in 1993, so the Ebasco name no longer exists. The company also designed Units 2 and 6 at Fukushima Daiichi. Toshiba designed Units 3 and 5. Hitachi designed Unit 4. Though these reactors had "architects," they are all based on GE's "Mark I" BWR, apart from Unit 6, which is a "Mark II." These systems are discussed below.

Fukushima became version 1.0, suffering all the setbacks one would expect as a result.[3]

A proposal to build a nuclear plant in Fukushima existed almost from the beginning. The prefecture was proud home to the largest coal mine on Japan's main island of Honshū: the Jōban Coalfield spanned from Tomioka in Fukushima, down to Hitachi in Ibaraki Prefecture. With the worldwide shift towards oil power, however, production at the mine – Fukushima's sole large-scale industrial employer – had been in decline since the late 1950s. Back in 1958, knowing of TEPCO's desire to build a plant in rural areas, Fukushima governor Zenichirō Sato became enticed by the economic prospects of hosting a nuclear power station and ordered a secret feasibility investigation, with the utility in mind.[b] In October 1960, a private landowner heard of the proposal and reserved an area of disused land near the small and sparsely populated coastal towns of Ōkuma and Futaba in Futaba County, 110 kilometres north of Tōkai.[4] The Imperial Japanese Army Air Service had used the space as a training airfield from 1938 until the United States bombed it in 1945, after which it found use as a salt farm from 1950 until its closure five years later. By-products from the salt production process rendered the earth unsuitable for farming and therefore almost worthless.[c]

The two towns ranked among the poorest areas in one of the poorest prefectures in Japan, so courting an enormous construction project of national importance was an easy decision. After settling on the land, the Ōkuma and Futaba town councils submitted a secret joint proposal to TEPCO and the national government in 1961, specifically hiding it from the public.[5] "The prefectural office and TEPCO were equally complicit in concealing the plan," claims historian Hiroshi Onitsuka, "even going to the extent of having young female TEPCO workers accompany engineers on inspections

---

3   Despite the delays, this Spanish plant did begin construction a few months before Fukushima but was commissioned after.

4   Exactly who this landowner was and how much he or she owned appears to be unknown to an extent. One source claims that former chairman of the House of Representatives Yasujirō Tsutsumi owned 90% of the land. Another source says that real estate mogul Kokudo Keikaku owned 30%. In still others, the majority owner is not known (odd, if the owner was a former high-ranking politician or large company president) and the percentages differ again. Without being able to confirm any of this, I have left the specifics out.

5   Whether this proposal was all top secret is also unclear. Some scholars, such as Daniel Aldrich, claim that an extremely rare local public vote was held in 1961 after the details were ironed out. Others, such as Hiroshi Onitsuka, claim the entire thing was kept on a strictly need-to-know basis until the entire deal was agreed. Given that Onitsuka provides a highly detailed account of how matters progressed, along with actual quotes from people involved in the secret deal, I am inclined to believe his version of events.

of potential locations so as to give the impression of being simply vacationers on a hiking trip."[d] The only other locals aware of the plan were representatives of the nearby fishing organisations, who agreed to relinquish a stretch of coastline 3.5 kilometres wide, and the land owners (the site exceeded the original plot, requiring eleven households to relocate).

Governor Sato passed away in 1964, but his successor's electoral district included the neglected town Futaba, and he supported the plan. For TEPCO, the proposed site was perfect: flat, dirt cheap, right beside the ocean and relatively earthquake-free. "In the past 700 years," TEPCO's 1966 government application stated, "Fukushima suffered almost no noteworthy damage from earthquakes except in the Aizu area [in the current Fukushima Prefecture].... Thus, the site can be described as an area of low-seismicity."[e] The report also noted that "it appears that the vicinity of the Fukushima [Daiichi] site has not previously experienced earthquake damage." This finding seems bizarre in hindsight. While 700 years is a long time and the report avoided definitive language, when compared to the epochal tectonic history of planet Earth – billions of years – those 700 are almost meaningless.

The available technology was indeed primitive when these early reactors were designed, but Japan was already a world leader in the field of seismological analysis. Humanity had known of the existence of earthquakes for thousands of years, of course, and conjecture on their cause ranged from the anger of the gods to pockets of trapped air deep below the ground, but the reliable measurement of seismic activity did not come about until the mid-20th century. A magnitude 6.8 earthquake struck the city of Fukui in 1948, killing over 3,700 people and injuring over 22,000 more as 36,000 buildings collapsed.[6]

Only basic seismographs existed at the time. So, after this event, Japanese scientists and the government joined forces to measure, record and classify the strength of earthquakes, particularly with respect to their ability to move and damage structures, while engineering firms concentrated on designing shake-resistant buildings. The government rescinded an earthquake law restricting buildings to 31 metres high in 1963. Interestingly, engineering firms were then free to invent their own approaches to earthquake resistance during this period – they merely needed to convince the government that

---

6   This 6.8 magnitude rating seems to have been retroactively applied, since precisely how strong the ground motion was remains unknown.

their design was sound. This paved the way for the Meiji-era Kajima Corporation to build Japan's first modern skyscraper: the 36-storey Kasumigaseki Building, which was completed in 1968 using cutting-edge techniques.[7] The building and its basement levels were covered in seismographs, pioneering a new approach to earthquake protection, and becoming a symbol of Japan's post-war economic resurgence in the process. Nonetheless, the more advanced technologies for detecting, predicting and (most importantly) resisting earthquakes did not appear until the 1980s, which greatly hindered TEPCO's efforts to estimate any genuine earthquake risks.

Accurate and reliable tsunami prediction science barely existed at all, which was problematic, given that Japan has endured more tsunami than any other nation. The Japanese even coined the term "tsunami," meaning "harbour wave."[8] Japanese tsunami scientific analysis began in 1896 after the magnitude 8.5 (on the "moment magnitude scale") Sanriku Earthquake.[9] The quake, originating 150 kilometres offshore, caused waves to reach heights of up to 32.2 metres (125 feet) on land and killed over 22,000 people along the Sanriku coast. This incident prompted the first-ever use of the word "tsunami" in English, in *National Geographic* magazine, and set a height record which remained unsurpassed until 2011. The Sanriku coastline is notorious for some of the world's worst and most frequent earthquakes. Running from the northern half of Fukushima-bordering Miyagi Prefecture (basically from about where the Onagawa Nuclear Plant was soon built), the coastline stretches north through Iwate Prefecture and up into Aomori Prefecture, which are the north-easternmost prefectures on Honshū. The 1896 earthquake produced extremely slow and mild tremors, which people ignored due to a recent spate of harmless minor quakes, hence the high death toll.

Emperor Meiji's governmental Council on Earthquake Disaster Prevention postulated that an earthquake could be a possible source for the devastation, to great scepticism. Tsunami were assumed to be the result of violent underwater geophysical events but, because this record-setting tsunami was not heralded by large-scale ground

---

7   In the year the height law was abolished (1963), Kajima "recorded the largest annual order value of any construction company in the world": https://www.kajima.co.jp/english/prof/overview/history_05.html

8   Tsu: harbour; nami: wave.

9   The moment magnitude scale is the successor scale to the well-known Richter scale. Though they calculate strength in a slightly different way, they both use the same number scale.

tremors, other underwater activities like a volcanic eruption or landslide were considered more likely sources. Tidal records from the event showed a long wave period which, by around 1910, became recognised as an indicator that earthquakes cause tsunami. People now knew to evacuate coastal regions after such an event.

After the Great Kantō Earthquake of 1923, discussed earlier, serious work began on developing earthquake and tsunami counter-measures, with the first requirement for earthquake resistance intro-duced the following year. A decade later, another earthquake rocked Japan's Sanriku coast on 3rd March 1933. This time, people felt it and fled towards high ground. Their quick actions worked: though the earthquake had the same magnitude as the Sanriku quake 37 years earlier, and the resulting tsunami flattened the exact same region as before, this time they were only 14% of the fatalities. Three months later, the Council on Earthquake Disaster Prevention recommend-ed the government adopt ten tsunami defensive measures.[f] These included erecting seawalls, moving homes to higher ground and building structures nearest the ocean out of stronger materials. In addition, every village would gain designated high-ground evacuation routes and tsunami watchers, since incoming tsunami are visible from great distances, with dense "tsunami control forests" helping to dissipate the incoming water. The central government adopted them all, while local governments produced and distributed booklets for the public, warning them of possible tsunami indicators.[g] People might encounter mild but long ground tremors, possibly accompa-nied by a loud rumbling like thunder, and the ocean tide withdrawing away from land. In such circumstances, citizens were advised to flee towards high ground and remain there for at least an hour. Those aboard ships near the shore were to sail far out into open waters.

By 1941, Japan had its first dedicated tsunami warning system. Waves of seismic energy transferred through rock were detected by seismometers at the Sendai District Meteorological Observatory, one of five district monitors still in use today. From there, the earth-quake's wave amplitude and estimated distance were compared on a chart against other recent earthquakes and tsunami in the region. If things seemed dicey, the observatory broadcast a public warning via radio, with an emphasis on caution over optimism. Within eleven years, the entire Japanese coastline operated a system based on the pioneering Sendai method. Japan's first proper seawalls and coastal embankments were put to the test when Super-Typhoon Vera swept across the Ise Bay in September 1959. The typhoon won, washing away embankments built of compacted soil, solid only on the seaward side. Over 5,000 people died. Coastal defences were redesigned again, this time covered in concrete on all sides, but by the time of Fukushima Daiichi's construction, tsunami height prediction remained mostly guesswork.

Due to the global inexperience of designing earthquake-resistant buildings and detection technology, no formal seismic safety regulations for nuclear plants existed until 1977.[10] TEPCO, after studying earthquake research papers and investigating known incidents in the area, decided that Fukushima Unit 1 needed to withstand a ground motion of 265 Gal, a unit of gravitational acceleration named after Galileo. One Gal is defined as one centimetre per second squared. In other words, if the ground on which a plant is built accelerates at one Gal, its speed increases at one centimetre per second, every second. Earth's gravity (one g) is equal to 980 Gal, so 265 Gal equals 0.27 g, or an acceleration equivalent to approximately one-quarter of Earth's gravity. Tertiary areas of Fukushima Daiichi's Unit 1 not critical to safety could only withstand 176 Gal or 0.18 g, but all plants are different. Tsuruga, for example, sits on an area of more recent activity and was therefore originally designed to withstand 368 Gal.

A seismic rating of around 1,000 Gal is not unusual today, with the lowest rating approved by the modern Nuclear Regulation Authority (NRA) being 700 Gal, at Fukui Prefecture's Takahama Nuclear Power Plant. The highest appears to be at Kashiwazaki-Kariwa, where proposed upgrades would bring the rating of some reactors up to 2,300 Gal. Prior to a 1981 revision of the Building Standard Law, however, earthquake-resistant design was still in its infancy and engineers were unsure how sturdy buildings ought to be. They determined the strength of ordinary buildings from a formula for calculating lateral seismic force using a building's height, weight and location (split into seismic zones), with a baseline seismic coefficient of 0.2.[h] For nuclear plants developed in the 1960s, when sophisticated computational analyses did not yet exist, Japan simply adopted a standard of three times the yielding at a given seismic zone, which, at Ōkuma, was 0.18, giving a coefficient of 0.54.[i] "There was no basis in deciding on three times," Tokyo University professor emeritus of

---

10 When the AEC created the Regulatory Guide for Reviewing Seismic Design of Nuclear Power Reactor Facilities.

structural engineering Hiroyuki Aoyama said in 2011. "They were shooting from the hip.... There was a vague target." [11, j]

Following the 2011 disaster, TEPCO repeated their official reasons for lowering the land, as included in their original 1,000-plus page application to the government in 1966: improved earthquake resistance from building the plant foundations straight onto bedrock, and aiding the delivery of heavy reactor components to the site via boat. [12] These were indeed factors, but public comments in 1969 from the then-deputy head of Daiichi's construction office indicated another motive: "We decided to build the plant at ground level after comparing the ground construction costs and operating costs of the circulation water pumps." [k] This approach was confirmed in a July 2011 interview with the *Wall Street Journal*, when a former TEPCO executive who had helped oversee the plant's construction conceded that "we figured it would have been a major endeavor to pump up seawater from a plateau 35 metres above sea level." [l] TEPCO and GE both knew the massive cooling pumps used to draw in Pacific Ocean water and push it upwards into the reactor would be cheaper to operate if they were as low as possible, relative to the ocean.

The Japan Atomic Energy Commission would not introduce the first tsunami guidelines for nuclear facilities until 1970, leaving TEPCO to choose a seawall height for themselves. Its engineers looked to the most powerful earthquake ever recorded: the May 1960, magnitude 9.5 Valdivia Earthquake off the coast of Chile, 17,000 kilometres away. They assumed ocean water would never surpass the peak height at which the resulting tsunami eventually reached Japan's coast, 50 kilometres south of Fukushima Daiichi: 3.122 metres, or about 10 feet. This assumption may seem ludicrous, since water easily exceeded ten metres high (32 feet) near the epicentre, but tsunami around Japan are *relatively* rare; typhoons are more common and are considered a bigger threat. As the plant application states, "Most large waves in this coastal area are the product of strong winds and low pressure weather patterns, such as Typhoon No. 28 in February of 1960, which produced peak waves measured at 7.94 metres." [m] Yet, even knowing that such intimidating waves were possible near Fukushima's coastline, the planning

---

11 Note that this approach was not unique to nuclear facilities but included hospitals and other vital structures.

12 In fact, strong evidence suggests that Fukushima Daiichi was not built onto hard argillite rock at all, but softer and porous sandstone, which is not great for earthquake resistance.

application did not include coastal defences designed specifically to protect against a tsunami.

Kajima Corporation's mighty construction vehicles began by removing most of the site's 35-metre-high sandstone plateau until the ground was only 10 metres (33 feet) above sea level. Next, they dug foundations and laid concrete for sub-levels reaching a further 14 metres down into the bedrock. Beginning in September, the first sections of Unit 1's 32-metre (105 feet) tall "Mark I" containment—comprised of two connected parts, a containment vessel (also called a "drywell") and a pressure suppression chamber ("wetwell") – were gently lowered into position via crane and welded together, about 80 metres (260 feet) from the coastline.[13]

The drywell is the upper portion, shaped like an inverted light bulb with its narrow end pointing upwards. The drywell's spherical section had a diameter of 17.7 metres (58 feet), while the upper cylindrical portion was 9.6 metres (31 feet) wide. The wetwell, a torus-shaped chamber lying beneath the drywell, had a cross-sectional diameter of 8.45 metres (28 feet), with a total width of almost thirty metres, and was two-thirds filled with water. These two sections were joined by pipes thick enough to stand in, called vent lines, which reached from the upper surfaces of the torus to the lower half of the drywell's light bulb. According to the design, if the pressure became too great during an accident, steam would be forced down through the vent lines and into the water below, where it would cool and condense before being discharged out of the reactor building.

The incredibly strong containment vessel was designed to survive a simultaneous worst-case earthquake and reactor accident. During construction, each slice was milled from three-centimetre-thick Japanese steel and resembled a giant bicycle wheel, spanned by taut spokes across its width to prevent buckling while in the air. Outside the steel vessel, two-metre-thick reinforced concrete encompassed the entire reactor assembly, to constrain the immense pressures and prevent any stray neutrons from escaping the core. This density of reinforced concrete was customised for Japan and not part of the original design, contributing to problems during maintenance down the line. A five-centimetre gap separated the two barriers, to

---

13 Units 1, 2, 3, 4 and 5 used "Mark I" containment as described here. Unit 6 uses "Mark II" containment, which is totally different; the containment is cone-shaped, and its pressure suppression chamber is beneath the pressure vessel, all in one space.

account for thermal expansion of steel during operation. The top cylindrical section of the light bulb housed the actual reactor "core": the 19-metre-tall pressure vessel.

A BWR's pressure vessel, shaped like an upright sausage (cylindrical with hemispheres at each end), contains the fuel and is where the fission reaction occurs. A specialist steelworks in Yokohama shaped high-quality steel 16 centimetres (6.3 inches) thick for the job, before welding sections together and adding a five-millimetre stainless steel lining to the inside to prevent corrosion. Once complete, the vessel underwent rigorous testing, including pressure tests and an ultrasound test to check for defects. In total, it took two years to make Fukushima Daiichi Unit 1's pressure vessel.

To generate electricity, water inside the vessel boils into 285°C superheated steam as it passes uranium dioxide fuel. Next, at the top of the vessel, this steam travels up through tube banks called "steam separators" to remove moisture, before passing through a hooded steam dryer, where any remaining moisture collects on rows of dryer vanes. This is done to improve thermal efficiency and prevent degradation of the electric turbine. From here, the dried steam exits the containment and blasts through one high-pressure and three low-pressure steam turbines inside an enormous (and loud) turbine hall. These turbines spin, creating electricity via a connected generator.

Beneath each low-pressure turbine sits a condenser tank, through which runs a cold-water pipe from a nearby source (the Pacific Ocean in Fukushima Daiichi's case) that condenses the reactor steam back into water. Pumps then force the cooled water along pipes penetrating the middle of the pressure vessel and down to its base, then back up through the centre, where the cycle begins anew. Operators control reactor power output by varying the amount and circulation speed of cooling water entering the reactor (the more coolant, the less fission), or by inserting control rods made of a fission-dampening material up into the core from below.

There are several pros and cons to choosing between pressurised and boiling water reactors. BWRs use a single loop system, where one unbroken chain of water/steam traverses the reactor and drives the turbines, making them relatively simple as far as reactors go (none of them are simple). They also operate at a lower fuel temperature and less than half the pressure of a PWR, around 75 atmospheres compared to 155, making them less prone to leaks or containment failure. Combined, these factors mean less radiation hits a BWR's pressure vessel than a PWR, so the steel is weakened less over time. This all generally makes BWRs smaller, cheaper, quicker and simpler to build and maintain than PWRs.[14] Prior to the Fukushima disaster, no water-moderated BWR had ever suffered a serious accident.

That said, BWRs and their single coolant loops have been controversial for decades (although not when they were first built) and have some critical disadvantages compared to PWRs, which use multiple loops. In a PWR, the primary loop flows cool water into the core, where it heats, leaves the core and passes down through a cold water tank within the containment vessel, from where it's fed back into the core. The tank is a steam generator – a component of the secondary loop. On contact with the primary loop's extremely hot water (around 320°C / 600°F), the generator's own water boils into steam that then exits the containment vessel and forces its way through a steam turbine. The steam then condenses back into water – using cold water pipes from a third loop that pulls in water from outside – before re-entering the reactor.

A BWR's single loop system means contaminated water exits the containment vessel and travels around the entire construct. This arrangement not only contaminates the turbines but also means any water leak is going to release radioactive water. Although that sounds rather risky, the bulk of the water's radiation comes from nitrogen-16, which has a half-life of seven seconds, meaning it halves its intensity every seven seconds and dissipates very quickly. PWRs also have the safety benefit of control rods inserting down into the core from above, meaning their own weight allows them to drop in the event of a power failure or other emergency. BWR rods insert from below and are powered by water pressure because the space above the core is occupied by steam separators and dryers, which aren't required in a PWR, leaving the area underneath as the only option.[15]

* * *

---

14 BWRs are smaller overall but the actual pressure vessel is larger than that of a PWR of equivalent power.

15 Multiple emergency redundancy systems are in place to guarantee the rods will work in all scenarios, but this lack of the gravity option is considered a negative.

Fukushima Daiichi gradually took shape. Its 441-ton pressure vessel arrived as a complete unit via cargo ship, before being hoisted onto land by an enormous dock crane. The vessel was then laid on its side in a cradle and pulled up the hill, inch by inch, on rollers. Finally, on a warm and sunny day, another crane gently lowered the core into a gaping hole in the reactor building.[16] A turbine hall, water processing plant, a 500,000-volt, 190-kilometre main transmission line and other support structures also emerged from the ground. Meanwhile, in the turbine building basement, four metres above sea level, construction workers installed emergency DC batteries, switchgear equipment and backup diesel generators (cooled with seawater) to drive the plant's enormous water pumps in case of a power failure.

Placing emergency backups in the basement came from General Electric's original designs for American plants, which are typically sited inland along riverbanks and are more vulnerable to tornadoes than sudden unpredictable water rises. Being underground made them safe from high winds, strengthened them against earthquake vibrations and granted easy access for repairs. They were not located in the reactor buildings because those require an airtight seal. TEPCO held a longstanding policy of not modifying GE's blueprints – a policy that continued after Japanese vendors took over the design work – but the layout wasn't ideal. "If an earthquake hits and destroys some of the pipes above, water could come down and hit the generators," said former GE employee Yukiteru Naka, who helped with the design. "DC batteries were also located too close to the diesel generators. It's not at all good in terms of safety. Many of the middle-ranking engineers at the plant shared the same concern."[n] A former TEPCO executive who oversaw construction of the plant would claim in 2011 that he was unaware any diesel generators were in the basements. Still, TEPCO or GE must have realised the diesels were vulnerable at some point, because Daiichi's eventual Units 2, 4 and 6 each had one of their two generators up on the first floor. The shared spent fuel building held the second generators for Units 2 and 4, while Unit 6's was in its own generator building. Unit 6 was also unique in that it had a third generator, however most of the electrical switching equipment for *all* backups remained below ground.

---

16 There's a film of Daiichi's construction on YouTube; it's an amazing sight. The old promotional video is called Reimei ("Dawn") and can be found here: https://www.youtube.com/watch?v=HZfhlu_Wy6c.

Fukushima Daiichi

Fukushima Daiichi

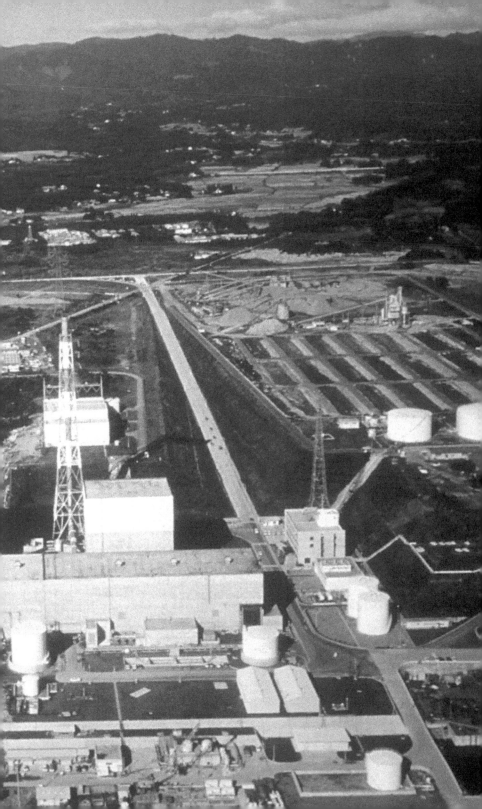

Truck after straining truck, laden with one of 10,000 concrete tetrapods, each weighing up to 20 tons, drove from the on-site concrete works down to the harbour, where Daiichi's massive ocean barrier took shape. From there, workers wrapped reinforced lines around each tetrapod and a floating sea-crane hauled them aloft before stacking them into what became the breakwater. This mainly existed to shelter boats and offer some protection against a typhoon – not tsunami. TEPCO determined that a height of five metres (16 feet) would suffice to halt the highest water nature could ever conjure, noting in their 1966 site application that "there is no recorded history of a severe earthquake in the immediate vicinity."[o] Just how "immediate" they meant is unclear, given that region's infamy for its earthquakes.

In 1970, late in Fukushima's construction, the JAEC added natural phenomena to the industry safety guidelines for the first time, writing that "systems and devices that, upon malfunction, could potentially become direct causes of accidents with serious safety ramifications must be designed to withstand the most severe forces of nature that may be anticipated based on the prevailing natural conditions at the plant site and in the surrounding regions." Not only would these safety systems have to survive the worst weather imaginable, they must do so during "the simultaneous impact of any accident conditions that may be present."[p] But how could they prepare without knowing the future? The nearby Onagawa Nuclear Power Plant may hold the answer.

ii. Predicting the Worst

The Miyagi prefectural government had announced their selection of a site for Miyagi and Tōhoku Electric Power Co's first nuclear plant beside the old fishing town of Onagawa back in January 1968.[17] The following month, a team of nine experts began a two-year investigation of past and possible future coastal incursions by scouring historical records, interviewing local residents and investigating the nearby Koyadori beach for tell-tale signs of tsunami deposits. They concluded that the maximum possible

---

17 The utility uses the disjointed abbreviation Tōhoku EPCo, also sometimes known as TEPCO, although different from Tokyo Electric Power Co. Very confusing, so I call them Tōhoku Electric with no abbreviation to make things easier.

tsunami height near where they planned to build would be three metres, but this estimate did not satisfy former Tōhoku Electric vice president Yanosuke Hirai.

Born in Miyagi Prefecture in 1902, Hirai studied civil engineering at what was then called the Tokyo Imperial University. He became a protégé of Yasuzaemon Matsunaga, who hired him to work for Tōhō Electric Power Co at the age of 24 in 1926, before moving to the mid-century megacompany Nippon Hassoden in 1941, where he became the head of its Civil Engineering Division. After playing a prominent role in the post-war reconstruction of destroyed electricity infrastructure, he joined Tōhoku Electric in 1951. Over time, he inherited his old mentor's disdain for the stereotypical grovelling bureaucrat and, in a society famous for its subservience, stood out with a lifelong belief in going above and beyond, unafraid to stick his neck out. In 1957, during construction of a thermal power station in Niigata, for instance, Hirai worried about a potential earthquake causing soil liquefaction beneath the plant, so he ordered specialist (and expensive) foundations be added, against much resistance. His precautions saved the plant when that exact thing happened seven years later.[q] After retiring from Tōhoku Electric in 1962, Hirai took charge of technology research at the non-profit Central Research Institute of the Electric Power Industry (CRIEPI).[r] CRIEPI was the country's first private think tank, founded by Matsunaga back in 1951 as a means of coordinating research resources between the utilities to help prevent the government from retaking control.

Six years later, in 1968, Hirai joined the aforementioned Coastal Facility Planning Committee – comprised of nine experts in civil engineering and geophysics, led by Tokyo University professor emeritus and Toho University professor Masashi Honma – conducting preparatory work for construction of the Onagawa nuclear plant.[18] Hirai argued against the committee's majority conclusion that no nearby tsunami could exceed three metres, believing they had overlooked ancient events like the 9th July, AD 869 Jōgan earthquake Hirai had learned about as a boy. The quake was mentioned in the True History of the Three Reigns of Japan, the official historical record of the time. "The sea began roaring like a big thunderstorm," the

---

18 Whenever the Coastal Facility Planning Committee is mentioned – and it sometimes has different names due to poor translation – Hirai is always said to be a member. While he was not the group's leader, he was involved in approving their work and presenting it to Tōhoku Electric in some capacity. Professor Honma's name is sometimes written as Hitoshi rather than Masashi Honma. Sources differ as to the number of experts on the committee, but Tōhoku Electric themselves say there were nine members.

dramatic entry states. "The sea surface suddenly rose up and the huge waves attacked the land. They raged like nightmares, and immediately reached the city centre. The waves spread thousands of yards from the beach, and we could not see how large the devastated area was. The fields and roads completely sank into the sea. About one thousand people drowned in the waves."

Hirai worried that such a colossal event could happen again. His colleagues reported to Tōhoku Electric's president that a 12-metre wall was beyond capable of repelling the worst tsunami imaginable but, still unsatisfied, Hirai pressed for it to be raised to 14.8 metres (almost 50 feet). This placed it five times higher than the maximum theorised tsunami – outrageous on the face of it – but his knowledge, experience and work philosophy were highly respected. The president reluctantly agreed to Hirai's request. Through strong leadership, therefore, Hirai helped instil a "safety culture" in the places he worked; an approach that reflected admirably when compared to TEPCO. He even considered what would happen at Onagawa's inlet pumps between tsunami waves, where the water level can often sink below normal sea level. The plant's design incorporated a special reservoir of water for just such an occasion, providing the reactor with an additional 40-minute supply.

The utility did lose some of his unwavering devotion to social responsibility after Hirai's full retirement from the industry in 1975, but it retained more than several of its peers.[19] When the time came to plan for a second reactor at Onagawa in the late 1980s, for instance, Tōhoku engineers used the most up-to-date simulations and estimated a 9.1-metre tsunami, after which they built a new 9.7-metre concrete wall to reinforce the ocean embankment. In 2002, they checked again with the latest method and predicted 13.6 metres.[s] Hirai passed away in 1986, but the 13-metre tsunami in 2011 proved him right once more. "I respect the law," he once said, "but engineers are responsible for the consequences beyond the standards and guidelines stipulated in laws and regulations." Tōhoku Electric

---

19 While designing the Noshiro Thermal Power Station in Akita Prefecture, for example, Tōhoku Electric decided that building it 4.2 metres above sea level would provide sufficient protection against tsunami. In 1983, during the early stages of construction, the Sea of Japan earthquake occurred 100 kilometres west of Akita Prefecture. The ensuing tsunami enveloped the construction site, killing 34 of the 306 people on site and injuring 58 more. Nevertheless, the utility chose not to raise the plant height, even though construction of the main power unit had not yet begun. Tsunami Disaster Prevention Research Institute Research Report No. 1, May 26, 1983: A paper and research report on the Tsunami of the Japan Sea Chubu Earthquake (in Japanese). Toshio Iwasaki, Takehiro Nakamura and Ito Il: http://tsunami-dl.jp/document/109.

quickly promoted Hirai as a poster boy and used his outstanding reputation for leadership as evidence of the company's trustworthiness after the disaster, despite his name barely existing prior to that, even in industry publications from his day.

* * *

By the time of Fukushima Daiichi's first criticality of Unit 1 on 10th October 1970, work was well underway on two more reactors on-site, plus a second reactor at Mihama and two entirely new nuclear plants. The first, Takahama NPP, utilised KEPCO's origins as Osaka Electric Light Co and its longstanding partnership with Westinghouse to install one of the American company's PWRs. The plant, Fukui Prefecture's third, was around 50 kilometres west of Tsuruga and Mihama. The second was Japan's westernmost thus far, just outside Matsue City in Shimane Prefecture and appropriately named Shimane Nuclear Power Station.

TEPCO spent six rigorous months testing Fukushima's first reactor before its commissioning and connection to the grid, beginning commercial operation on Friday, 26th March 1971. Just one day earlier, the *Asahi Shimbun* ran a story declaring that the new plant would soon be "solving the summer power squeeze in a snap."[t] Optimism at atomic energy's incredible potential reigned supreme.

And yet, doubts emerged from the United States in 1972 about General Electric's BWR Mark I containment and exactly how it would perform in an extreme accident scenario. Nuclear power stations are built to survive what is termed a "design-basis accident," which the US Nuclear Regulatory Commission (NRC) defines as "a postulated accident that a nuclear facility must be designed and built to withstand without loss to the systems, structures, and components necessary to ensure public health and safety." Nuclear plants contain multiple layers of redundancy to ensure safety even in this worst-case scenario, which is one reason they have such high up-front costs compared to their thermal counterparts: they're designed on the assumption that anything that can go wrong will. A team of engineers in 1975 described the NRC's reactor backup requirements: in an accident that "might result in an uncooled core, two emergency cooling systems must be available, either of which could by itself cool the shutdown core, and both of which have considerable internal redundancy."[u]

BWR containment buildings were not designed to contain a core meltdown in a prolonged loss-of-coolant accident like what happened in 2011. During such an accident, the reactor core contents – fuel, cladding, water pipes etc – get so hot from a lack of water coolant that they melt together into a radioactive sludge and, in the worst-case scenario, burn downwards out of the pressure vessel. Instead, the BWR design-basis accident is a temporary loss of primary coolant for a few hours, where the pressures from overheated fuel are vented safely into the wetwell and emergency water is applied, but so much attention had been paid to preventing serious accidents by this time that few considered them to be possible. The number of things that would have to go wrong at once, including a total loss of power while somehow simultaneously losing all on-site and off-site emergency measures... it just couldn't happen.

Still, it isn't unfathomable to imagine Murphy's Law rearing its inevitable head and creating just such a cataclysmic scenario, where a loss of coolant wasn't temporary and everything – *everything* – did go wrong. A US Atomic Energy Commission (AEC) advisory committee highlighted these concerns in 1966 as they were first certifying GE's and Westinghouse's designs. In response, the companies were instructed to expend "a small-scale, tempered effort on [the] problems ... associated with systems whose objective is to cope with the consequences of core meltdown."[v] Essentially, the committee chose to overlook the problem because of politics. Nuclear power was taking off across the world, and American firms had already spent years and untold billions on designing and testing reactor configurations, pushing technology to the brink of human ingenuity. Redesigning them to withstand such an implausible accident, the commission's advisory committee determined, "might require both considerable fundamental research and practical engineering application."[w] It could set them back years. The Tsuruga and Mihama plants were well underway, never mind all those in the United States. Sooner or later, the AEC agreed, they had to draw the line. The established safety measures were extensive, reliable and, "for the time being, assurance can be placed on existing types of reactor safeguards, principally emergency core-cooling."[x]

Though government research on improving existing containment designs continued, the AEC did not view the matter as critical, and work eventually stopped. In 1972, a senior commission staffer named Steven Hanauer brought his concerns about pressure suppression containment methods to the AEC's Deputy Director for Technical Review, Joseph Hendrie. Hendrie, who would later become the US Nuclear Regulatory Commission's third chairman, responded with a memo on 25th September 1972. "Steve's idea to ban pressure suppression containment schemes is an attractive one in some ways," he wrote. He continued:

> Dry containments [as with most PWRs] have the notable advantage of brute simplicity in dealing with a primary blowdown, and are thereby free of the perils of bypass leakage. However, the acceptance of pressure suppression containment concepts by all elements of the nuclear field … is firmly embedded in the conventional wisdom. Reversal of this hallowed policy, particularly at this time, could well be the end of nuclear power. It would throw into question the operation of licensed plants, would make unlicensable the GE and Westinghouse ice condenser plants now in review, and would generally create more turmoil than I can stand.[y]

Oil power in Japan achieved near-total domination during this period with an astonishing 80% market share, yet few Japanese companies succeeded in the profitable business of oil exploration and extraction, instead continuing the tradition of relying on foreign imports. Hydropower capacity remained stagnant but steady at the level it had for over a decade, while coal power collapsed. The latter proved to be too uneconomical, even in areas with coalfields close to power stations and minimal distribution costs.

In May 1972, America returned the island of Okinawa to Japan. The Okinawa Electric Power Company was formed and took over electricity generation and distribution duties there, transforming the industry from nine to ten regional power companies. Then the 1973 oil crisis happened, and the world changed once more.

iii. Energy Diversity

Humanity now depended on oil. Plastics; asphalt for roads; photographic film; machine lubrication; fuel for cars, ships, planes and most trains; and the principal means of generating electricity in most industrialised nations.

Japan sustained an economic boom over the 13 years preceding 1973 in large part due to the influx of cheap foreign oil. At the time, global oil trades used US dollars as one unifying currency, giving the dollar tremendous stability and value, both real and political. After American oil production peaked around 1969 and 1970, the US balanced its increasing shortfall by importing oil from the Middle East in agreement with the Organization of Petroleum Exporting Countries (OPEC).[20] President Richard Nixon disrupted this arrangement on 15th August 1971 by removing the US from the international gold standard, which had tied the dollar's value to that of gold, in effect making the dollar the standard exchange currency against which all other currencies are valued.[21] The Japanese yen had been held at ¥360 to the dollar since the end of World War II, but now, in the new system, the yen's value declined. Other industrialised nations responded by increasing their money reserves to protect their own currencies, causing the dollar's value to drop. OPEC, in turn, began pricing oil based on the value of gold instead of the dollar, prompting the inflation of American oil prices.

Two years later, Egypt and Syria launched a joint surprise attack against Israel-occupied territories on Yom Kippur on 6th October 1973 – the holiest day on the Jewish calendar.[22] Unprepared for the offensive despite a warning from the king of Jordan, Israel lost ground, troops and equipment to superior numbers. Nixon authorised a $2.2 billion resupply of Israeli material losses after three days of fighting, partly because Israel began to ready its nuclear weapons in the face of a total defeat. Furious, OPEC raised oil prices by 21% and decreased production by 5% per month starting 16th October, to continue indefinitely until Israeli occupation of Arab territory ceased. On the 18th, a week before work began on Fukushima Daiichi's sixth and final reactor, OPEC member Libya proposed an oil embargo against the United States and its principal allies, Japan included, effectively weaponising oil production.[23] Saudi Arabia and others agreed, and so it began.

---

20 OPEC's Middle Eastern member states were Iraq, Saudi Arabia, Iran, Kuwait, Libya and Qatar. Other members are located farther west, in North Africa.

21 This action, which became known as the 'Nixon shock,' was part an of effort to curb rising inflation.

22 These territories were areas Israel had taken during the 1967 Six-Day War. Most fighting took place in or around the Egyptian peninsula of Sinai and the Golan Heights on Syria's eastern border.

23 This entire explanation of the embargo and how and why it happened is grossly oversimplified because the event was rather complicated and not the focus of this book.

Rather than unburden their home nation by increasing production and decreasing prices for the duration of the embargo, the major US oil companies exploited the situation with a 30% *increase* in price to all customers, including Japan. Days later, Arabian and other global oil companies announced a 10% reduction in oil exports to Japan. OPEC oil production dropped further in November, as promised, from −5% down to −30% beneath their pre-Yom Kippur War output. The resulting surge in oil price made the 1957 Suez Canal incident pale into insignificance, almost quadrupling from $3 to $11 per barrel in four months.

Japan found itself in a bind. "Japanese diplomatic policy is based on cooperative relations with the United States," remarked Vice Minister of the Ministry of Foreign Affairs Shinsaku Hogen at the time. "Japanese trade cannot exist without the United States. If we support Arabian countries by violating pro-US policies, the supply of oil by the Majors [the collective term for the biggest oil companies] might be stopped, and Japanese exports to the United States might also be damaged. Such results would have a tremendous effect on Japan."[z] On the other hand, Japan *needed* oil – its economy and electricity supply depended on it. This reliance on foreign oil suddenly became a national security issue, so Prime Minister Kakuei Tanaka asked if America would compensate Japan for the estimated loss in economic growth if he were to side with them. Japan had the world's second-highest GDP in 1973 and losing even a few percentage points could trigger a recession. The definitive response: no. Faced with little alternative amid heavy pressure from business leaders, Tanaka's government officially declared a pro-Arab, anti-Israel stance on 22nd November to regain unfettered access to oil. This was the first time since World War II that Japan had openly defied the United States in international diplomacy. Even then, however, it took months for the flow of oil to return to pre-crisis levels. In January the following year, Japan made arrangements to lend Iraq $1 billion in exchange for 160 million tons of crude oil over ten years, one of several such deals in the Persian Gulf. Yasuhiro Nakasone, now the minister for MITI, played a key role in negotiations with Saudi Arabia.

The cabinet's Urgent Policy Committee, with Prime Minister Tanaka as its chairman, immediately established a new national policy framework for Japanese oil use. In addition to implementing new energy conservation guidelines to reduce demand, such as using cars only when necessary, the government sought to diversify Japan's energy supply and decrease oil consumption by 15% by 1985.[24] Liquid natural gas (LNG) made a good medium-term choice, being relatively cheap and not requiring advanced technologies to burn it for power generation, with the added bonus of almost zero environmental harm. Nuclear power, together with LNG and energy conservation, became the long-term solution, with nuclear expected to be the main baseload power source by the year 2000.[aa]

The oil crisis caused enormous damage to Japan's economy and principal industries, despite lasting just a few months.[25] The petro-chemical industry – exploding from the late fifties and into the sixties – recorded no growth for the first time in 25 years as pre-approved petrochemical plants were cancelled and other funding decreased. The aluminium and, to a lesser extent, steel industries also suffered, with much-reduced orders and growth. Japan's shipbuilding industry, a global leader producing over half the weight of *all* ships in 1973, suffered a huge drop in orders, down from 35 million tons in 1973 to just 12.8 million in 1974. Diversifying away from heavy industry in the wake of the 1973 oil crisis is also one reason why Japan pivoted to the development of electronic goods and eventually became a world leader in that sector. The crisis caused one other change: the power companies lost their spirit of independence and henceforth increasingly relied on governmental guidance.

* * *

Government bureaucrats had introduced an electricity tax at the point of consumption back in the late 1960s. Only the Tōkai NPP benefitted at first, but the tax meant that money from large cities filtered back to more sparsely populated regions where nuclear plants were being built, with their rates initially set at 4% above cost.[ab] The more electricity consumed and the more power stations built, the richer small host communities became. This tax expanded

---

24 The goal was originally going to be 20% before Japan reached an agreement with OPEC to resume oil imports on Christmas Day 1973. Industrial businesses were offered incentives (low-interest loans and the like) to buy more energy-efficient equipment, for example.

25 At the same time, the prices for oil products went way up. The taxes on petroleum products alone increased dramatically over the five years after 1973: petrol/gasoline for cars went from 34.5 yen/litre in 1974 to 53.8; fuel oil went from 15 to 24.3; aviation fuel from 13 to 26. Source: Fuels Paradise: Seeking Energy Security in Europe, Japan, and the United States. John S. Duffield, JHU Press, 2015.

to include other sites in 1971 before becoming more defined in June 1974, when Tanaka's government enacted three laws, known as the Three Power Source Development Laws (*Dengen Sanpō*). Tanaka – who had long supported a proposed nuclear plant and its economy-boosting prospects near his birthplace of Kashiwazaki, in Niigata Prefecture – hoped the tax would bolster public opinion of nuclear power.[26]

These three new laws – the Law for the Neighbouring Area Preparation for Power Generating Facilities, the Electric Power Development Promotion Law, and the Electric Power Development Promotion Special Accounting Law – institutionalised and expanded the existing localised portfolio of subsidies by adding a national tax to all electricity bills across Japan, the funds from which were split into two areas.[27] Some went on research for alternative energy sources (mainly nuclear) with the remainder funnelled towards communities living near power stations of all types (though, again, nuclear subsidies were double those of all others).[28, ac] A property tax granted small towns millions of dollars a year (in 1970s money) for the first few years of a plant's life: tax money which could be equivalent to, or even greater than, that town's ordinary annual budget.[ad] On average, the government passed along taxes totalling around $370 million per gigawatt of installed capacity to these rural areas.[29] In addition to discounted electricity rates, nearby towns or cities received government grants and the utilities directly donated enormous amounts of money on top of the taxes to their respective prefectural governments and smaller local community projects. These donations counted towards a nuclear plant's construction and running costs, and therefore the utilities' profits – incredibly – rose in line with their spending.[ae]

---

26 Tanaka was technically from Nishiyama-machi, but the town was merged into the expanded city of Kashiwazaki in 2005 and no longer exists.

27 The tax is called the Electric Power Resources Development Promotion Tax. In 1980 it increased dramatically, from ¥85 per 1,000 kilowatt-hours in 1979 to ¥300 per 1,000 kilowatt-hours.

28 Similar measures were also used to "persuade" people living near new airports or dams, although these apparently required far less extensive and sophisticated manipulation due to less resistance than to nuclear sites.

29 In 2020 money, but this figure is simplified. The amount varied depending on the proximity to different generating capacities. By the 1980s, living near a plant generating 1,000 MW granted individuals ¥300 per month. Living near a plant generating over 6,000 MW entitled people to a personal grant of ¥900 per month. Business grants ranged from ¥75 to ¥225 per month. Prefectures exporting their nuclear energy also received money, and again the amount varied. Figures found in *Fuels Paradise: Seeking Energy Security in Europe, Japan, and the United States*, John S. Duffield. Johns Hopkins University Press, 2015. References cited: IEA 1979, 94; Suttmeier 1981, 121–23; IEA 1981, 185; Oshima et al. 1982, 97; Shibata 1983, 141; Samuels 1987, 246; Kohalyk 2008, 72.

All this money, as stipulated by the three laws, provided new or improved schools, new roads, hospitals, sports and welfare facilities, and all manner of other infrastructure in neglected areas that usually struggled to obtain meaningful government funding.[30] Because the compensation system was set up to be invisible and does not appear on the national budget, according to Aldrich, "few citizens know that a portion of their electricity payment is funding the distribution of side payments to communities hosting nuclear power plants; politicians, also ignorant about the instrument, have little power to affect this institution."[31, af]

In December 1976, power stations were designated an "Important Electric Power Resource Requiring Special Measures," entitling host communities to even more subsidies. Their once dying economies thrived. In time, communities came to benefit just for *considering* hosting a plant, and areas with a history of prioritising government subsidies over private-sector investment were especially likely to seek out nuclear plants.[ag] Thus, many citizens in rural Japan, from whence young people traditionally moved to cities in search of work, came to support nuclear power, which became a quick, reliable and easy way of injecting money into poor and neglected regions – hence the reason such funds came to be known as "cooperation money."

The effect is most pronounced during times of annual reactor inspections, when thousands of workers descend on a plant's local town for several weeks, spending money in hotels, bars and tourist attractions during their free time. Depopulation stops; the local economy grows. With money flowing in, local citizens disregarded their apprehension at the lack of public information on the safety of nuclear facilities and any environmental effects.[ah] This positive reinforcement, persuasion, or "soft social control" – as opposed to "hard" control like land expropriation and police force – proved to be excellent at silencing opposition and encouraging communities to embrace nuclear power in the late 1970s and into the 1980s, in stark contrast to the approach by other nuclear-heavy governments like France.

Dengen Sanpō has been altered many times since its inception, but all this tax funding originally had a five-year limit: the shorter

---

30 A 2003 revision allowed the money to be spent on less tangible things such as tourism promotion, social welfare and job training.

31 Aldrich has spent much of his career building on the work of others to analyse how Japanese civil society influences the decisions of policy-makers, particularly in relation to nuclear power stations, and I draw on his work numerous times in this book.

of either the period from the beginning to the end of a unit's construction, or from the beginning until five years later. This inevitably meant communities sought out additional reactors at a given power plant to bring a fresh influx of money. It came at a price. The flow of money, while plentiful at first, soon ran low and proved to be corrosive in the long term. Each area received a modest figure of a few million dollars per year while the utilities conducted their initial plans and environmental impact studies, then far more during a plant's construction, with the exact amount dependent on its size. The figure then fluctuated and declined further as operation commenced.

Between the introduction of the three power laws in 1974 and 2011, Fukushima Prefecture received around $3.8 billion (in 2020 money) in subsidies from its power stations, with most coming from Daiichi and its soon-to-be-built sister Daini. For a more detailed example, the village of Kariwa (next to Tanaka's Kashiwazaki plant) received ¥1.7 billion worth of subsidies from the plant in 1997, the commissioning year of its seventh and final reactor. This amount increased to ¥5.7 billion the following year before plummeting to ¥25.5 million by 2000 and zero in 2001 and 2002.[ai] Soon the towns wanted – and often needed, due to the ongoing costs of everything they had built – their lavish funding back, so they granted the utilities permission to build another reactor as the national economy expanded. Then another and another, until almost all of a host community's income depended on their nuclear plant. This phenomenon is referred to as *genpatsu izon-shō*, or "nuclear power plant addiction." Such an addiction suited the utilities, because finding new willing rural communities became increasingly difficult.

Fear of radiation was and is to be expected, regardless of its rational veracity, but that fear has a very real impact on the social fabric of a small town or village: research shows young families generally won't move to an area with a nuclear plant, and existing families move away.[aj] Fewer families means less local talent and reduced prospects for the local economy to grow. Fishing and farming communities also worried about the potential loss of livelihood. They believed, rightly or wrongly, that people visiting markets in larger towns or cities wouldn't buy produce from areas near nuclear plants in case they were contaminated somehow. Even if contamination did not occur, fishermen were concerned that nuclear plants would upset the ecological balance and drive fish

away. Boiling water reactors expel a phenomenal volume of water
– around 160,000 tons per hour – which, because it has been used
to cool a reactor, is a few degrees warmer than the water it took in.[32]

To combat these concerns, MITI's public relations budget
ballooned. Promoting plant safety and assuaging general fears of
nuclear power in potential host communities became a significant
task.[33] The ministry handed out information booklets, sent speakers
from utilities and the government to regale people in town halls,
and flew locals to existing nuclear facilities to speak with others
in the same jobs. Children joined in too. Aldrich again writes that
the government "provides a number of programs for middle and
high school students ranging from educational materials (with
syllabi on nuclear power) to field trips to visit functioning nuclear
power plants."[ak] MITI even published articles in business magazines
reassuring farmers and fishermen of the lack of ill effects on surrounding habitats.

To divide up the hotly contested fishing grounds along Japan's
coastline, back in 1948 the government had organised local
fishermen into large groups called cooperatives (*gyogyō rōdō kumiai*)
and gave each cooperative a stretch of coast to fish in.[34] Any
company wishing to take control of a given section of water had to,
by law, gain permission from a two-thirds majority among a given
cooperative's members. The agriculture industry, which included
fishing, still employed around 50% of the Japanese workforce by
the end of World War II, and, though in steep decline, remained
and remains a powerful force in rural and coastal regions. Getting
individual fishermen's support, therefore, was often key to quelling
local resistance.

Local governments, tasked with representing the population in
siting procedures, often faced the problem of unknown feelings
from the "silent majority," whose opinion could not be easily sought
or anticipated in the early stages. Even as proposals solidified, input
from ordinary men and women was deliberately limited and, in any
event, local governments lacked the knowledge or resources to rig-
orously evaluate the utilities' technical and environmental plans, nor

---

32 This 160,000 m³ of water produces around 7,500 tons of steam per hour. Source: https://www.euro-nuclear.org/info/encyclopedia/boilingwaterreactor.htm.

33 PR promotion had been ongoing since the beginning, but MITI stepped it up several notches at this stage.

34 This explanation is somewhat oversimplified. The cooperatives had been around for decades, but a proper law establishing their legal rights and boundaries was enacted in 1948.

their long-term socioeconomic effects. In the early days, during the fifties and sixties, when nuclear optimism prevailed, local officials consistently sided with the national government and utilities. By the eighties and beyond, this trend reversed.

iv. Japan's First Radiation Accident

Japan's fleet of nuclear reactors expanded after the oil crisis, further propelled by a second oil crisis in 1976. These twin crises, paired with stricter regulation of (and dormant public attention to) industrial pollution, meant opposition to nuclear power softened as the need for a self-sufficient energy source became clear to all. Between 1974 and 1990, construction work commenced on 29 nuclear reactors at 18 different plants. [35] And, as more utilities announced more reactors, a pattern emerged that mirrored the 50 / 60 hertz boundary line and its nearly century-old allegiances: power companies in eastern Japan adopted General Electric's BWR, while those in western areas chose Westinghouse's PWR. Over time, Toshiba and Hitachi became the principal contractors on BWRs, while Mitsubishi Heavy Industries built PWRs. This divide occasionally caused some unexpected issues, such as the broken turbine incident at Hokkaido Electric Power Co's Tomari plant in 1991; the plant was the first PWR built in a 50 hertz area but used a 60 hertz design. The number of turbine revolutions differs with each frequency, but Mitsubishi didn't modify its stationary blades to compensate, resulting in hundreds of cracks.

* * *

In addition to harnessing fission at power stations, the Japan Nuclear Ship Development Agency (JNSDA) and its semi-precursor, the private Japan Nuclear Powered Ship Research Association, worked from 1963 to join engineering giants the Soviet Union, the United States and Germany in developing a nuclear-powered ship. In 1959, Soviet engineers had completed the world's first non-military nuclear ship, the icebreaker *Lenin*, while the US and Germany built the cargo freighter *Savannah* and bulk carrier *Otto Hahn* respec-

---

35 Twenty-seven, if not counting two that began in 1990.

tively, the latter named after Germany's pioneering scientist in the field of radioactivity.[36] Foregoing a conventional combustion engine and its increasingly expensive oil for nuclear power offered a few theoretical but tantalising benefits. First, for large, high shaft-horse-power (shp) cargo vessels between 50,000–100,000 shp and capable of 25–30 knots, a nuclear reactor would be more economical.[al] (For comparison, the *OOCL Hong Kong*, the largest container ship ever built as of the time of writing, uses an 84,000 shp oil-fuelled engine, right in the nuclear sweet spot, though the vessel has a top speed of only 21 knots.) A nuclear ship could also remain at sea almost indefinitely without the need for refuelling, boasted zero carbon emissions and greater hull space for cargo because it didn't need a colossal fuel tank (the *OOCL Hong Kong* carries almost 15,000 cubic metres of fuel). Keen-eyed observers around the world, including in Britain, which was still considering a nuclear vessel of its own, watched to see how the Japanese ship would perform.

After a year without bids on the construction contract from any of the seven approved shipbuilders or five reactor manufacturers, the Japan Shipbuilding Industry Association negotiated an agreement with Ishikawajima-Harima Heavy Industries for the hull and Mitsubishi Atomic Power Industries for the powerplant in March 1965, at almost double the original budget of ¥3.6 billion ($157 million in 2020). The inflated price stalled momentum for a couple of years, during which time politicians and bureaucrats argued over whether to simply import American technology again, or to proceed with their own design. In the end, those favouring a domestic reactor won, soon backed by an optimistic projection from the Japan Atomic Industrial Forum of almost 300 nuclear container ships worldwide within the next three decades.[am]

Workers of Ishikawajima-Harima, originally founded by Daimyō Nariaki Tokugawa in the same year Matthew Perry arrived at Japan, laid down the keel on the prototype nuclear merchant ship at their Tokyo shipyard on 27th November 1968.[37, an] Construction proceeded at an impressive pace and the new ship, christened *Mutsu* after what was to be her home city, was launched before a large crowd on 12th June 1969, accompanied by streamers, balloons

---

36 The *Lenin* had its own colourful history that is worth reading about. The ship faced far worse problems than Japan's ever did: a partial meltdown that required the removal of its entire reactor compartment, including steam generators and pumps.

37 Engineers designed the vessel around nuclear shielding research for shipping conducted at JAERI's JRR-4 reactor.

and confetti. Those gathered included the prime minister and the future Emperor Akihito, son of Hirohito. The *Mutsu* then sailed, sans reactor, to her new home port at Mutsu city in the far north of Aomori Prefecture using a secondary oil-powered, non-nuclear boiler. Officials chose the remote location after other, more centralised ports such as Yokohama "declined on the basis of sea traffic congestion or fear of nuclear contamination."[ao] Mitsubishi soon delivered the completed 36-MW pressurised water reactor in a self-contained unit to Mutsu from their own Kobe shipyard, where it was then lowered into the hull by crane on 16th August. The reactor supplied steam to a 10,000 shp turbine capable of pushing the 8,240-ton, single-screw ship up to a cruising speed of 16.5 knots. One batch of uranium fuel provided an impressive estimated range of 145,000 nautical miles.

Unlike her American counterpart, Eisenhower's baby, the NS *Savannah* – 181 metres (596 feet) of 1950s-style beauty; sleek, with a swept-back superstructure all in brilliant white – *Mutsu* was a utilitarian and unremarkable-looking ship. She was first intended to be an oceanographic survey vessel, but a requirement that at least some of her budget be recovered through commercial operations saw her redesigned into a nuclear fuel transporter that could best be described as functional. Her main hull was painted in a faded cantaloupe colour with fern-green decks and none of *Savanna*'s flair or size (*Mutsu* was deliberately small, at 130 meters). She was not lacking in technology, however. In addition to clever designs to cope with abrupt load changes from the constant pitching and rolling not encountered on land, she employed cross-flooding ducts to right the ship and ensure that its fuel remained covered in the event of an impact.[ap] Two huge balance valves under her hull, similar to those on a household radiator, would prevent external water pressure from crushing the containment even if the ship were to sink.[aq] A maze of pumps, valves and pipes ensured that no liquid or solid waste ever left the vessel while she was at sea, and her crew and the surrounding environment were protected from gamma rays by a metre-thick concrete containment shell backed up by 19 centimetres of lead and 15 centimetres of polyethylene slabs.[38, ar] She could also survive a larger direct collision to her reactor compartment than the *Savannah* could.[as]

---

38 *Mutsu*'s waste stores only had a six-month capacity, presumably meaning she would be forced to return to her home harbour every six months for decontamination.

Work to outfit *Mutsu* concluded in July 1972. Her schedule called for a low-power criticality test while moored in the harbour, but opposition from local residents and unsympathetic communications by government officials scuppered the plan.[at] Sea trials were instead changed to commence in September, followed by two years of experimental voyages and then a permanent role as a fuel transporter, but threats to block the harbour exit by fishing cooperatives – afraid the experimental ship might contaminate their fishing grounds – repeatedly prevented her from putting to sea. Negotiations between the cooperatives, local officials and the government dragged on for two whole years before the latter agreed to establish a central compensation fund in case anything went wrong. Even then, hundreds of the more determined fishermen swarmed *Mutsu* in the harbour on her eventual departure date of 26th August 1974, with around a hundred of their small 30-foot fishing boats lashed together.[39] One man even dove into the water wearing nothing but his underwear and a traditional Japanese *hachimaki* headband to symbolise courage, then lashed his and dozens of other boats to *Mutsu*'s anchor. That evening, an ominous storm forced the dissenters and their little craft to retreat, whereupon the ship finally departed under cover of darkness.

Mutsu's 80-strong crew originally planned to test the reactor nearby in Ōminato Bay after a mooring test, but, per the government's deal with the fishing community, they instead sailed east, far out into the Sea of Japan under auxiliary power. *Mutsu*'s reactor achieved basic low-power criticality on the 28th, but three days later, as the crew brought her up to 1.4% of full power, radiation sensors detected neutrons escaping the core. Alarms blared across the ship, but poor contingency planning meant the crew had almost no tools or materials to affect any kind of temporary repairs at sea. When they identified the source of the breach, all they could do was plug it using an impromptu sealing paste of boiled rice (carbohydrates slow neutrons) and boron (to absorb them). This containment dumpling, measuring 70 centimetres in diameter and 55 millimetres thick, was spread across the gap in the shielding by hand and proved somewhat

---

39 Around 500 boats attended according to the Japanese Maritime Safety Agency, with reports of "over 300" boats lashed together, although photographs imply less of both. Images can be seen here: https://www.gettyimages.co.uk/detail/news-photo/in-this-aerial-image-local-fishboats-surround-to-disturb-news-photo/185553319. The ship sat here: 41°16'41.5"N 141°10'41.6"E at the time. This isn't Ōminato harbour, which is four kilometres south-west. Presumably the vessel was moved ahead of her maiden nuclear voyage, but I'm unsure why there's a discrepancy in the location.

– and surprisingly – effective; the onboard scientists recorded a two-thirds reduction in escaping thermal neutrons, though the fast neutrons sped right past.[au]

Nevertheless, the escaping neutrons were essentially harmless and the crew shut down the reactor without further incident.[40] The Japanese news media, on the other hand, flew into a frenzy over their first nuclear accident.[41] Widespread reports on *Mutsu*'s "leaking radioactivity" failed to highlight the distinction between radiation and radioactivity when no nuclear material had leaked at all, and soon sparked the country's second panic over contaminated fish.[42] Derisive articles mocking the improvised rice-versus-reactor containment modifications (instead of applauding the crew's ingenuity) were just the icing on the cake. Fishermen immediately threatened to blockade any port *Mutsu* sailed towards, stranding her out at sea. As the ship floated adrift and supplies of fuel oil and food for its crew ran low, Aomori's fishing cooperatives – furious that their worst fears appeared to be coming true and their prized scallop beds might be contaminated – negotiated a multi-million dollar agreement with the national government which not only guaranteed compensation if the price of their catch fell, even if it was due to unwarranted panic over the radiation risk and not to actual contamination, but also provided money for lost income while demonstrating against *Mutsu*'s return.[av] In the end, the ship made a temporary return to her home harbour under auxiliary power with a skeleton crew on 15th October after 50 days at sea, before eventually moving to the port of Sasebo in the south (again swarmed by fishing boats) for repairs in October 1978.[aw] The large shipbuilder Sasebo Heavy Industries faced bankruptcy at the time, hence why local officials accepted *Mutsu* there.[ax]

Westinghouse had reviewed the reactor design and highlighted weaknesses in its lead containment back in the 1960s, while other researchers ran shielding experiments using JRR-4 and a vessel mock-up prior to *Mutsu*'s completion. Combined with the more

---

40 The crew were also rumoured to have used socks in the containment gap, but unlike the rice story, I have never seen anything to corroborate this detail.

41 The incident was most likely not the first accidental nuclear release of any kind in Japan, but it seems to have been the first to be widely reported.

42 The media likely failed to explain the difference because it's impossible to provide an adequate explanation in a sentence or two, but here goes anyway. Radiation, at its most basic, is the word given to a travelling particle. Radioactivity is the process of atoms splitting apart in radioactive decay (fission is radioactivity, for example). By reporting that 'radioactivity' had leaked, they were saying that some material had leaked and would continue to pollute the environment, which was not the case.

advanced simulations than had been initially available when it was first designed, the results pointed towards a clear problem with leaking neutrons, but things were already in motion and the construction plans had gone ahead unchanged.[ay] The government's accident investigation committee struggled to assess what happened because they couldn't bring the ship's reactor online for testing while in port, but they did ultimately identify various causes for the incident. Perhaps most fundamental was the fact that – though managed by JAERI and JNSDA – the reactor and *Mutsu* herself were designed and built by two separate private companies based on research by three organisations. Each company designed different sections of her outer containment shielding, causing a gap in the paraffin wax–sealed boundary between shielding pieces and the pressure vessel.[az] This design flaw partly arose from overconfidence that a small ship reactor came with small problems compared to a large commercial type. According to Japan's Association for the Study of Failure, the engineers who designed *Mutsu*'s prototype reactor had "made poor judgments about the capacity shielding with hard to calculate complex shapes."[ba]

Then there was what would become a recurring problem: the people responsible for nuclear safety did not take it seriously. The report noted that the Nuclear Reactor Safety Review Panel's members worked only part-time, were not experts in radiation containment and only conducted written rather than practical safety tests. Most damning of all, the report concluded that the NRSR panel "tended to produce passable conclusions, blurring the demarcation between their responsibility for the outcomes and their roles. There seemed to remain engineering and technical gaps between the examinations and the real design."[bb] With all that said, it's important to note that this phenomenon is not limited to Japan's nuclear industry but, as we shall see, appears to be ingrained to some extent in Japan's culture, manifesting as strong deference to authority and general unwillingness to "create problems" by highlighting deficiencies.

The Prime Minister's Office conducted a national survey on attitudes towards nuclear power two years after the incident. The results showed a sharp increase in public opposition, up to 15% from only 5% in the same survey in 1969, as cells of insubstantial local discontent coagulated into proper organisations during the 1970s.[bc] This negative sentiment was dwarfed by the 49% of positive responses, but it had dropped from 62% in the 1969 survey. The gap was beginning to close.

Mutsu spent almost all of the next 16 years in port, first at Sasebo and then, from September 1982, at a small purpose-built harbour at Sekinehama.[43] Her reactor compartment was redesigned and rebuilt with an enhanced concrete containment at a cost of $90 million: $36.5 million more than the original cost of the whole vessel. She then operated as an experimental ship for 18 months from June 1990, sailing 80,000 kilometres while her crew recorded and studied the effects of a ship's motions in the water on its reactor. She was decommissioned in 1992, for a total combined cost over the ship's lifetime of almost ¥120 billion at the time, or around $1.2 billion in 2020. In August 2020, the Japan Society of Naval Architects designated *Mutsu* a 'Heritage Ship' in recognition of its technical and historical value.

v. Reprocessing and New Regulations

Japan's long-term goal of energy independence hinged on the ability to enrich and reprocess nuclear fuel, thus creating the prophesised closed fuel cycle. The country has limited uranium deposits of its own, and its water reactors relied on imported enriched uranium from the West. In theory, however, Japan could purchase natural uranium for enrichment directly from any of dozens of countries with uranium mines, including the Soviet Union and parts of Africa.

To understand enrichment and reprocessing, it helps to have a basic knowledge of the atomic structure. Atoms are comprised of three different types of subatomic particles: a cluster of protons and neutrons – the nucleus – and its orbiting electrons. The number of protons in an atom define the element, its atomic number and its weight, be it hydrogen (1 proton, hence an atomic number of 1), oxygen (8), sodium (11), copper (29), gold (79), uranium (92), etc. Within each of these elements are variants, known as isotopes, which are defined by their respective number of neutrons. When you buy a car, say a Honda Civic, each model comes with several different trim options. The basic option is the cheapest and least well equipped, but with more money comes sportier bodywork, more advanced driver assistance and navigation options, larger wheel

---

43 Here are the harbour's coordinates on a map: 41°21'54.4"N 141°14'19.7"E.

alloys, and a more powerful engine. Each is still a Civic, but they are demonstrably different from one another. The same principle applies to isotopes, which don't usually get their own name but are instead written using the element name plus the total number of protons and neutrons, known as the atomic mass number. Uranium-235 has 92 protons and 143 neutrons, while uranium-238 has the same 92 protons but 146 neutrons. Uranium-238 is non-fissionable, so is useless in a reactor, but the fissionable uranium-235 makes up just 0.7% of natural uranium.

Reprocessing spent fuel offers several potential benefits over simply burying it, mainly the ability to extract plutonium (for use in Fast Breeder Reactors), shorten the toxicity period by thousands of years and reduce the volume of hazardous waste by up to 95%. The original import contracts with the United Kingdom, United States and Canada, however, forbade Japan from reprocessing its spent fuel and instead required all fuel to be returned to its country of origin, with safeguards in place to ensure this condition was honoured. Delays in sharing the technology arose from the concern that both enrichment and reprocessing equipment could be used to make weapons-grade materials for nuclear bombs. This could be done in two ways, using two different elements. First, and earliest in the fuel cycle, natural uranium could be enriched far beyond the level necessary for a power reactor (typically 3–4% enrichment of uranium-235) up to around 95%. Second, Japan's existing reactors formed the artificial element plutonium as a fission by-product.[44] Britain's Magnox reactor was originally designed to make weapons-grade plutonium – where around 93% or more of the plutonium is the isotope plutonium-239 – for Britain's nuclear arsenal, but regular BWRs and PWRs also produced small amounts (around 0.9%). Yet, without access to spent fuel and the ability to chemically extract and recycle its individual elements, Japan was prevented from ever making such weapons, should it decide to do so.[45]

---

44 Fission is initiated by a process of neutron bombardment, where neutrons pummel the nucleus of the fuel's fissionable isotope, usually uranium-235. When the non-fissionable isotope uranium-238 (made of 92 protons and 146 neutrons) absorbs (or "captures") a neutron, it becomes uranium-239 (92 protons, 147 neutrons). This new isotope has a half-life of 34 minutes and decays into neptunium-239 (93 protons, 146 neutrons), which has a half-life of 2.3 days. This then decays into plutonium-239 (94 protons, 145 neutrons).

45 Whether Japan was hedging its bets and working towards creating nuclear weapons in the shadows during the fifties, sixties and seventies seems to be open to debate. Some evidence suggests that they were doing so under Prime Minister Eisaku Sato, and there were (and are) certainly some outspoken Japanese politicians in favour of the idea, often with the feeling that they should not actually make any – but should be allowed the ability to do so.

Japan pushed back against the terms of its agreement with the US in the mid-1960s, while the JAEC temporarily switched focus away from creating its own fuel reprocessing system towards importing technology from abroad to save money and time.[bd] Concurrent to the primary negotiations for obtaining an American reprocessing plant, however, the commission continued background discussions towards developing a non-US facility. This initially meant partnering with British company Nuclear Chemical Plant to develop a pilot plant in 1963, but a year later Japan opened the design work up to a bid, which French company Saint-Gobain Nucléaire won in 1966. This protracted design period meant the Tōkai Reprocessing Facility, which had been scheduled for completion in 1970, did not even begin construction until June 1971. The United States reached an agreement to permit reprocessing in Japan in the meantime but kept a veto power to withdraw permission.

Fuel fabrication began in 1972 at the sprawling Tōkai-mura campus's new Plutonium Fuel Fabrication Facility (different from the facility discussed above) for Japan's experimental 50-MW fast breeder reactor: Jōyō. Utilising a Mitsubishi-designed, liquid metal–cooled reactor core, Jōyō began construction 20 kilometres away at another new facility, the Ōarai Research and Development Institute, back in the spring of 1970.[46] Being an experimental reactor, it had no means of generating electricity but would serve as a proof of concept and testbed for future fast breeder technology.

America began to regret sharing its nuclear technologies by the mid-1970s. Japanese energy consumption had tripled between 1960 and 1975, far outstripping the United States' own energy growth of just over a third, with Japanese industry accounting for a massive 50% of that amount. The subject of reusing nuclear fuel had become a hot topic. JAERI had their own fuel reprocessing research facility and an untested reprocessing plant, but progress on forming a domestic reprocessing industry hit a brick wall on 28th October 1976. In response to fears about an unstable Iran, suspicious activity in Pakistan and Indian exploitation of its own nuclear power industry, which it used to develop and test a nuclear weapon, President Gerald Ford suspended spent fuel reprocessing in the United States and all reprocessing-related exports.[47]

---

46 The Ōarai Research and Development Institute is its current name, but it has changed names a few times over the decades. I've kept this name for simplicity.

47 According to a report in the National Security Archive, "In 1974 Department of State officials wrote

In a statement, Ford acknowledged that "the 1973 energy crisis dramatically demonstrated to all nations not only the dangers of excessive reliance on oil imports but also the reality that the world's supply of fossil fuels is running out. As a result, nuclear energy is now properly seen by many nations as an indispensable way to satisfy rising energy demand." Despite this assessment, however, he believed "that the reprocessing and recycling of plutonium should not proceed unless there is sound reason to conclude that the world community can effectively overcome the associated risks of proliferation."[be] The US paused reprocessing technology development and encouraged other "supplier" nations to follow its lead, while at the same time shifting focus onto the problem of long-term storage of spent fuel.[48] Japanese hopes for an American reprocessing facility collapsed. Consequently, the Japanese government reached agreements with British and French firms and began to send their spent fuel to Europe.[49]

* * *

Concerns about General Electric's boiling water reactors appeared once more in 1975 and '76, this time beginning in an internal report written for GE. The report's authors found that, "the PWR design [as opposed to BWR] is inherently more seismic resistant because of lower reactor vessel placement and the need to design for larger [loss of coolant accident] loadings." It continued, "because of

---

that if the Shah's dictatorship collapsed and Iran became unstable, "domestic dissidents or foreign terrorists might easily be able to seize any special nuclear material stored in Iran for use in bombs." Moreover, 'an aggressive successor to the Shah might consider nuclear weapons the final item needed to establish Iran's complete military dominance of the region.'" The 1979 Iranian Revolution illustrates the foresight of these concerns. Source: https://nsarchive2.gwu.edu/nukevault/ebb268/.

48 Federal defence reprocessing (i.e. for nuclear weapons) continued.

49 There is some debate on whether "reactor-grade" plutonium can be used in nuclear weapons. Apparently it can, but such plutonium won't produce the yield of ordinary nuclear weapons because of the existence of too much of the non-fissionable isotope plutonium-140 and insufficient plutonium-239. The more of the former and the less of the latter, the harder it is to build a weapon. Still, any kind of nuclear explosion is a bad thing, hence the strict international safeguards to protect spent fuel. The United States conducted a nuclear weapon test in 1962 using British plutonium from the Calder Hall reactor, and it did explode, but since Calder Hall had been designed with the specific purpose of producing weapons-grade plutonium, this test was arguably not indicative of the overwhelming majority of reactors in operation today, which produce less useable plutonium. Note that at the time and for three decades after, the UK government claimed Calder Hall's plutonium was not weapons-grade and could not be used in a nuclear bomb. The use of this plutonium in an American test weapon was not disclosed until 1994. Of course, using spent nuclear fuel in any kind of bomb is going to produce a "dirty bomb," which is arguably even worse in some ways.

phenomena recently discovered, all BWR containment types are undergoing extensive additional analyses to evaluate structural adequacy," and concluded that the Mark I and Mark II containments "are likely to be redesigned and retrofitted."[bf]

On 2nd February 1976, three senior engineers with a combined 50 years of experience resigned from GE in disillusioned protest against the Mark I design, partially in response to a fire at the Browns Ferry nuclear plant in Alabama the year before. In a cable room just below the plant's control room at noon on 22nd March, an electrician and an electrical inspector were working to seal air leaks. The room routed electrical cables that carried signals to the plant's two reactors, which were both operating at full capacity. The pair stuffed foam sheets into cable penetrations in the walls, which were around 75 centimetres (30 inches) thick, then checked for air flow using a naked candle. A strong draft pulled the flame across the foam, which ignited, then pushed flames through the building. Panicked confusion followed before other personnel managed to activate the plant's CO_2 fire suppression system, which the electricians had deliberately disabled, but by then the fire had taken hold. When they finally did get it working, the CO_2 not only failed to extinguish the fire but also drove smoke and CO_2 from the cable room up into the control room, choking and blinding the operators.

Even with alarms sounding and warning lights flashing across their control panels, the Unit 1 control room operators did not shut down the reactor for 16 minutes after the fire alarm sounded, and did so only once the emergency pumps had repeatedly started on their own because the operator kept terminating them. Suddenly, a plant worker's worst nightmare: the electronics died. All their controls became unresponsive, including the emergency pumps and diesel generators. Even the control room phones, radiation monitors and temperature monitors stopped responding.[50] Unit 2's equipment began to fail several minutes later.

The incident was a laundry list of problems waiting to happen: flammable materials used when they shouldn't be (not to mention live flame present in their vicinity); procedures not followed; the

---

50 According to a shift engineer, "Panel lights were changing color, going on and off. I noticed the annunciators on all four diesel generator control circuits showed ground alarms. I notified the shift engineer of this condition and said I didn't think they would start." The Unit 1 operator said, "I checked and found that the only water supply to the reactor at this time was the control rod drive pump, so I increased its output to maximum." Doing so isn't remotely sufficient to keep a reactor cool, but it was better than nothing.

proper people not alerted; firefighter breathing apparatuses not fully charged; and, most foolish of all, electrical cables for backup systems all routed through a single room. The NRC rewrote the rulebook.[51]

One of the three men who subsequently resigned was told by his superiors that if GE "had to shut down all these Mark I plants it would probably mean the end of GE's nuclear business – forever."[bg] In the end, however, GE did address many of the problems, retrofitting existing plants and implementing new designs at great expense. These refits were carried out on all Japanese Mark I reactors over ten years beginning in 1975 as part of MITI's Improvement and Standardisation program, with fixes including structural reinforcement of some pipe penetration points and the pressure suppression chamber, which had been identified as a weak spot that "could be damaged in a very short time in an extreme case," but they still did not anticipate the extreme pressures experienced during the Fukushima disaster.[bh,][bi] Construction costs increased dramatically – sometimes tripling – during the program as the designs integrated new safety measures.[bj] Fukushima Daiichi Unit 1's construction cost, for example, increased from ¥98.6 billion to ¥272.8 billion by 1976.

Daiichi gained a sister plant in November 1975, when Kajima Corporation began construction of the first of four reactors at a site 12 kilometres (7.5 miles) away. Fukushima Daini's reactors are all Hitachi- or Toshiba-designed BWRs, the last of which began commercial operation in August 1987.

* * *

The government commissioned a review of existing governing bodies, laws and practices after the *Mutsu* incident. The resulting July 1976 report recommended a number of fundamental changes that brought the first phase of Japanese nuclear regulation to a close.[52] The new regulations, which were enshrined into law in

---

51 The electricians seemed to be the only group who knew both that the foam rubber was flammable and that candles were being used as the testing method. As one electrician later recounted, "The electrical engineer called the group [of electricians] together and warned us how hazardous this method was. 'Why just the other day,' the electrical engineer said, 'I caught some of that foam on fire and put it out with my bare hands, burning them in the process.'" One of the electricians who started the fire said that candles had been used for more than two years but said, "I thought that everybody knew that the material we were using to seal our leaks in penetrations would burn.... I never did like it."

52 The 1976 report was known as the Arisawa Report after Hiromi Arisawa, an emeritus professor of Tokyo University, who was appointed chairman of the Advisory Committee on Atomic Energy Administration.

1978, beginning phase two, spanned a range of safety considerations, including natural hazards such as floods, tsunami, wind and
typhoons, freezing, snow and landslides. [bk] Yet, among nature's
threats, the guidelines still only provided details for earthquake
protection. Perhaps more importantly, provisions were made only
for accident prevention, not accident mitigation. They stated, for
example, that power stations should be able to halt a fission reaction
and keep a reactor cool in the event of a loss of power "for a short
period of time" – a period not defined but assumed to be a matter
of hours. It was "not necessary to simultaneously consider the loss
of function of power supply facilities that can expect high reliability," such as the kinds of facilities that supply outside electricity to
a given power station.[bl] Long-term power loss, as would happen at
Fukushima in 2011, was neither mentioned nor ever given serious
consideration by utilities.

Responsibility for reactor licensing changed hands, from the
Science and Technology Agency (STA) to the Ministry of International Trade and Industry (MITI). Licensing for research reactors
remained with the prime minister's office and STA, while the
Ministry of Transport would handle any future commercial marine
reactors (though there were none). In addition to its new licensing
power, MITI remained responsible for the general oversight of
nuclear power stations as a group, though it was the STA's responsibility to conduct routine inspections of individual parts and components at each site. The scope of these plant inspections increased
dramatically over their pre-1978 levels.

Most significant was the separation of the Nuclear Safety Commission (NSC) from the Japan Atomic Energy Commission (JAEC),
with which it had previously been an arm of and shared some safety
responsibilities. The new NSC comprised five expert commissioners
(increasing to 20 over the next two decades) appointed by the prime
minister and responsible for overseeing and cross-examining *all*
nuclear safety matters in the so-called "double-check" system. Administration of the NSC fell to the Science and Technology Agency.
While MITI and other relevant government ministries provided
direct regulation of nuclear utilities, they were supervised by and
required to listen to the NSC's views at all stages of a facility's life,
from conceptual planning to decommissioning. Nevertheless, the
NSC remained an advisory group only, causing some controversy,
while MITI actually wrote the safety regulations for commercial

reactors.[53] MITI staff had to respect the NSC's recommendations but were not required to write them into law – an odd setup justified because the NSC having law-making powers would inherently make them a part of the government, and thus neither independent nor neutral. Others voiced concerns about the scope and overlap of authority. As a result, the regulations themselves were not as stringent as they perhaps ought to have been. Roger W. Gale, former director of external affairs for the US Federal Energy Regulatory Commission, summarised the relationship best when he wrote that "MITI does not give TEPCO the freedom to do as it pleases, but it does guarantee it a broad freedom *from* unwanted regulatory controls." (Emphasis in original.)[bm]

If you find all this confusing, you're not alone: the entire system was a tangled spaghetti-mess of baffling complexity. They could not see the forest for the trees at the government level because, while there were people and departments responsible for the ongoing safety of every last detail, down to individual pipes, nobody managed the high-level safety of a power station as a whole.[bn] The overemphasis on component inspections meant an extreme volume of paperwork, causing two main problems. First, inspectors became less attentive to a given item because of the sheer number of things to check. Second, and more problematic, was the long paper trail of the smallest faults, which made utilities uneasy. Beginning in the 1980s, all of them – TEPCO in particular – began to falsify failed inspection reports and continued doing so for two decades.

If that wasn't enough, the government possessed a serious shortage of advanced nuclear physics and engineering knowledge due to another peculiarity of Japanese bureaucracy: regular job rotation. Managers typically undergo regular job changes every two to three years in both the public and private sectors. This phenomenon extends to the judiciary, where even Supreme Court justices rarely sit for more than six or seven years before retiring.[54] The logic is that a succession of different jobs, combined with continual on- and off-the-job training (sometimes including training within

---

53 People were not blind to the apparent problems this arrangement would cause, and each opposition party in Parliament, along with the Federation of Electric Power Related Industry Workers' Unions of Japan, called for the NSC to have legal authority, but arguments against these more tangible powers prevailed.

54 I use the Supreme Court example to illustrate the court's frequent rotation of members, but they don't leave because they're forced to move onto another job; they are simply required to retire at age 70 yet are almost never granted membership before age 64.

the private sector for government employees), creates a workforce armed with comprehensive knowledge which can handle almost anything. In many ways, the system makes sense. Western men and women at the highest levels of business and government sometimes face accusations of being out of touch and not understanding that which they control. In Japan, managers can gain first-hand experience of different aspects of that business, giving them a true appreciation of the work and the people doing it. TEPCO's own grovelling report into the Fukushima disaster mentions personnel rotations almost a dozen times as a partial justification for their company-culture problems.

Japan's basic structure of government differs from that of many advanced nations, too, in that its central body is quite small and relies on outsourcing all sorts of things to branches of local governments (which are comparatively large) as well as various industry leaders. As a result, government officials – sometimes woefully unqualified because of rotation – seek advice and input from technical specialists in private industry. But a quandary surfaces when government ministries monitor and regulate nuclear facilities of labyrinthine complexity: in a land of generalists, experts are scarce, so how do you know if you've been told everything you need to know? Before anyone had time to experience the breadth of their new jobs and attain the standards required of nuclear oversight, staff rotated out into other positions. "Since there were personnel rotations within every administrative organ," wrote the Rebuild Japan Initiative Foundation's Independent Investigation Commission in their report, "each encountered constant difficulty in finding and allocating officials possessing the required expertise.... Before long, relative amateurs were tasked with creating proposals and planning safety regulation policy."[bo] Another unfortunate consequence of rotation saw each new batch of employees repeat the same mistakes over and over.[bp] What transpired, in effect, was the regulated dictating to the regulators.[55] A related example invited disbelieving derision in November 2018, when 68-year-old deputy chief of cybersecurity strategy Yoshitaka Sakurada admitted to never having used a computer in his entire professional life.

---

55 A similar role reversal spilled into public discourse in October 2012, when, after three weeks on the job, Japan's new justice minister, Keishu Tanaka (not to be confused with former prime minister Kakuei Tanaka), was forced to resign after journalists exposed his ties to organised crime.

Despite the drawbacks, eliminating the old centralised system in favour of each party writing their own rules meant more consistent safety regulations in theory. The STA, for instance, as the body responsible for research reactors, now focused on scientific research reactors, just as MITI now wrote regulations specific to the commercial power industry. The side effect, however, was that MITI now had two competing roles: promoting the commercial industry while also keeping it in check with safety regulations – regulations that cost businesses time and money. This system could have remained distanced and ethical, but it did not.

CHAPTER 4

Deviation:
the Nuclear Village,
Safety Lapses and 1980s

i. Descent from Heaven

The right-wing Liberal Democratic Party (LDP) has been in constant control of Japan from the party's foundation in 1955 until the present day, save two brief interruptions in 1993–94 and 2009–12, making its political elite an influential cornerstone of society. The power companies and government came under intense criticism from consumers in 1974 for the latter agreeing to the former's request for a 57% rate increase, brought on by the oil crisis. A further 21% rise two years later placed Japanese electricity among the most expensive in the world. The utilities were known for their lofty donations to conservative politicians, seen as a form of bribery, so opposition politicians made such donations a campaign issue. As a result, the practice was banned in 1977, but individual utility executives still regularly accounted for over 50% of political contributions to the LDP prior to 2011. Exceeding 70% wasn't unheard of (it was 72% in 2009, for example), making it unwise for career politicians to cross them.

The top people at TEPCO, especially, enjoyed immeasurable influence. In 2011, as the first wave breached Fukushima Daiichi's coastal defences, its president, Masataka Shimizu, headed not only the largest privately owned utility on the planet, with over 30% of the nation's population and industry (not to mention the central government) under its umbrella, but he also chaired the Federation of Electric Power Companies (FEPC), Japan Nuclear Fuel Limited (JNFL), presided over the Japan Society for Corporate Communication Studies (JSCCS) and was vice chairman of the Japan Business Federation (known as Keidanren, an amalgamation of its Japanese name, Nippon Keizai-dantai Rengōkai). The FEPC is a forum for discussion and cooperation among the utilities, and, as such, lobbied their own national policy to government bureaucrats. JNFL own the enormous Rokkasho Reprocessing Plant, the country's only spent fuel recycling facility, which will be discussed later. The JSCCS, according to the society's own website, "is engaged in research activities related to public relations and communication of management entities such as companies and government."[a] Its management at the time included a smorgasbord of leading university professors and current and former top executives from, among many others, the country's largest alcohol producer (Asahi Breweries), its largest (and the world's second-largest) car manufacturer (Toyota) and the world's largest advertising agency (Dentsu), all vying for control over their portrayal in the media. Keidanren, meanwhile, is Japan's most powerful business lobbying group and is strongly pro-nuclear.

With influence concentrated in a few small areas, institutionalising the siting, approval, regulation, promotion and host-community subsidising of nuclear power stations, in addition to the general government planning and oversight role, created foreseeable problems. Beginning in the 1950s but at full strength by the 1970s, an unofficial group of individuals and organisations emerged through their shared vested interest in the success of nuclear power, collectively nicknamed the "nuclear village" (Genshi-ryoku-mura) by those outside it. The nuclear village contained electric utilities, academics, media groups, financial institutions, construction companies, the big five manufacturing and nuclear vendor groups (including Mitsui [Toshiba], Mitsubishi, Hitachi and Sumitomo), bureaucrats and, inevitably, local and national politicians from both major parties. Together, as some of the country's most powerful and influential entities, they ate away at the fundamental principles of Japanese nuclear energy policy: democratic transparency with an emphasis on safety.

That isn't to say that these groups all synchronise in perfect harmony when out of the public eye – they do not. It is no accident that Japan stands as one of the few first-world countries with an energy industry that has never been nationalised (except during World War II), instead remaining private precisely because that industry fought against government control at every turn. Rather, they have evolved what political scientist Richard J. Samuels calls a give-and-take system of "reciprocal consent" in which "firms give the state jurisdiction over markets in return for their continuing control over those markets."[b]

Yet, collectively, the village's goal was and is to utilise their combined resources to secure energy self-sufficiency via the proliferation of nuclear power, all while fiercely defending one another. The core of the nuclear village – and one main reason this whole system arose and continues to exist – is the inbred relationship between senior government bureaucrats and electric utility executives. This symbiosis breaks down into two distinct but related components, both of which may seem far-fetched to a foreigner: *amakudari* and *gakubatsu*.

Of the two, *gakubatsu* is more abstract and difficult to explain. Translated into English the term means "alma mater clique," but that doesn't adequately explain the concept. The university from which one graduates is seen as incredibly important in Japanese society, stemming from feudal traditions, and graduates of the same university share a foundational bond for their entire lives. When two office workers – almost irrespective of age and stature – meet in a bar after hours and discover they are both Kyoto University alumni, their connection is sealed. When recruiting for a job, employers show favouritism towards candidates from the same university as themselves, even if they graduated decades apart. Former Osaka University attendees can find common ground on which to cooperate during inter-business negotiations, which they may otherwise lack without that shared background. Even the everyday Japanese equivalent of office politics revolves around these graduate cliques in the workplace, rather than the more typical Western style where individuals are in it purely for themselves.

Tokyo University, founded in 1886, is the oldest and most prestigious of all. It's also where the *gakubatsu* kinship is strongest, and Tokyo U's elite graduates – not coincidentally – have occupied the nation's top business, political and bureaucratic positions for over a century. "Japanese [political] parties are coalitions of factions which, in turn, are based on career, school and family ties," wrote Dr James R. Soukup, Professor of Political Science and International Relations at the University of Texas, in 1963. "The loyalties of most politicians are directed more toward these factional cliques than their parties as a whole."[c] It's difficult to overemphasise how significant this intangible connection is, but it is felt throughout Japanese business culture. During the Fukushima disaster, *gakubatsu* would influence the behaviour of the prime minister himself.

Amakudari, or "descent from heaven," is a form of bureaucratic compensation that can be explained using a relevant example. A few weeks after the Fukushima disaster, it came to light that, since the 1960s, almost 70 top-ranking MITI bureaucrats (who controlled regulation of the nuclear industry) had landed in executive-level or advisory positions at all of Japan's electric utilities upon retiring from their public-service roles.[1] Top bureaucrats joined the most influential utilities (such as TEPCO, which gained five of the 70), while less senior figures settled for smaller firms. Toru Ishida, the head of MITI's Agency for Natural Resources and Energy (ANRE), the organisation specifically tasked with promoting nuclear power and the man who drew up a plan for 14 new reactors in 2010, was one of the five. Ishida resigned from ANRE in August that year before joining TEPCO as a senior advisor the following January. He and others like him then used ties with their former colleagues to negotiate more favourable terms with the government. Between 1959 and 2010, four TEPCO vice presidents were former top-ranking members of nuclear regulatory agencies, and the position was considered reserved for whomever held the top MITI job when a vice president retired – assuming, of course, they had been kind to TEPCO during their tenure.[d, e] These retirees weren't gifted plush, high-paying industry jobs by chance: the positions were lined up for them, on the understanding of generosity towards their future benefactor. At the time of the Fukushima disaster, 13 former top ministry bureaucrats were employed at TEPCO.[f] Ishida was compelled to resign ("for personal reasons," he said) in April 2011 amid public outrage.[g]

Controversial transfers are not exclusive to the top echelons. In the late 1990s, the government nuclear licensing panel's principal seismologist and advisor to the Nuclear and Industrial Safety Agency, named Yoshihiro Kinugasa, approved a license for a second reactor at the Shika Nuclear Power Plant based on a study of nearby faults. The report noted three fault lines just below the ten-kilometre-long threshold that would prohibit construction. By 2005, no longer working on the panel, he co-wrote a paper with staff from Hokuriku Electric Power Co (HEPCO, the owner of Shika NPP) refuting claims in a lawsuit from local residents that the site was unsafe.

---

1   Later METI, the Ministry of Economy, Trade and Industry.

HEPCO lost the case. Two years later, on 25th March, the Noto Peninsula Earthquake shook the region around Shika, killing one person, injuring 350 others, damaging over 25,000 buildings, and causing widespread power outages. The HEPCO study Kinugasa approved as the government's expert had been wrong, and the three faults were in fact a single large fault 18–21 kilometres long that could shake the plant beyond its limits. Thankfully, the reactor was offline at the time. "The same people are making the rules, doing the surveys and signing off on the inspections," said Takashi Nakata, Hiroshima Institute of Technology seismologist and previous member of the government's earthquake survey committee. "The regulators just rubber-stamp the utilities' reports."[h]

While *amakudari* prevails in the nuclear power industry, the system is widespread and not exclusive to any one sector; sociologist Richard A. Colignon calls it "the hidden fabric of Japan's economy."[i] It was and is not an isolated or unknown phenomenon at all – ministries openly endorse, encourage and often themselves engage in the practice. Naoto Kan, prime minister during the Fukushima disaster, was informed of 4,240 instances in the year after his party won the election.[j] Another example was the 2017 Education Ministry scandal. Sixty-two former officials "retired" from the ministry into lucrative positions at universities across the country, despite a 2007 law making this practice illegal. Why? Japanese universities rely on government funding rather than endowments, as in the United States, so employing ex-high-ranking government officials can increase the flow of Education Ministry money that they want and need. In fact, schools and businesses which hire these individuals are usually looked upon favourably by the ministry officials that rule on regulations, award contracts or grant state funding. This also relates to the power companies, which give generously to universities and think tanks supportive of the industry. On the flip side, "If you are a critic of nuclear power," says Minister of Foreign Affairs (in 2011) and current (2020) Minister of Defence Tarō Kōno, "you are not promoted, you don't even become a professor, and you are certainly not appointed to key commissions."[k] While this observation is true as a general matter, in more recent times (particularly since the Fukushima disaster), a sprinkling of contentious academics have sat on important committees, though this strategy is generally seen as a cynical exercise in impartiality. Such people are almost always stonewalled.

Career movement between the public and private sectors in Japan is usually a one-way street, but there's a lesser-known sibling to *amakudari* – *amaagari*, or "ascent to heaven."[2] A relevant example is that of Tokio Kano.[3] Kano joined TEPCO aged 22 in 1957, straight out of university. After three decades of service, he became director general of TEPCO's Nuclear Power Division and nine years later (in 1998) was elected to the Diet's upper house. He served in Parliament for two terms, or 12 years, and to say his career comes across as suspicious is an understatement. Kano enjoyed the backing of Keidanren, the powerful business lobbying group, of which TEPCO was a principal member, and one *New York Times* article quotes a member of his own party complaining, "He rewrote everything in favor of the power companies."[1] Kano became Minister for Education, Culture, Sports, Science and Technology in 2001 and focused on expanding Japanese reliance on nuclear power before retiring from government and returning to TEPCO as an advisor eight months before the Fukushima disaster.[4] And from which institution did he graduate as a young man? Tokyo University.[5]

Once again, this is not to say that disagreements between government and industry don't happen (they do and always will) and, to be clear, a revolving door between industry and regulators is not inherently bad and can be achieved in an ethical manner. Retirement age is a factor. Most public servants must retire at 60, and the higher you rise in government, the fewer positions are available, so many retire early.[m] When combined with low pensions, you can understand why people in late middle age with decades of experience would want to continue to work, and private businesses are usually their only option. But the high-level problem is obvious: privileged people, appointed to act with good faith in overseeing a vital sector of modern society, in reality cosy up to the companies and institutions they are tasked with monitoring. Similar practices occur elsewhere, but they have more effective checks in place. In the US, former Nuclear Regulatory Commission members cannot represent

---

2   Apparently, the term *amaagari* was initially considered a joke, but it seems to have become accepted (albeit uncommon) parlance.

3   His name is 加納　時男, which may be translated with a few different spellings. Tokio Kano seems to be the most common spelling, although Tachio Kano is also common.

4   The "Minister for Education, Culture, Sports, Science and Technology" title sounds like he was in charge of every ministry under the sun, but it is actually a single title.

5   He graduated with a law degree and was not an engineer or technical specialist. He also attended Keio University, the same institution as Yasuzaemon Matsunaga.

nuclear businesses for a year after leaving government service and are permanently barred from doing so to the government on any topic they had previously been involved with. In Japan, however, an atrophy of honesty developed at the upper echelons because there were few checks and balances, and the entire system devolved into an ingrained, institutional means of collusion and corruption.

* * *

The Japanese press does a reasonable job of holding businesses and politicians to account yet was strangely reluctant when it came to the nuclear industry prior to 2011. Conservative outlets like Shōriki's *Yomiuri* went all-in on nuclear power from the start, but one or two, like the traditionally liberal and anti-nuclear *Asahi Shimbun*, were somewhat critical of the nation's safety regulations in the early years. After the 1973 oil crisis, however, even they went quiet; advertising revenues had subsided, and power companies stepped in to exploit the deficit.[n] A major push began in August 1974 on the 29th anniversary of the Hiroshima bomb, when the FEPC took out a two-thirds page ad on radiation's impact on the environment. "Since *Asahi* has many intellectual readers," recalled former FEPC publicity director Tatsuru Suzuki in his 1983 memoir, "we mobilized scholars and researchers to create an ad as a PR campaign from an independent source."[o] After the ad ran every month for two years, the *Yomiuri* came knocking. "'Nuclear power generation was introduced by our late President, Shōriki Matsutaro,'" said the paper's representatives, according to Suzuki "'With our rival *Asahi* exclusively carrying the ads, we are losing face.'" Soon, the *Mainichi* – Japan's third-largest and another anti-nuclear paper – requested an ad. "Your company is campaigning against nuclear power generation," came Suzuki's derisive response. "You are free to do that. If you think you are opposing [nuclear energy] for the benefit of society, go ahead and stick to your position. You shouldn't care about a petty thing like revenues from advertisements." The *Mainichi* changed their position.[p]

Thus, the largest media organisations came to rely on this money to stay afloat. Condemning negligent accidents was allowed, but if you criticised the wider nuclear industry or Japan's reliance on it, your money dried up. On an individual level, journalists or politicians daring to criticise the industry could find their careers ruined, while support was rewarded. On the day of the 2011 earthquake, for example, a flock of journalists were accompanying TEPCO chairman Tsunehisa Katsumata on a luxurious, nigh-all-expenses-paid trip to Beijing, China.[q]

The nuclear village has, above all else and since the days of Shōriki's public exhibitions, exerted a singular message to prevent popular opinion from succumbing to the cries of protestors: perfected absolute safety. This message became known as *anzen shinwa*, or the "safety myth," for the simple reason that nuclear power – like any industrial endeavour – is not absolutely, inherently safe. Absolute safety is the ideal, forever out of reach, the intangible peak which regulators strive to conquer with the implicit understanding that no such summit exists. Despite their monopoly status, utilities throughout Japan have spent a fortune to push this myth and others, totalling over $17.5 billion on advertising and PR between 1970 and 2011, of which TEPCO was by far the largest contributor.[r] The figure now exceeds $2.4 billion annually when combined with government promotional spending.[s] To put this in perspective, Panasonic topped the list of Japanese advertising expenditures during fiscal year 2010, just prior to the Fukushima disaster, at over $880 million, while TEPCO was the only utility to reach the top ten, at $324 million in tenth place.[6] Ad spending from the nuclear industry has declined since 2011, but industry members remain influential clients of the major media organisations.

Safety preparations and procedures can *always* be improved, thinking otherwise carries fundamental danger. And yet, that's what happened in Japan, as conscientious individuals like Professor Eizo Tajima discovered. A prolific nuclear physicist specialising in radio-activity, he helped survey the damage to Hiroshima in 1945, lectured at St Paul's University in Tokyo for decades and authored countless papers up to the 1980s. He became a member of the Nuclear Energy Safety Expert Working Group of the Japan Atomic Energy Commission in 1961 and served in this capacity for 13 years. During his tenure, Tajima argued to the JAEC's new chairman that Japan's safety regulations were poor, but his concerns were dismissed. It was

---

6   The figures are ¥73,356,000,000 for Panasonic and ¥26,903,000,000 for TEPCO. Source: Advertising Expenditure of Leading Corporations (fiscal year 2010). Nikkei Advertising Research Institute: http://www.nikkei-koken.com/surveys/survey14.html. I have also seen ¥11.6 billion (over $100 million) for TEPCO's fiscal year 2010 advertising expenditure, which is significantly less, but this is a more reliable source. Either way, the figure is enormous.

not the commission's place, he was told, to question the perfection of existing precautions.[t] In any case, the commission had no legal power: it existed as an advisory group and nothing more.

ii. American Influence

Japan completed the pilot fuel reprocessing facility at Tōkai-mura, located one mile south of JAERI's campus, just as President Gerald Ford's successor, Jimmy Carter, announced an outright ban on nuclear reprocessing in the United States on 7th April 1977. After a meticulous review, the Carter administration rejected the original fuel cycle concept embodied in light water reactors and "concluded that the serious consequences of [nuclear] proliferation and direct implications for peace and security – as well as strong scientific and economic evidence – require: a major change in US domestic nuclear energy policies and programs."[u] Instead of recycling their nuclear waste, the US would bury it. Abroad, America would "continue to embargo the export of equipment or technology that would permit uranium enrichment and chemical reprocessing." Two weeks later, on 24th April, Japan's Jōyō experimental fast breeder reactor achieved its first criticality, so the timing could not have been worse.[7]

This decision understandably upset the Japanese government, as the country already relied on Britain and France to reprocess and return much of its spent fuel.[8] Government bureaucrats spent months lobbying for permission to use the new Tōkai facility to make plutonium fuel for their planned breeder reactors. A US State Department memo dated 30th July said the Japanese negotiators highlighted that "the reprocessing issue is extremely important to Japan in the context of their long-range energy program, because of their extraordinary energy dependence." Prime Minister Takeo Fukuda even publicly declared reprocessing a matter of "life and death."[v] They were not alone in rejecting the "bury it" solution – France became a major proponent of reprocessing and re-using spent fuel, and remains so today.[9]

---

7   It took another six years for Jōyō to attain its full power of 100 megawatts-thermal.

8   Japan used British Nuclear Fuels Ltd (BFNL) in the UK and Compagnie Générale des Matières Nucléaire (COGEMA) in France, both government-owned.

9   The Ford administration had blocked the planned sale of French reprocessing plants to Pakistan and South Korea the previous year.

Subsequent to three rounds of talks and fourteen proposed "technical alternatives," the Carter administration agreed on 1st September 1977 to permit the opening of Japan's new reprocessing facility on a "very limited" and "experimental" basis for two years.[10, w, x] In exchange, Japan agreed to reprocess less than 100 tons per year and to indefinitely postpone the construction of a much larger facility at Rokkasho-mura in the country's far north, which would have converted the materials extracted from spent fuel back into new fuel. This postponement was "expected to set back by several years Japan's ambitious nuclear energy development program," reported the *New York Times*, which meant that "Tōkai-mura, in effect, may be turning out a nuclear product without the ability to make practical use of it." In any event, multiple equipment setbacks meant the Tōkai Reprocessing Facility did not begin full-scale operation until January 1981.[11]

On 20th March 1979, officials commissioned the home-grown 165 MW-electric Fugen prototype Advanced Thermal Reactor (ATR) at the Tsuruga site after almost a decade of construction and testing.[12] The site was developed and operated by the Power Reactor and Nuclear Fuel Development Corporation (Dōryokuro Kakunenryō Kaihatsu Jigyōdan, or Dōnen for short, abbreviated to PNC in English), a quasi-government research organisation created in 1967 to invent new nuclear technologies, controlled by the Science and Technology Agency. JAERI originally controlled Fugen, but labour disputes over poor pay and other matters, legal entanglements, and general mismanagement in the early 1960s led to a loss of credibility and the government sidelining the institute for all but its long-term theoretical and scientific research. PNC, formed in October 1967, also absorbed the STA's other organisation, the Atomic Fuel Corporation, thus inheriting four main responsibilities judged too risky and expensive for an ordinary private company: fuel enrichment commercialisation, fuel reprocessing, the breeder program and ATRs.

---

10 As cited in the US State Department memo, these alternatives were "evaluated against such criteria as technological feasibility, non-proliferation advantages, safety and regulatory features, and implications for the Japanese research and development program."

11 The first test run after the September 1977 agreement failed because of a damaged heat transfer tube. Further problems followed with the acid recovery evaporator, two dissolving tanks and various other pieces of machinery.

12 Construction began in December 1970; testing began in early 1978. The JAEA website includes an extensive history of the Fugen project: https://www.jaea.go.jp/04/fugen/en/guide/history/.

As a demonstrator ATR, Fugen formed the test-bed for Japan's long-term goal of power stations fuelled by recycled plutonium and unenriched uranium. Advanced Thermal Reactors use an unusual combination of heavy-water moderation and light-water cooling, light water being plain old water, albeit filtered and purified so as not to damage any machinery. Traditional light water reactors such as a PWR and BWR both use light water as their moderator, which is required to slow down flying neutrons in the fuel, giving them a greater chance of colliding with one another. Heavy water, named for its higher density than light water, allows a reactor to use unenriched uranium fuel, much like graphite. Some excitable scientists in the early 1930s opted to name the first discovered hydrogen isotope, rather than simply use its atomic number, even though this radical endeavour "threatened to become acrimonious" during a 1933 gathering of the American Physical Society.[y] In the end, hydrogen became the naming-convention exception, with this first known isotope bestowed the name deuterium, and it is deuterium that makes water "heavy." See footnote for a more detailed explanation.[13]

The creation of a new type of fuel known as mixed-oxide (MOX) used plutonium separated from spent fuel out of – and to be re-used in – Japanese light water reactors, which provided an interim solution to the growing stockpile of the lethal by-product. The fuel recycling program had hit some bumps in the road along the way but was proceeding well overall, and Fugen was the world's first reactor to be fully loaded with MOX fuel – a matter of national scientific pride.[14] Fugen's secondary role was in furthering Japanese scientific understanding and technological prowess in the field of

---

13 Three hydrogen isotopes were named in time. Hydrogen-1 is protium (the normal type of hydrogen); hydrogen-2 is deuterium; hydrogen-3 is tritium. Water is H_2O: two hydrogen atoms and one water atom. Light water uses ordinary hydrogen, with a nucleus of one lone proton and zero neutrons. In heavy water, those two hydrogen atoms are actually deuterium, which has one proton and one neutron, making it twice as heavy and 10.6% more dense than regular water. The heavy water used in Japan's ATR had a lower neutron absorption rate than light water, allowing more neutrons to fission and thus making enrichment unnecessary. In addition, some of these extra neutrons are absorbed by the non-fissionable uranium-238, which converts it into plutonium-239. Much of the furore of Barack Obama's Iran deal (nulled by the Trump administration in 2018) surrounded Iran's ability to make deuterium, because it can be used to make plutonium for nuclear bombs.

14 France and Britain were also building their own, but Japan was the first to operate with a full MOX core. The US Nuclear Regulatory Commission says MOX fuel was tested in "several commercial reactors in the 1970s," and the first recorded use of MOX in a reactor of any kind was in 1963, so the Japanese were not the first to experiment with it. The use of a full MOX core is still unusual today. An interesting background on MOX fuel can be found at this "Overview of MOX Fuel Fabrication Achievements": https://inis.iaea.org/collection/NCLCollectionStore/_Public/31/062/31062329.pdf.

reactor design and manufacturing, and later in life acting as a sort of transition reactor from ordinary light water to fast breeder reactor technologies.

Even at this early stage, however, the economic viability of plutonium recycling and MOX looked questionable, and several nuclear nations began to scale back their ambitions. Japan, in contrast, amended the law in 1979 to allow private companies to begin their own reprocessing work, then went out of its way to coerce them into doing so by dropping support for the use of foreign reprocessors.[z] As a result, the following year the country's utilities and dozens of other companies established Japan Nuclear Fuel Service Limited, what is now known as Japan Nuclear Fuel Limited (JNFL), and began preliminary work towards building three nuclear fuel cycle facilities at Rokkasho-mura.

PNC continued to reprocess its spent MOX fuel at their Tōkai Reprocessing Facility, then refabricated it into new MOX fuel assemblies which were loaded back into Fugen in May 1988, thus completing the fuel cycle and demonstrating that the JAEC's long-term energy plan could work. Despite this success, for around 25 years Fugen was the only Japanese reactor to use MOX fuel, due mainly to cost, delays at Rokkasho and strong public opposition.

Still, the JAEC considered the Fugen prototype a success at the time and, in 1979, decided to build a large-scale, 600-MW demonstration ATR to take the project one step further. With the northern town of Ōma – just 34 kilometres (20 miles) from Mutsu – selected as the build site and a proposed cost split of 30% from the government, 30% from the utilities and 40% from the Electric Power Development Co, things were ready to move forward. But the power companies had doubts: the reactor was just too uneconomical for the low output it offered. They resisted and the project collapsed in July 1995. Fugen itself continued until March 2003 but remains the only heavy water plant ever built in Japan.

* * *

On 1st December 1978, former prime minister Kakuei Tanaka's wish came true when Kajima Construction workers began digging up coastal sand at a site that would become the world's largest nuclear power station, TEPCO's Kashiwazaki-Kariwa.[aa] (Tanaka himself had, by this time, been arrested and would soon be found guilty of accepting $1.8 million in the famous Lockheed bribery scandal.) The plant gained its name from the nearby town of Kashiwazaki and a much closer village called Kariwa, in Niigata Prefecture, 215 kilometres north-west of Tokyo. TEPCO had, years earlier, bought the land off a real estate company controlled by Tanaka for an astronomical fee and announced a plan in 1969 to build a plant there.[15] Because it was built in the catchment area of Tōhoku Electric Power Co, TEPCO reached an agreement whereby the plant would supply power to the Tokyo region *and* Niigata.

Groundwork began after a decade of bitter resistance from nearby residents and a fishing cooperative, who worried about possible health and environmental risks and were angered by what they saw as placing money above all else.[ab] During this time, MITI was forced to adopt new methods of subduing dissenters. First, they tried convincing locals of their national duty to accept the power station. When that didn't work, officials agreed to hold the public hearings promised by law back in 1959, with the purported purpose of allowing people to question attending experts and bureaucrats. Despite this apparent willingness to listen, however, questions were vetted ahead of time and each person had a limited time to speak before being cut off. The hearings "lasted just one day," according to social research professor Akira Suzuki, "and did not have any authority on the licensing process or on the granting of construction permits."[ac] It was all for show, and, by the time Kashiwazaki-Kariwa was completed almost two decades later, it boasted a whopping seven boiling water reactors. Units 6 and 7 were a brand-new type, the first Advanced Boiling Water Reactors (ABWRs) ever built, which are safer and more powerful than the older versions. All seven reactors output a mammoth total capacity of 8,212 MW (8.2 gigawatts) and can supply power to 16 million homes.[16]

---

15 The story behind this sale is somewhat convoluted. The land was originally owned by the Hokuetsu Paper Company, which sold it to Mayor Hiroyasu Kimura of the nearby village of Kariwa in 1966. Kimura then sold it to Tanaka's company, Muromachi Sangyo, which sold it back to the mayor, who then sold it – on Tanaka's instruction – to TEPCO at 1,600% of the value he'd originally paid for it. It was for dealings like this that Tanaka ultimately became notorious as one of Japan's most corrupt politicians. He controlled his own shady network of private companies, of which Muromachi Sangyo was only one, and is said to have personally made $13 million (in 2020-inflated money) from the sale. *Kakuei Tanaka: A Political Biography of Modern Japan*, Steven Hunziker and Ikuro Kamimura, Times Editions, 1996: http://www.rcrinc.com/tanaka/ch4-2.html.

16 Five reactors outputting 1,100 MW, and two at 1,356 MW.

Local citizens' inability to oppose nuclear plants meant that, though many were indefinitely postponed, almost none had ever been outright cancelled prior to the Fukushima disaster. One of the first cases occurred at the small Niigata town of Maki-machi (*machi* being Japanese for "town"), just 40 kilometres from Kashiwaza-ki-Kariwa.[17] The area was famed for its rice but saw its prosperity decline during the mid-20th century. Maki-machi was first thought to be considered for a nuclear plant by Tōhoku Electric Power Co when they bought a stretch of coastal land back in the mid-1960s, with the stated – and, one assumes, deliberately misleading – intention of developing it into a resort. The utility confirmed plans for a nuclear plant in 1971 but did not submit their official application until 1982. A year later, they learned that ownership of several parts of the proposed build site were in dispute, with one stretch held by an anti-nuclear group, so mothballed the entire plan.

Fast forward to the early 1990s, when most land disputes were settled and ownership had passed to the town. After Tōhoku Electric modified their plan to exclude the area of land owned by the anti-nuclear group, Maki-machi's incumbent (and ostensibly anti-nuclear) mayor Kanji Sato flipped and called for the almost-for-gotten plant to be reborn to revitalise the area's lagging economy. Sato won re-election on a campaign of support for the proposed plant in 1994, despite his two anti-nuclear opponents receiving more combined votes – potentially influenced by the Chernobyl disaster several years earlier. Several prominent local citizens, including local sake brewery owner Takaaki Sasaguchi, were unsatisfied with the apparent contradiction in results on the matter, so they requested a referendum to clarify public opinion. Though they took great pains to ensure voting would be fair and unbiased, Sato rejected the idea, so Sasaguchi's group took the extremely unusual step of holding a monitored but unofficial "citizens' referendum" themselves in January 1995. Pro-nuclear voters boycotted the referendum, but the tally against the plant again exceeded Sato's winning campaign from the year before. Sato refused to acknowledge this and further efforts, culminating in a signature campaign to remove him as mayor days before a fire at the experimental Monju facility – which is detailed

---

17 This is the first cancellation, depending on at which point you consider the proposed plant to have been cancelled. Strictly speaking, if you take the 2003 Supreme Court decision as the moment of cancellation, then it was beaten by the Ashihama nuclear plant, first proposed in 1963 and cancelled in February 2000. See: "Japan cancels nuclear plant." BBC, 22nd February 2000: http://news.bbc.co.uk/1/hi/world/asia-pacific/652169.stm.

in chapter 5 – caused a national scandal. The ensuing outrage galvanised local misgivings about nuclear power, and the apparently impossible task of collecting the required number of signatures was achieved within two weeks.[ad]

Sato was replaced by none other than Takaaki Sasaguchi, who held another referendum to settle the matter. An unheard of 88% turnout voted by over 60% against the proposed plant.[ae] Sasaguchi held on to the land for the duration of his term; then, to ensure no future mayor could overrule the vote, he sold it in secret in September 1999 to a group of private citizens using a law that permitted small, unapproved private sales. This move sparked a national debate around whether a few thousand people should be able to disrupt national energy needs, and the case eventually reached the Supreme Court in December 2003. Tōhoku Electric, remarkably, lost and abandoned their 40-year-old plans on Christmas Eve, though they retained ownership of most of the relevant land.

Two weeks earlier, the Kansai, Chubu and Hokuriku Electric power companies announced that they, too, were freezing joint plans for a nuclear plant at Suzu City in Ishikawa Prefecture. Having faced the trifecta of entrenched local resistance, stiff competition brought on by commercial electricity deregulation in 1995 and an economic downturn, the three utilities felt that "the notion that the construction of power plants is connected to [a] stable energy supply has changed."[af] Sasaguchi commented on his town's fight in the wake of the Fukushima disaster and, though Maki-machi's economy continued its downward spiral during the intervening years, he harboured no regrets. "We planted a seed of doubt about the nuclear industry's assurances of safety that resonated deeply," he said. "Our heart goes out to the Fukushima people. That could have been us if a plant had been built here."[ag]

The entire episode is notable as an outlier. "I think it can't be denied that a psychology favoring the safer path comes into play," said one former district court judge who ruled against Hokuriku Electric in 2006, speaking to the *New York Times*. "Judges are less likely to invite criticism by siding and erring with the government than by sympathizing and erring with a small group of experts."[ah]

* * *

Eight days after Fugen's commissioning in March 1979, the impeccable safety reputation of US civilian nuclear reactors ended with an event that frightened the world and perfectly illustrated how a string of small mechanical failures, compounded by human error, could cascade into a serious accident. During the early-morning hours of 28th March 1979 at the Three Mile Island plant in Pennsylvania, water from a previous effort to clean a stubborn blockage in one of eight condensate polishers seeped its way into an electric circuit.[18] This caused the condensate pumps to fail, then the polisher bypass valve – which would have allowed the continued flow of water – failed to open. The main coolant pumps tripped, completely halting the flow of water and prompting three backup water pumps to activate. The entire reactor had automatically shut down eight seconds after that first failure.

The safety systems functioned as intended thus far, but there was another problem: these backup pumps' valves were closed for maintenance (in violation of regulations, which require sequential work), thus preventing them from working. All these failures were anticipated in the plant's design and did not cause a reactor-destroying event by themselves, but then one final valve tipped the balance. The primary coolant loop's pilot-operated relief valve (PORV) opened for ten seconds to offset the rapid build-up of pressure caused by the lack of water in the secondary loop (again, as designed) but then failed to close. The overly complex control room interface indicated that the PORV was closed, when in fact the instruments only showed that a "close" signal had been sent, not that the valve had received and complied with it. Built-up steam tricked sensors into showing high levels of water in the system, confusing the plant's operators. To them, there appeared to be *too much* water, when water was actually pouring out. Overwhelmed by deafening alarms and experiencing sensory overload from their instruments, they ignored clear signs the reactor was losing coolant. Radioactive fission products from the damaged core mixed with water and then left the containment boundary via the PORV, soon flooding the auxiliary building where vented liquids were stored in relief tanks.

By 7:00 a.m., as coolant boiled away in the core, exposing the fuel, the plant declared an emergency. An hour later, around half the fuel had melted and slumped down to the pressure vessel's base, but in the

---

18 A condensate polisher is basically a water filter, used to remove impurities after steam from the turbine has condensed back into water – part of the secondary coolant circuit in a PWR.

end, few radioactive particles left the vicinity of the plant. Nobody died, but 140,000 nearby residents fled the area over the subsequent days, while operators struggled to keep the reactor under control. The Three Mile Island accident rated a 5 on the 7-point International Nuclear and Radiological Event Scale, making it tied for the fourth worst nuclear accident in history – roughly equivalent to the Windscale fire of 1957, the 1987 Brazilian Goiânia accident and an accident in 1952 at Canada's Chalk River Laboratories research facility, in which multiple hydrogen explosions badly damaged a reactor, leading to a radioactive release.[19] (Future US president Jimmy Carter, a former submariner, was part of the clean-up crew for the Chalk River event.)

The accident caused widespread apprehension among the surrounding population and alarm throughout the rest of the world, including Japan, where it made front-page news. Press coverage there was "predominantly sensationalist, concentrating on alleged catastrophic potential of the events at Three Mile Island, the negative environmental impact of nuclear power generation in general, and calling for a thorough-going reassessment of Japan's nuclear power program," according to a US State Department cable dated 5th April, which also noted that most coverage was "devoid of factual information / analysis."[ai] A post-TMI national survey by the *Asahi Shimbun* in June 1979 showed a small uptick in opposition to nuclear power from 23% to 29%, mirrored by a decrease in support from 55% to 50%, but this trend soon reversed and, by December, a follow-up *Asahi* poll placed supporters several percent higher than before the accident. Public interest in (and/or opposition to) nuclear power inevitably spiked during large-scale incidents like Three Mile Island, but the numbers often settled to reflect broad support in less tumultuous times. Still, surveys by *Jiji Press* indicated that public support reached a historic high of 54% in 1978, then dropped consistently until before bottoming out in 1989 at 19%.[aj]

Japan's Nuclear Safety Commission produced two reports in May and September 1979, listing dozens of failings in the United States and numerous recommendations for the Japanese industry to avoid a similar accident. Unfortunately, almost all these recommendations were ignored; Japan had just overhauled its entire regulatory

---

19 I covered the Goiânia accident quite extensively in my first book, *Chernobyl 01:23:40*. Thieves stole a radioactive component from an abandoned medical facility, then cracked it open and spread highly dangerous materials among themselves and around the city. The IAEA's 157-page report can be downloaded here: https://www-pub.iaea.org/mtcd/publications/pdf/pub815_web.pdf.

framework the previous year and was now focused on making that work. Critical safety issues relating to a plant's ability to withstand a design-basis accident (the theoretical worst-case scenario) were addressed, but the rest were forgotten, including fixes derived from studies into how multiple small faults could cascade into major accidents. It didn't help when the NSC's chairman announced that Japan could never experience its own Three Mile Island, despite his agency recognising a range of deficiencies equally applicable to identical Japanese facilities, prompting the public to rename his organisation the Nuclear Energy Safety Advertising Agency.[ak] He and much of the political establishment displayed classic traits of a phenomenon called "distancing through differencing," where people search for reasons why something couldn't happen to them. In focusing on the differences instead of the similarities, they over-looked many pertinent red flags.[al]

The official US government report on Three Mile Island offered a more pragmatic commentary on the world's nuclear industries and the relentless, unarguable portrayal of safety – one which Japanese utilities would have done well to consider. "The belief that nuclear power plants are sufficiently safe grew into a conviction," the report read. "One must recognize this to understand why many key steps that could have prevented the accident at Three Mile Island were not taken. The Commission is convinced that this attitude must be changed to one that says nuclear power is by its very nature potentially dangerous, and therefore one must continually question whether the safeguards already in place are sufficient to prevent major accidents."[am]

iii. The Labourers' Plight

The Japanese government joined other industrialised nations in de-regulating after the twin oil shocks, spurring a second round of massive growth. By 1980, electricity generating capacity reached 140 gigawatts – behind only the US and Soviet Union. Between 1980 and 1989, construction work began on twenty-one reactors while sixteen began commercial operation, four of these at brand-new plants (five, counting Monju). TEPCO now served 31% of the population, 35% of its industry and had become the world's largest privately owned power company.[an]

Japanese nuclear power technology surpassed that of their benefactor – the Americans – during this period, accounting for 17% of all Japanese energy use at the start of the decade, with a planned increase to 30% by the year 2000.[ao, ap] Oil was in decline but still provided 46% of electricity from a huge capacity of 70%, but nuclear would briefly surpass it for the first time in 1986, with oil dropping to a low in 1987 before retaking the lead. Of the remaining slices of pie, hydro matched nuclear at 17%, and liquid natural gas (LNG) crept up with a 15% share.[20] Coal, the source of almost all electricity only 40 years earlier, plunged to a mere 5%, virtually all of it imported. Government research and development spending on nuclear technologies now accounted for 18% of the science and technology budget, and annual subsidies for nuclear energy had grown from ¥250 million in 1954 to ¥1.6 trillion by the early 1980s ($14.6 million to $19.4 billion respectively in 2020).[aq] The number of individual subsidies eventually reached 28 by 2003 before being reigned in, but for now included – and I'm not kidding – a subsidy-management subsidy.

Though these subsidies meant nuclear power created vast communal wealth wherever it went, a MITI official conceded in late 1980 that there now existed "a lot of opposition by local residents at the sites where we plan to construct nuclear power generation stations."[ar] An STA report in December acknowledged the problem, conceding that public fear about inadequate safety precautions had become the biggest obstacle to expansion. Plants were still being built but lead times increased as local opposition grew more organised, despite the ever-increasing sums brought to bear on convincing (or coercing, depending on your point of view) rural communities to embrace them. During the 1970s, average lead times – from first public proposal to finished facility – were around seven to ten years, but by the 1990s this had risen to over twenty-five years.[21]

Public perception of the industry soured after newspapers

---

20 TEPCO was the first Japanese utility to import LNG, from Alaska in 1969, to power the world's first exclusively LNG plant in Yokohama. This was mainly done to combat the environmental pollution concerns at the time, because LNG contains almost no sulphur or nitrogen oxides, which were making people sick. This pioneering and controversial use of the pollution-free fuel came at the sole discretion of TEPCO president Kazutaka Kikawada – another former protégé of Matsunaga – against a cacophony of criticism because he stood among the few large business executives who did not consider emissions irrelevant. TEPCO eventually became the world's largest importer of LNG in 1980.

21 The power companies deliberately kept their plans secret to a certain extent in the early stages to prevent landowners inflating their prices and other similar reasons.

published a string of scandals in late April 1981 about accident cover-ups at the Tsuruga plant. The first revealed that 56 labourers had been exposed to radiation starting 8th March after 40 tons of radioactive sludge leaked from a waste processing plant when someone forgot to close a tank valve.[22, as] They worked around the clock with "buckets and rags" for over a month (until 15th April) to prevent it leaking from the building, but 16 tons flowed down drains and into the nearby Wakasa Bay.[at] Routine seaweed testing by the Fukui Prefectural Institute of Public Health uncovered the leak, leading to fish caught in the area being withdrawn from sale in a scene reminiscent of the *Lucky Dragon* incident 25 years earlier.[au] Local fishing cooperatives faced a boycott of their catch and were awarded damages as a result.

A similar accident from late January came to light mere days later.[av] (Even though this incident was made public second, the 8th March accident was discovered by government inspectors while looking into this one.) On this occasion, between 29 and 45 labourers were contaminated with radiation and many more were exposed during clean-up of another waste spillage, which had leaked out of a "drain from a feed heater" while undergoing repairs for a cracked pipe.[23] This was Japan's worst nuclear accident at the time. The *Fukui Shimbun* reported that utility subcontractors had deliberately recruited poor, unemployed and even homeless men from common lodging houses to act as "suicide squads" to clean up the mess.[aw] "We knew it was dangerous to their health but figured they would not be used twice," said a spokesperson for one of the two subcontractor companies involved. "We usually picked these kamikaze workers from Osaka and housed them in inns near the power plant."[ax] It was typical for labourers to lodge in hotels and taverns local to Tsuruga (or any other nuclear plant) for the duration of their contract, with the cost subtracted from their wages.

Days after the last story broke, another bombshell: Japan Atomic Power Co (JAPC), which operated Tsuruga, admitted the plant had been leaking radioactive water into the local town sewer and out into the bay for *four years*.[ay, az] As bad as this all sounds, the individual

---

22 I've seen as many as 138 reported, but 56 is the most common number.

23 The number was definitely at least 29, but the utility thinks the higher number may have been a mistake. As a company spokesman told the *New York Times*, "Our record shows up to 45 workers were exposed to radiation but there may be double counting as any one of the workers could have been exposed twice." https://www.nytimes.com/1981/04/26/world/45-japanese-workers-are-reported-exposed-to-nuclear-radiation.html.

doses received in each incident were insignificant. Nobody required medical treatment or was exposed to more than one millisievert – far beneath the government's safe daily limit for nuclear industry personnel.[24] Also, though Tsuruga released radioactivity into the environment, it was a mercifully small amount. This didn't stop the newspapers misleading people by reporting that cobalt-60 levels in the bay were 5,000 times higher than normal. Cobalt-60, used in radiotherapy machines and radiation sterilisation devices, is indeed quite a harmful isotope. But, being a synthetic isotope made inside nuclear reactors, it is not present in nature, and 5,000 times higher than basically nothing is still a minute amount.[25]

Still, people were outraged that JAPC hid these accidents from view despite a potential risk to public and environmental health. "There is no room for excuse," conceded JAPC president Shunichi Suzuki, days before resigning. "I deeply apologise that these accidents and cover-ups took place at a time when the government puts emphasis on power generation by nuclear energy and undermined the people's trust in nuclear power."[ba] Tsuruga's manager, Yonesuke Iwagoshi, also lost his job, but reports of up to 21 other undisclosed incidents at the plant soon forced the government to step in. On 19th May, in the first instance of its kind, MITI ordered a six-month cessation of operations at Tsuruga for failing to report the accidents. Local residents encouraged the Tsuruga city public prosecutor to charge JAPC with criminal negligence but were unsuccessful because the law did not define "excessive" contamination.[bb] MITI also opted against further repercussions once the six months were up, setting a bad precedent that would arguably invite many of the mishaps ahead. Recurring public surveys by the Prime Minister's Office showed a drop from just over 50% to 24% in the belief that "appropriate safety measures are adopted at the nuclear power stations."[bc]

The JAEC commented in early 1982 that "public trust in the safety of nuclear power stations must be won by taking all conceivable measures for their safe operating and by convincing the inhabitants … through accident-free records."[bd] The commission stressed that the incidents "show that the managers of these nuclear

---

24 The maximum amount received by an individual was reported as 92 millirems. Source: https://www.upi.com/Archives/1981/04/27/More-exposed-to-radiation-in-nuclear-accident/4310357192000/.

25 To be clear, I am not suggesting that the lack of serious consequences made any of this remotely acceptable – it did not. The situation just was not as desperate as the reporting implied.

power stations must take extra care to win public trust by keeping the public informed of any and all developments that have a bearing on the safety of their operation." Despite the series of failings, an *Asahi Shimbun* survey in December 1981 indicated that national opinion of nuclear power remained the same, with only a 1% increase in opposition. Nevertheless, elected officials in Japan's more remote regions came under mounting pressure from their constituents to reject proposed nuclear facilities in their areas, lest they be voted out. To counter this pressure and incentivise bureaucrats to toe the line, the central government created an annual program called the Dengen ricchi sokushin kōrōsha hyōshō, or "Citation Ceremony for Electric Power Sources Siting Promoters," to reward compliant officials with free publicity and a meeting with the prime minister.[be]

The string of accidents further highlighted a dark underbelly of the Japanese nuclear industry which first came to the public's attention in 1979. An undercover investigative journalist named Kunio Horie had released a book on his experiences at several plants posing as a day labourer – or, as he termed them, "nuclear gypsies" who rove constantly from plant to plant. He was hired within minutes in December 1978 and shipped off to Fukushima Daiichi. When he later fell and broke a rib, his recruiters drove him to hospital in a van to avoid alerting TEPCO to the accident, even offering to pay his expenses in exchange for silence. These workers were all exploited by a long-standing labour system, the modern iteration of which stems from the military-industrial complex of World War II, but its origins lie hundreds of years ago.[bf]

* * *

Power stations have their own core staff of highly educated and well-trained operators, engineers and other specialists, but maintenance and basic manual labour – the bulk of day-to-day work, including almost all work done around radioactive materials – is outsourced in a pyramid structure. When a power station utility like TEPCO needs something done, it usually gives the job to one of the five prime contractors (*moto-uke*) mentioned in chapter 2: massive conglomerates such as Hitachi, Toshiba and Mitsubishi that design, build and maintain reactors. This prime contractor doesn't physically build whole power stations by themselves, they subcontract much of the work out to one or more of Japan's major subcontractors (*shita-uke*), such as construction firms like the Kajima Corporation. Kajima, in turn, requires a constantly fluctuating number of labourers as old projects end and new ones begin, so they employ one or more sub-subcontractors – *mago-uke*, sometimes called subordinate subcontractors in English – to procure them. Each of these *mago-uke*, depending on project size, often then employs more sub-subcontractors, each smaller than the last, which, once you reach the bottom rung of the ladder, employ individual labour brokers (*tehaishi/ ninpu-dashi*). These brokers venture out in vans or minibuses to a given city's *yoseba* – an unofficial day-labour auction site, such as a city park, and the surrounding slums crammed with legions of unskilled workers. There, along with competing brokers from other companies seeking labour, they negotiate a day's work and depart with their cattle. Each company takes a percentage of the money assigned by the electric utility for a given labourer, leaving the person doing the actual work with as little as one-quarter of the full amount, handed over with an uncaring disregard for their well-being.

Every major Japanese city has at least one *yoseba* employment hotspot, often in run-down areas such as Kamagasaki in Osaka or Yokohama's Kotobuki district. One of the oldest and most famous covered an entire district of eastern Tokyo's Taitō ward, called San'ya. San'ya became an assembly point for men seeking temporary employment after World War II, and during the sixties, seventies and eighties was regarded as a slum, holding the country's largest concentration of day labourers, but it is much older than that. Established in 1848 during the Shogunate period, the area was home to outcasts, criminals and the *Burakumin* ("village people"): the much-maligned and ostracised lowest caste of Japanese society and the victims of discrimination for hundreds of years.[26] They are often branded with the epithet *eta* – "filth." Japan officially abolished the feudal caste system in 1871 but, despite government reforms during the 1970s designed to prevent third parties from searching for *Buraku* ancestry among potential marriage partners, for example,

---

26 The origins of the word *yoseba* lie in late-18th-century Japan. High unemployment in rural areas drove peasants into the cities, which could not cope with them. The first *yoseba* was a non-criminal homeless labour camp on an island just outside Tokyo, but it gradually became a dumping ground for unwanted citizens and criminals alike.

the United Nations estimates that between one and three million still live in disadvantaged communities such as the *yoseba* districts.[bg] Not all subcontracted day labourers are from recognised ethnic minorities like the *Burakumin*, but many are.

Problems with poor mental and physical health in these slums were typically far worse than the national average, with each man confined to the space on his one-by-two-metre *tatami* mat in crowded, claustrophobic dormitories. Ordinary citizens at the time viewed San'ya with slightly scared contempt and its residents in a manner not dissimilar to how immigrants or racial minorities can sometimes be seen in other countries. One doctor who worked in a nearby blood bank recalled, "Whenever [San'ya's workers] get desperate for money, they come in and sell their blood. Most of them do it far too often. You know there are weight requirements for blood donors. Well, we'd be putting these guys on a scale and the next thing we know we'd find they'd put rocks in their pockets to get them up to standard weight."[bh] Officials wiped San'ya's blighted name off city maps during the mid-1970s as part of a local government beautification effort to clean up the district's image, but many of the workers remain to this day.[27]

Back in 1973, the year of the first oil crisis, temporary and day labourers comprised 17% of Japan's workforce, but two years later a sharp increase in unemployment saw this figure decline to 14.4%. Men aged 45 and over, many of them ex-coal miners who could not compete with their younger peers, formed most of the drop.[bi] Nuclear power, as one of the few sectors still enjoying rapid growth, presented those older and more desperate workers with an employment opportunity as subcontracted labourers.

Maintaining nuclear plants was where the real health concerns lay, as, of course, not all subcontractors dealt with hazardous material. Security guards, for instance, were subcontracted, but unlike business sectors that lack the need for such things as legal radiation exposure limits, the ordinary supply of itinerant workers can run low at nuclear plants each time the reactors go down for their annual maintenance. During these periods, labourers go *inside* the clammy reactors and other hot zones to scrub them clean, absorbing large doses in a short space of time. The brokers, therefore, are forced to branch out onto

---

27 I have seen several different dates for when the name San'ya disappeared from official records and maps, ranging from 1966 to 1978. Almost always the dates are somewhere around the mid-70s, so that is what I've gone with.

the streets, into homeless hostels or wherever else potential labourers might be found. Recruiting the mentally handicapped and young people below the age of 18 was not unheard of but, as numbers dwindled, brokers travelled further and further afield.[bj, bk] The *Fukui Shimbun* quoted subcontracted workers during the Tsuruga incidents claiming some had travelled from as far as Osaka, 140 kilometres away. Even then, the flow of bodies can run dry.

In steps the *yakuza* – the Japanese mafia. It is well known that yakuza syndicates operate some of these subcontracting companies – an estimated 2,500-plus construction businesses were affiliated with yakuza gangs in the mid-1970s – and are funded by the wages they skim off workers they contract out to nuclear plants. The Yamaguchi Syndicate, which boasted around 40,000 members and operated several branches in Tsuruga during this time, were not above intimidating people into working around radiation.

At night, after an exhausting day's work, labourers spent their evenings gambling in "illegal, but tolerated" dens, fuelled by alcohol and drugs that were often deliberately supplied by the very same yakuza who recruited them each morning, ensuring they remained shackled to that lifestyle.[bl] Oddly, many yakuza themselves are also uneducated *Burakumin* outcasts, with most of the rest comprised of Korean-Japanese immigrants.

In total, subcontracted workers received over fifteen times the dose of utility employees in 1981, the year of the Tsuruga spills, and subcontractors outnumbered regular employees by almost ten to one.[28] Inspections of the most contaminated areas meant people could only enter for minutes or even seconds before hitting their safe dose limit, hence the disproportionate number of subcontracted workers. At the end of each shift, each labourer read his or her radiation dose – as recorded on their pocket radiation counter, carried at all times – back to the shift foreman (often from the subcontractor that had hired them), who wrote it down in pencil. This system led to a recurring situation, as would soon happen

---

28 Based on data compiled by historian, political scientist and *Asia-Pacific Journal* editor Yuki Tanaka, see below. The data, published at the direction of the Agency for Natural Resources and Energy (ANRE), indicates that in 1970 subcontracted workers were exposed to an almost equal dose of radiation as regular company employees (236 vs 326 rems in total for all subcontractors, rems being a now-obsolete unit for measuring the effects of radiation on a person). By 1980, the total dose for all company employees was 828 rems, while subcontractors received 11,105 rems, and the gap continued to widen. Having said that, the average dose per worker was higher for company employees prior to 1975 but dropped below that of subcontractors after this. Source: Tanaka, Yuki, "Nuclear power plant gypsies in high-tech society." *Bulletin of Concerned Asian Scholars*, vol. 18, no. 1, January–March 1986. http://criticalasianstudies.org/assets/files/bcas/v18n01.pdf

at Chernobyl, where workers were incentivised to lie about their radiation dose and sometimes even altered the recorded logbook readings after the fact.[29] Anyone reaching their legal limit was forced to leave for at least a few days, if not longer, thereby losing income. Conversely, if labourers were not working, then the subcontractor that hired them wasn't being paid either, so they were known to pressure their employees to lie. If workers gave a low number, on the other hand, then the men continued to work, everyone got paid and jobs were done sooner. The doses they reported were double-checked against a secondary wearable radiation monitor (a film badge with a thermoluminescent dosimeter) roughly once a month. Because this badge could not be altered, it was not unusual for workers to switch film badges with colleagues, or even not wear them at all in extreme cases.[bm]

Other monitoring systems and procedures were in place for subcontracted workers, but none were immune to wilful alteration in some form. Proper company employees of a utility or prime contractor were subject to more thorough and frequent checks. At Fukushima Daiichi until the mid-1980s, for example, "subcontracted workers used only hand-foot monitors and a simple small Geiger counter for the rest of the body," according to historian Yuki Tanaka, while "employees of [TEPCO] used the 'gate monitor,' which checks the entire body surface at once."[bn] Subs were only checked with a full-body detector once every three months and also had a higher permitted safe dose than their utility counterparts. By law, the limit was set at 5,000 millirems (5 rems) per year at the time, but a utility's own employees were often restricted to 1,500 by company policy. (As noted earlier in a footnote, rems are a now-obsolete unit for measuring the effects of radiation.)

The health of these anonymous labourers, in other words, was not a high priority – quite the opposite, in fact. "If you voice concerns about the welfare of temporary workers at the plants," admitted a TEPCO executive decades later in an interview with the *Atlantic*, "you're labelled a troublemaker, or a potential liability. It's a taboo to even discuss it."[bo] Industry regulators did not attempt to verify the accuracy of labourers' doses, which were provided by the subcontractor, nor did the utility.[bp] "Guys needed the work,

---

29 Photographer Kenji Higuchi has published images where the original written dose was clearly erased and a new, lower number was written over it: http://klikdotsystems.jp/stronapolska/files/rok2015/KenjiHiguchi.html.

so they cut corners," said one 56-year-old subcontractor in a 2011 interview with the *Los Angeles Times*. "The plant bosses knew it but looked the other way."[bq] In March 1977, the *Asahi Shimbun* published the results of research by Diet Lower House member Yanosuke Narazaki on subcontractor deaths at Japanese nuclear facilities over the previous eleven years.[30] The statistics included 15 deaths from falls and 16 "other" non-radiation related deaths. Of the suspected "deaths resulting from irradiation," 32 were from cancer, 23 from brain damage, 12 from heart disease, 3 from leukaemia and 5 from "other." But poor record keeping of radiation doses and a lack of tangible proof or funds for legal support mean that, historically, almost no unwell former labourers have ever won legal cases against their former employers.

Sick of the mistreatment, workers at nuclear plants in Fukui Prefecture formed their own union called Gempatsu Bunkai in July 1981 which, while pro-nuclear, demanded safer and fairer working conditions.[br] After some early resistance from JAPC and the subcontractors themselves, matters did improve, but Gempatsu Bunkai remained a fringe organisation throughout its existence.[31] Readings began to be recorded in pen, workers gained access to their own records and their average exposure to radiation declined significantly over the decade, but the organised exploitation of weak and vulnerable labourers did not. Nor did the callous attitudes of government officials. "There is work that exposes people to radiation that has to be done so long as you want to sustain the current energy supply," one unnamed Labour Ministry representative told the *Los Angeles Times* after a worker died from acute radiation exposure in 1999. "They say it's discrimination, but there is freedom of work in our country, and if people don't want to do these jobs they can quit. If nobody wants to do the work, eventually the industry will have to be shut down."[bs] Dosimeters were digital by that time and hence could not be altered, so workers simply switched them off once they reached their limit. "We were sweeping up dust and had bleepers which went off when the radiation levels were too high, but the

---

30 Narazaki held a reputation for uncovering high-level misdeeds, the most famous probably being the Lockheed scandal, where Prime Minister Kakuei Tanaka personally received $3 million for his involvement in All Nippon Airways' purchase of Lockheed Tristar aircraft, mentioned earlier in the chapter. Source: *Asahi Shimbun*, 17th March 1977.

31 "Resistance" is something of an understatement. JAPC ignored the union's demands for a long time and ordered its subcontractors to destroy it, resulting in threats from yakuza members and other unpleasantries. Only once Gempatsu Bunkai had expanded and gained national sympathy were their demands listened to.

supervisors told us not to worry, even though they were bleeping," one homeless man told the *BBC* in the 1990s. "I got out when I started to feel ill."[bt] JAPC refused to compensate him, claiming he couldn't prove his illness arose from his work.

* * *

The yakuza played one other role in the nuclear industry, at the earliest stages of a plant's life. "Nuclear plants-in-planning," writes J. Mark Ramseyer, Harvard University's Mitsubishi Professor of Japanese Legal Studies, "invite extortion and protection."[bu] Utilities invest billions of dollars, decades of time and untold man-hours on planning, research, acquisitions, political manoeuvring, legal challenges, construction and testing on any given power station: investments which, overwhelmingly, cannot be transferred to another site. Situations like this are unmissable goldmines for organised crime, which can either threaten to fan the flames of fear among those able and motivated to mount an effective opposition, thus scuttling the utilities' plans, or offer to smother any resistance by greasing the right palms. If that proves ineffective, they resort to threatening – and inflicting – violence. "Why did we make so much money when a nuclear plant came to town?" asked one yakuza of an undercover journalist. "It's obvious. We made money because we could make anyone who threatened to complain shut up."[bv] The same principle applies to any mega-sized construction project, like an airport. The cost of dishonesty is far cheaper than decades-long court fights, so they lavish the local mayor or prefectural governor with gifts and pay for thugs, through innocuous shell companies, to crack down on any dissent. When this specific practice began is unclear, but it stretches at least as far back as the early 1980s yet flew under the radar of all but the most observant until 2011.

* * *

October 1981 brought another shift in American nuclear policy with the election of President Ronald Reagan, who lifted the ban on nuclear reprocessing and directed his government to "proceed with the demonstration of breeder reactor technology." [bw] The following June, Reagan approved a foreign policy change that gave consent for "certain reprocessing, retransfers for reprocessing and use of US-controlled material."[bx] These changes included removing limits on "reprocessing at the Tōkai-mura plant for its operative life [and the] use of plutonium in Japan's breeder, advanced reactor and thermal recycle programs." Exports of "US sensitive reprocessing technology and equipment for facilities in Japan" would commence, providing a boost for Japan's long-term nuclear goals. Things were getting back on track.

iv. Chernobyl

On 28th April 1986, the world learned of shocking news: a powerful 1,000-MW reactor at the Soviet Union's Chernobyl Nuclear Power Station in northern Ukraine had exploded two days earlier, disgorging tons of highly radioactive debris across a vast area. The world's most famous industrial disaster manifested people's worst fears about nuclear energy as radiation hysteria spread across Eastern Europe, fanned by inaccurate newspaper reports of "thousands" dead. Authorities declared a 2,600-square-kilometre region around the plant a no-go zone of immense radioactivity, while over 600,000 soldiers and draftees spent years decontaminating the area as part of the largest peacetime mobilisation in history. The accident was (and, in many ways, remains) unprecedented.[32]

Surprisingly, some initial Japanese public surveys indicated a general indifference toward Chernobyl, though a national opinion poll by the *Asahi Shimbun* four months later showed opposition to nuclear power surpassing support for the first time, with far greater opposition among women than men (a common trend all over the world). Editorials in Japan's biggest newspapers, such as the *Yomiuri*, distanced the domestic industry from that of the Soviets, claiming that their technology and safety practices were superior and such an accident could not occur in Japan or the West. One month later, the newspaper sought to clamp down on anti-nuclear dissidents, declaring that Japan needed to "ensure nuclear safety rather than to

---

32 None came close to Chernobyl, but the Soviet Union actually had countless accidents resulting in radiation release, particularly in the 1950s. The Mayak nuclear weapons facility, for example, released radiation over a large area after a waste storage tank exploded. What appears to be a comprehensive list of radiation incidents is shown here: http://www.johnstonsarchive.net/nuclear/radevents/radaccidents.html.

give up nuclear power through … fear."[by] Other papers like the *Asahi* took a similar approach, stating the country should look closely at what had happened and enhance existing safety policies, but that was about as critical as the mainstream media went.[bz]

The JAEC issued a white paper in late 1986, expressing their belief that Japan could act as the global standard for exemplary nuclear safety efforts.[ca] The press lapped it up, writing prideful prose like, "given that Japan is the only country that was affected by atomic bombs, Japanese nationals have kept a strict and critical watch over its safety management of nuclear power."[cb] This fallacy of "we were victims of atomic bombs, therefore we have the best nuclear safety practices" appeared in a range of media outlets (though, oddly, not the *Yomiuri*). As concerns regarding Chernobyl's continued global impact grew, the press sought to downplay fears that a similar incident might occur at home.[33] Their efforts weren't entirely successful, as the number of protests grew from 12 in 1987 to 83 in 1988 during a brief union of anti-nuclear weapons and anti-nuclear power activists, with one Tokyo protest numbering almost 20,000 people.[cc]

The Nuclear Safety Commission spent almost a year studying the accident. Their final report commented on the stark differences between Chernobyl's RBMK-type reactor and Japan's own reactors while echoing comments from an August 1986 MITI report that existing personnel training and accident-prevention countermeasures were sufficient.[cd] If any improvements were to be made, the NSC argued, they were in the areas of human error prevention, control room user-interfaces, emergency procedures manuals and accident clean-up safety. The commission stressed the importance of safety training for reactor operators but considered Japan's reactor technology itself to be flawless. That being said, it is important to note that this report appeared long before the revelation that Chernobyl's reactors possessed serious design flaws and that the disaster was not caused by a deliberate violation of operating procedures alone, as was believed at the time.

The NSC also touched upon the topic of mass evacuations, as had happened at Chernobyl and would eventually happen in Fukushima. Their report confidently stated:

---

33 The press's respective approaches differed, though the outcome was the same. Some called for more dialogue with the government, some dismissed fears as irrational and some tried to explain away those fears.

It is extremely difficult to think of a situation similar to the accident at the Chernobyl nuclear power plant [happening] in Japan ... and it is unlikely that it would lead to such a situation where a wide range of inhabitants must evacuate.... Also, even if it becomes necessary to take measures against disasters, such as evacuation, it is believed that there will be time to spare. Therefore, it is considered possible to carry out withdrawal and evacuation measures without social confusion.[ce]

Still, they acknowledged the wisdom of conducting further research into how people might react under such circumstances and what could be done to ensure a smooth evacuation. Japan's prime minister during this time, incidentally, was none other than Yasuhiro Nakasone, who led the country from 1982 to 1987 – an unusually long time for a Japanese premier.

* * *

On 9th June 1986, Harold Denton, director of the US Nuclear Regulatory Commission's Office of Nuclear Reactor Regulation, raised fresh concerns about BWR reactors like those at Fukushima Daiichi, telling utility officials "There has been a lot of work done on those containments, but Mark I containments, especially being smaller with lower design pressure – and in spite of the suppression pool – ... you will find something like a 90% probability of that containment failing [in a serious accident]."[cf] He reaffirmed his stance on the Mark I during a government hearing after the Chernobyl disaster, adding that "they are all designed for the design-basis event, but they vary with regard to their capability to withstand the core meltdown."[cg] In 1989, GE received NRC approval to begin installing a new hardened venting system to supplement the existing reactor venting, which could not function under the extreme pressures expected during a theoretical emergency. The new vents, installed on Daiichi's reactors between 1999 and 2001, could bypass the normal route's particle scrubbers and direct radioactive steam and gases straight into the atmosphere to prevent an explosion. The suite of upgrades applied to the Mark I since its inception made it arguably safer than many PWRs, even in a design-basis accident, but its ability to withstand an extreme scenario remained in doubt.

Japan published new, more thorough earthquake-specific guidelines for nuclear plants in 1987, which utilities had to apply to all existing plants.[ch] They extended the window in the search for historic tectonic fault activity to 50,000 years and recommended applying modern theoretical prediction models to better anticipate tremors for each site, with less shared generic data and more random outliers thrown into the calculations.

Tsunami received a cursory mention. The word "earthquake" appears 1,675 times, "tsunami" a mere 25, several of which are in references and the index. Actual tsunami content occupied little more than a page in the almost 1,000-page document, and, though it stressed that "when the site is selected and designed, the influence of a tsunami must be fully taken into consideration," the accompanying information amounted to little more than two mathematical formulae for calculating tsunami scale and velocity.[ci] The document contained nothing about potential measures to protect against water damage, demonstrating how little consideration was given to secondary hazards arising from an earthquake. Most warnings from outside the nuclear industry were dismissed as obstructionism, causing extensive debate on the extent to which it had ignored advice from the academic community following the Fukushima disaster.

On 8th December 1988, Unit 3 at Fukushima Daini shut down automatically after a strange fluctuation in one of its recirculating water pumps caused an abrupt power spike. The unit resumed service but began exhibiting signs of faint vibration three weeks later, on 1st January; operators reduced power by 3% to just below the warning level. On the morning of 6th January, several bolts, washers and a 100-kilogram ring attached to the pump bearing dislocated, shattering the vanes and sending hundreds of metal fragments – ranging from powder size to pieces up to 10.5 centimetres long – flying off into the core, with some getting as far as the turbines.[cj, ck] Unaware of this accident but apparently unconcerned by the ensuing abnormal vibration alarm, the operators slowed the pump but continued on for another 14 hours before shutting down the reactor. After months of recovery and investigation work and around 30 kilograms of metal recovered from the pressure vessel, TEPCO announced in October 1989 that they needed to replace all

770 fuel rods. The reactor remained offline until 5th October 1990, while the government revised safety regulations to require a full stop upon any pump vibration alarm.[cl]

While Daini was offline, Fukushima Daiichi closed out the decade with the commissioning of its sixth and final reactor in October 1989, a 1,100-megawatt Series 5 BWR with Mark II containment. (Unit 1 is Series 3, Units 2–5 are Series 4, each new version has certain differences which will be covered later.) While Units 1 to 4 lay in Ōkuma's district, 5 and 6 were in neighbouring Futaba, whose town council members were so insistent on receiving government subsidies that Unit 5 began operation six months before Unit 4. TEPCO opted to build Units 5 and 6 on earth three metres (ten feet) higher than the ten-metre height of Units 1 to 4, in a sign that they were potentially now aware of the tsunami risk. They similarly built all of Fukushima Daini's reactors 12 metres above sea level.

As the 1980s came to an end, Japanese research and development spending on energy – and especially nuclear energy – outstripped that of every other country, equivalent to around 1% of gross national product, having risen by over 4% each year.[cm] Improvements made during the 1980s meant the average radiation exposure of workers halved while the number of accidents, failures and automatic shutdowns also declined, until up-time of Japanese reactors surpassed those of the United Kingdom, the United States, West Germany and France.[cn] Japanese utilities, meanwhile, came to believe – perhaps correctly, with an average of 0.4 unplanned reactor shutdowns nationwide in 1988, compared to 4.0 in the US and 5.3 in France – that their streamlined preventative maintenance programs and safety standards were the world's best.[co] The future seemed bright.

CHAPTER 5
The Lost Decade: 1990s

i. Downturn

By 1990, 38 nuclear reactors provided 27% of Japanese electricity, with a capacity surpassed only by France and the United States. The economy continued to soar following its recovery from the oil shock and a new target share of 40% seemed achievable, but in late 1991 the bubble burst, the stock market crashed, unemployment rose, wages fell, and the economy collapsed to the cumulative tune of $1 trillion off the national GDP by the year 2000. The "lost decade," as it came to be known, introduced fifteen new reactors at eight plants (not including Monju), one of which was a brand-new facility, though almost all of them began construction during the 1980s. Despite an ever-increasing demand for electricity, the recession saw numerous projects cancelled, meaning work began on just five new reactors during the 1990s – a significant downturn from 19 new builds in the 1980s and 21 during the 1970s. Japan's original commercial reactor, the British Magnox at Tōkai, reached its natural end of life and shut down for the final time in March 1998. It had generated around 29 billion kilowatt-hours of electricity during 277,752 hours of operation. While the 1990s were quiet in terms of notable headway, the millennium's final decade gave birth to truly organised nationwide opposition to nuclear power, fuelled by a string of high-profile accidents and cover-ups. This movement nevertheless remained largely ineffective at influencing government policy until the beginning of the next decade.

* * *

In early 1991, Japan experienced its first-ever use of a reactor's Emergency Core Cooling System (ECCS), the principal emergency safety mechanism, when Unit 2 at Mihama sprung a leak in one of its two steam generators. Inside the steam generator on this model of PWR, 3,260 water pipes, each shaped like a lowercase "n" with an inside diameter of 20 millimetres and an outer wall thickness of 1.27 millimetres, carry hot reactor water from the primary coolant circuit. Heat exchange occurs when cooler water from the secondary circuit, with a water pressure one-third that of the primary circuit, boils and turns to steam as it flows around the thin-walled pipes.[1] This steam then spins a series of turbines to generate electricity, as in a BWR. An anti-vibration bar (AVB) used to secure the web of pipes was misaligned in one instance, allowing a single pipe to flex and rub against the AVB as water rushed through and around it, weakening the metal. A routine inspection in 1990 had failed to notice the misplacement.

At 12:40 p.m. on 9th February, the pipe flexed and tore a small hole, providing the high-pressure radioactive water an avenue to escape into the lower-pressure secondary circuit.[2,a] The leaking water then triggered four distinct control room radiation alarms over the next 80 minutes. A sampling analysis taken at 1:20 p.m. indicated a rise in radioactivity inside the steam generator, but the operators – who had not been trained for an event that was considered impossible in Japan – took no action while they debated whether to take the drastic step of shutting down the reactor, or if the readings might be incorrect.[b] "We never thought that the radiation could increase so quickly because such a thing hadn't happened before," remarked one safety official quoted by the *Associated Press*. "It is so inconvenient to stop the whole thing if the monitor was showing the increase because it was broken." As pressure and coolant levels fell, water began to boil in the core and radiation levels rose until, at 1:45 p.m., the pipe flexed one final time then snapped clean off. The operators, still confused, then finally activated a backup water pump to correct the imbalance and commenced a slow power reduction at 1:48 p.m.

---

1 To clarify, none of the water in this system is cold. It is all highly pressurised and very hot, but water outside the reactor is colder than inside. Primary loop pressure is 150 centimetres squared per gram, while secondary loop pressure is 50 cm²/g.

2 The rupture is often said to have occurred at around 1:40–1:50 p.m., which was the time given to reporters on the day of the accident. The correct time was provided the following day, but the original time often remained stuck in subsequent reporting. Note also that the timings in the US NRC report and some others do not match the JAEA report perfectly. I've attempted to reconstruct events by matching them up as best I can while using the JAEA as a basis.

Sensors detected a sharp drop in primary loop pressure two minutes later, subsequently "scramming" (emergency insertion of all control rods into) the reactor. This, in turn, caused a further sudden pressure drop that activated the ECCS's twin high-pressure coolant injection pumps after seven seconds. They began forcing water into the reactor compartment. Operators now realised the steam generator was the source of the problem and tried to isolate it, only to discover, to their mounting unease, that the main steam isolation valve – which is kept open during normal operation but closable in an emergency – was stuck open. Someone would have to locate and close the valve by hand.

The ECCS pumps quickly over-pressurised the primary loop and began struggling to function. The operators first tried to open two pressuriser relief valves at around 2:15 p.m., but neither moved – an inspector had disabled them by mistake the previous summer and none of the three people who reviewed his checklist had noticed the error – forcing them to attempt depressurisation using the auxiliary water spray system inside the pressuriser itself.[c, d] The situation was resolved by switching the ECCS *off* at 2:37 p.m., thereby allowing pressure to equalise between the primary and secondary loops, stopping the leak.[e] This bold but slightly risky move was not in the operating manual.[f] Though water surrounding the fuel temporarily boiled again upon deactivation of the ECCS, the reactor's trio of make-up pumps[3] could now insert water, and the incident was brought under control.[g]

Fission ceased and Unit 2 achieved a "cold shutdown," where the coolant temperature falls below 100°C, but not before 20 tons of primary coolant leaked into the secondary non-radioactive circuit. Officials from Kansai Electric Power Company (KEPCO) initially claimed no radioactivity had escaped and that the ECCS functioned as intended, saving the day, before revising and increasing their statistics several times over the following weeks. Though the amount released was insignificant overall, it struck another blow to public faith in nuclear safety, aggravated by the two-day delay in admitting radiation had escaped the site. The situation was most pronounced for residents of the nearby towns of Tsuruga and Mihama, who,

---

3   According to Mitsubishi Heavy Industries, a make-up pump (sometimes called a filling pump) "supplies make-up water to maintain the volume when the primary reactor coolant has contracted due to decrease of load. Also, it supplies sealing water to the sealed part of the primary reactor coolant pump." Source: https://www.mhi.com/products/energy/pumps_for_nuclear_power_plant.html.

despite their local government officials being informed after one hour, only learned of the accident from newspapers and TV reports rather than from KEPCO, who were required to keep the nearby public informed. Nevertheless, KEPCO took the unusual step of establishing the Institute of Nuclear Safety System (INSS) [sic] in March 1992 as a direct result of what happened. The INSS's admirable goal is to conduct "a wide range of investigations and researches to improve the safety and reliability of nuclear power generation, and share the results within Japan and abroad."[h] It continues to publish an annual journal of all research conclusions.

* * *

After Chernobyl, the Nuclear Safety Commission began a second, longer-term study into severe-accident countermeasures and how to cope with potentially disastrous events at nuclear plants, but the pace was sluggish. Both France and the United States – the Western Hemisphere's two largest nuclear consumers – refreshed their safety policies within five years of the disaster, many of which were written into law. Japan, by comparison, took until late May 1992 just to release a report. Though it declared there was a "low risk" of a severe accident in Japan, the report conceded that improvements were possible and prudent and suggested numerous changes to both the reactors and safety procedures.[4] First and foremost, the NSC said, "it is strongly encouraged that reactor [utilities] voluntarily develop effective accident management [preparations] and be able to properly implement these in the unlikely event" of a severe accident.[i] The report stated that accident countermeasures should be based on the latest research and made numerous excellent suggestions – one section reads like a laundry list of system failures that would occur at Fukushima 20 years later – but it was all undermined by one critical word: "encouraged." The NSC still lacked legal authority, allowing utilities like TEPCO to take their sweet time preparing for extreme circumstances. All according to

---

4   I think most reasonable people would accept that there was a low risk on the technological side of things. The human element is another matter: that same low risk caused utilities to not take the more severe risks seriously. An element of risk is always involved, but until this point Japan had never suffered a "severe" accident, so the report's comment should not be dismissed out of hand. That said, I think it's fair to suggest that Japan's natural vulnerability to earthquakes and associated phenomena makes the country more at risk than is the case with nuclear facilities in many other countries.

plan: the utilities lobbied hard against such regulation, at least in part because admitting that safety measures could be improved meant acknowledging further risk.[j] The NSC justified this voluntary implementation as a way of maximising reactors' up-time while avoiding costly and time-wasting litigation.

Kobe University professor and seismologist Katsuhiko Ishibashi released a book in 1994 titled *An Era of Underground Convulsions: A Seismologist Warns* in which he dared to highlight deficiencies in Japanese buildings' outdated earthquake protections. His employer ordered Ishibashi to the Construction Ministry to apologise for the work, but just a few months later, on 17th January 1995, the magnitude 6.9 Great Hanshin Earthquake struck his home of Kobe, a city of 1.5 million people.[k] Though the quake was "only" a 6.9, it was the first to directly affect such a dense, modern city.[l] As such, the shaking caused extensive infrastructure damage – a long stretch of the city's overhead freeway tipped over on its side, obliterating the vehicles below – and almost 6,500 people lost their lives. In response, the government set up the Headquarters for Earthquake Research Promotion later that year, to "promote research of earthquakes in order to strengthen earthquake disaster prevention measures."[m] The organisation predicted a 99% chance of a magnitude 7.5 or higher earthquake striking the coast of Miyagi Prefecture, a region just 60 kilometres north of Fukushima, within the next 30 years.[n] Ishibashi's book became a bestseller.

ii. Monju

Most industrialised nations were now walking away from nuclear power; some had decided the cost of making reactors safe made them uneconomical even before Chernobyl. The United States, which started the nuclear power craze, commissioned three reactors in 1990, but as of 2020 only three more have been built, two of those in the 1990s. Prior to that, the US proudly built 127 reactors over four industrious decades.

Japan pushed on through the downturn, determined as ever to chase the dream of energy independence. Jōyō performed well as an experimental fast breeder reactor (FBR) and, since 1968, the Japan Atomic Energy Commission had thrown money at the next evolutionary step: Monju. Construction finally began in October

1985 at a site nestled in the north-east corner of the same small, hilly Fukui peninsula as Fugen and the Mihama and Tsuruga power stations.[5] It was an incredible feat of advanced engineering, designed to show that a full-scale FBR could work as a commercial power plant. Even so, critics at home and abroad increasingly questioned Monju's long-term rationale.

During the 1950s and '60s, the global supply of uranium was expected to last just a few decades, and the breeder program came about, in part, because of that concern. By the early 1990s, however, geologists had discovered plentiful reserves across the world, new reactor builds had slowed and uranium was cheap, therefore many Japanese taxpayers wanted to know why they were still funding such an expensive project. Debate on the matter regularly made the news as the public opposed not just Monju's extraordinary cost but also its use of plutonium – which is many times more expensive for creating electricity and infinitely more toxic – as fuel. Japanese officials did the breeder program no favours by demanding that North Korea close the nuclear weapons plutonium reprocessing plant at its Yongbyon facility, while simultaneously hailing the magnificence of their own plutonium prowess (the irony of which did not go unnoticed). The story held another awkward comparison: that the reactor at Yongbyon, which began operation in 1986, is a Magnox based on the earthquake-reinforced Tōkai plant. Some Asian neighbours even suspected the Japanese government of surreptitiously planning to build a nuclear weapon of their own.[6] "It is almost inconceivable that such a good idea could have turned this bad," one senior official told the *New York Times* in December 1992.° "We spent the last 20 years building this project," he continued, with ominous foresight, "and we'll probably spend the next 20 killing it."

The first of many planned shipments retrieving Japanese plutonium from Europe (France, in this case) neared the end of its two-month voyage at the time, to enormous public dissent and strained international relations, stoked by fears the ship might sink or be attacked by terrorists. Matters had not improved by August 1995, when the Massachusetts Institute of Technology published a report on global concerns surrounding Japan's plutonium program:

---

5 Preparation for construction began in May 1983, but actual construction started in 1985.

6 The Japanese government tried to assuage these concerns by becoming the first signatory of the IAEA's 1997 Model Additional Protocol, which allowed IAEA inspectors to widen their safeguards and conduct more thorough site inspections with shorter notice than before.

Circumstances have changed since the basic contours of Japanese reprocessing and breeder plans were first formulated more than 30 years ago. All the claimed advantages – security, political, and economic – that appeared initially to favor plutonium use have changed. Yet the rationales and main elements of the Japanese plutonium program have not changed.... Whatever arguments Japan has for proceeding toward a "plutonium economy" within Japan, many responsible observers believe it would be very dangerous if the world at large accepted the widespread commercial use of plutonium in nuclear power programs.[p]

The United Kingdom – one of the few countries to build a functional breeder prototype – had decommissioned its Dounreay facility the previous year, and both France's Phénix and Superphénix breeders encountered continual problems. The technology, it seemed, was on its way out for one reason or another. Regardless, Japan's long-in-development program approached another milestone, and nothing was going to stop it.

Nuclear energy generation in 1995 accounted for 34% of Japan's total (from a capacity of 21%), which was the highest it would ever go.[q] The rate of new reactor builds had slowed considerably since the 1970s, but it's interesting to note that 1995 was both the year nuclear power achieved dominance in Japan, overtaking oil and everything else, and also the year that arguably signalled its downfall. The year also marked another paradigm shift, albeit one that would not matter much for another 16 years: deregulation.

Much had changed in the 43 years since the country formed its vertically integrated regional energy monopolies. In addition to the ten major utilities wielding 76% of the national generating capacity, a further 14% came from 56 electricity wholesalers (Japan Atomic Power Co is a wholesaler, for example), which provided additional supply to the utilities but did not sell to the public. Most of the rest came from heavy industry's own private plants, but the entire system was plagued with high prices and inefficiencies. Why, for instance, were TEPCO and others spending obscene amounts on advertising when they had no competitors? Because, with a guaranteed profit, they had no incentives for higher productivity or lower prices.[r] "Only after the populist rise of the Hosokawa cabinet," argues social scientist Dr Paul Scalise, referring to the first time in its history that the Liberal Democratic Party lost power, "did MITI reluctantly agree to investigate reform, changing their emphasis from 'energy security' to 'energy efficiency.'"[8] Complaints from Japan's industrial sector, too, pressured legislators to usher in change by loosening the restrictions on who could enter the market for high-voltage customers and forcing utilities to set efficiency targets in advance. The Electricity Utility Industry Reform Bill passed on 14th April 1995. It did cause a small decline in rates, but prices in general remained high and few ordinary people took notice.

* * *

Monju, like Fugen before it, was operated by the Power Reactor and Nuclear Fuel Development Corporation (PNC) and was commissioned to industry fanfare on 18th May 1991. The facility was the culmination of decades of research and design, and eight years of construction work by Japan's heavy hitters Hitachi, Toshiba, Mitsubishi and Fuji Electric, plus a shoal of smaller companies. The construction bill alone reached a gigantic $6.3 billion.[7] A further three years of tests, months of loading fuel and several faulty false starts followed before the reactor's first criticality on 5th April 1994. Japan's first FBR-generated electricity followed, in August 1995. The dream was within reach, but Monju only ran for four months before disaster struck.

The reactor's unique design employed a highly unusual and complex liquid metal–cooled, triple loop system, each with primary and secondary heat transport and auxiliary cooling functions. Graphite was the original moderator of choice for breeder reactors but using the ultra-efficient coolant sodium allowed for highly enriched uranium and plutonium fuel (which can fission by themselves), removing the need for a moderator. At its most basic, Monju was similar to a PWR in that the sodium primary coolant circuit passed through the core, where it reached temperatures between 400°C and 530°C, then continued through a heat exchanger with the secondary loop, all within the main reactor containment structure.[1] The secondary, non-radioactive coolant loop left the containment, but instead of passing through a turbine (which wasn't

---

7   The cost is listed as ¥593.4 billion. The facility opened in 1991, so I've used the 1991 inflation start date to calculate this dollar value for 2020.

possible because this secondary circuit also used liquid sodium), it then passed through a heat exchanger connected to the third loop. This third loop contained water which boiled into steam and blasted through a turbine to generate 280 megawatts of electricity. The design was far, far more complicated than that, to the point where I can't explain it without the aid of a diagram, but that's the essential structure.

Workers spent much of 1994 and 1995 conducting power tests, gradually raising the output and checking their systems, confident they had overcome the difficulties experienced with other nations' sodium reactors. On the evening of 8th December 1995, they were again increasing the power for another round of tests and had reached 43% of thermal capacity when a control room alarm at 7:47 p.m. made everyone jump. A temperature sensor deep inside one of Monju's serpentine piping rooms flickered, then leapt off the scale where the secondary circuit exited the reactor containment. The room's smoke and sodium leak detectors tripped seconds later, moments before the temperature sensor succumbed to the extreme heat.

There was no radiation risk because the secondary loop didn't enter the reactor core, but a dearth of remote cameras forced a lone operator to rush to the room, which he found filled with smoke.[8] Nuclear reactors are not like hydroelectric power stations, where you can close the water intakes and stop everything within minutes – they take time to reduce power (unless there's a real emergency), to avoid sudden temperature changes that might damage their complicated components. As such, the shift manager ordered a normal shutdown at 8:00 p.m. Confusingly, this decision violated guidelines which required an emergency shutdown when fire alarms activate but followed the procedure for a small sodium leak.[u] A second visit to the piping room at 8:50 p.m., prompted by another fire alarm, revealed a space opaque with thick, white fumes.[v] The plant's volatile sodium coolant (which, under ordinary circumstances outside a reactor, must be isolated from its surroundings under oil) instantly and spectacularly reacted with oxygen as it escaped, burning with a

---

8 Such a leak sounds as if it would be urgent, but all other systems were operating normally, and the primary loop had backup systems to cool itself, plus an emergency shutdown "burdens" the system, which is why the reactor was not scrammed immediately. More information about what was happening is available at this link and the other links from this page: "Present status of cause investigation." Japan Atomic Energy Agency, http://www.jaea.go.jp/04/monju/category05/mj_accirep/mj_accirep05.html. Archived link: https://web.archive.org/web/20070222035737/http://www.jaea.go.jp/04/monju/category05/mj_accirep/mj_accirep05.html.

fierce orange glow. This reaction changed the molten sodium into a white sodium-oxide powder that billowed around before settling on the walls, floor and machinery, triggering 66 separate smoke detectors in the process. Upon realising they had a serious problem on their hands, the plant staff decided to forego normal power-down procedures and scrammed the reactor at 9:20 p.m. Secondary loop flow ceased and was drained, while the plant's backup systems provided reactor coolant.[9, w]

The compromised pipe leaked around 640 kilograms of sodium into the piping room, where it melted a nearby ventilation duct and access walkway.[10] The damage left one corner resembling the scene from *Aliens* where Ripley's colonial marines first enter the "Hadley's Hope" outpost to discover melted walkways and ruined equipment. Sodium pooled by the wall in a mass three metres wide and thirty centimetres high, where it burned as hot as 700°C (1,292°F), but the steel-lined floor was built to withstand situations like this. They were lucky the leak did not occur in the primary radioactive loop, with its higher temperature and flow speed, but apart from a large mess, the plant otherwise contained the accident undamaged.

Had PNC officials been honest about what happened, they may have avoided serious public scrutiny. Yes, a leak had occurred, but less than 1% of what the plant was designed to cope with, and leaks in all kinds of pressurised systems are common throughout the world. Even sodium leaks have happened at other fast breeder reactors. Nobody was hurt, the operators responded more or less correctly, and the facility sustained minimal damage – officials should have owned the problem, fixed it, identified any procedural flaws and moved on. Instead, they tried to cover it up and stumbled right into the "Streisand effect," wherein an attempt to bury information has the opposite result, creating another national scandal in the process.

They began by taking almost an hour to report the accident to local government officials, something they were required to do immediately.[11] An hour was far quicker than past delays, but not quick

---

9   Draining began between 23:40 and 23:55 (reports vary) and lasted around 1.5 hours. All three loops were drained, but the secondary is the most relevant.

10   The JAEA claim the amount was 640 kg ± 42 kg in total, "of which 410 kg was collected in the building." This claim is backed up by several other technical reports. Early speculation of various numbers up to five tons, then the widely reported two to three tons, were overestimates. Source: "Leakage status and recovery of sodium." Japan Atomic Energy Agency. https://archive.is/20130501081831/http://www.jaea.go.jp/04/monju/category05/mj_accipanf/accipamph02.html.

11   STA and the Fukui Prefectural government were notified by PNC at 8:40 p.m., but Governor Kurita himself was not informed until 9:25 p.m.

enough for Fukui's unimpressed governor, Yukio Kurita. "Along with the delay in the advisory bulletin, subsequent reports were insufficient," he complained. "This is an extremely regrettable loss of the prefecture's trust in the overall safety of Monju."[x] Clearly fed up with the industry's lethargic attitude towards informing anyone of accidents, he also demanded a "change of consciousness" in public relations.

At around 2:00 a.m., six hours after the leak and soon after draining the secondary loop of sodium, men in hazard gear entered the space to take photographs and video footage. Copies of this "two o'clock video," as it came to be known, found their way to PNC headquarters at 9 p.m. the following night, where its existence remained a secret for some time. The company published a few ambiguous photographs of unscathed walls and equipment but were soon pressured to release more after Fukui prefectural officials visited the damaged piping room and anti-nuclear groups complained about the lack of transparency. The resulting one-minute-long, heavily edited footage released by PNC came from a second video, which they claimed had been recorded at 10:00 a.m. on the morning after the accident, but which was actually recorded at around 4:00 p.m. – 14 hours after the first tape and 19 hours after the accident. This "four o'clock video" also reveals little: in each ten-second cut, the camera operator is stationary, the camera doesn't pan around much and there's no sign of smoke or fumes. PNC released a longer four-minute version of the second video soon afterwards, which they claimed was all available footage. None of it showed the damaged piping.

Beginning on 18th December, news reports emerged of faults with the way operators had handled the incident. Upon activation of the smoke alarms, regulations required the plant's air-conditioning system to be sealed off immediately because sodium compounds could leak out of the building (which they did), and a sodium fire cannot be extinguished by conventional means. Hitting it with water creates an explosive hydrogen reaction; halon and carbon dioxide extinguishers would have much the same result.[12] The only real way to combat burning sodium is to starve it of oxygen, hence the rule about the air circulation, but Monju's operators did not seal the air vents for three and a half hours. Perhaps worse, officials admitted

---

12 Halon fire extinguishers have been banned since the early 2000s because they use CFCs, which harm the atmosphere.

that the operators were afraid to immediately scram the reactor because they did not have permission, hence the initial slow decline in power.[y]

Fukui officials, meanwhile, noticed large discrepancies between a tape they'd recorded on their visit to Monju and the footage released by PNC. On 22nd December, STA investigators confirmed that PNC lied about how long plant workers took to actually enter the piping room and record their original videos. The edited version was produced, according to what I imagine was a sheepish PNC spokesperson, on the instructions of company director Yasuhito Ohmori, "in order to make the tape easier to understand." Isao Sato, the deputy chief of PNC's Monju office, was comfortable admitting at a press conference on 23rd December that he "was determined to cover up the case."[z] A day later, three senior officials responsible for the edited video, including Sato, were relieved of their posts, though they were reassigned and did not lose their jobs. The unedited footage aired on national television, with a soundtrack of heavy breathing from the camera operator's respirator. Workers dressed in bulky head-to-toe silver hazard gear enter the piping room, with its blank concrete walls and machinery everywhere. You can't see more than twenty or thirty feet because of the thick sodium fumes, and for the longest time you assume the footage is in black and white because *everything* is white. Eventually, you catch hints of colour and realise it's just because sodium powder covers every surface. At one point, the camera pans down to reveal footprints cutting through sodium on the floor like snow. Some damage is evident where sodium erupted from the compromised pipe, but the only surprise is just how much of it coats everything.

In addition to employee-misconduct investigations by the STA, the NSC and Fukui Prefecture's local government office, PNC launched their own internal probe, placing 49-year-old deputy chief of general affairs Shigeo Nishimura in charge. He committed suicide three weeks later, on 12th January, by jumping off a Tokyo hotel roof the morning after revealing his findings in a press conference. He had lied about which date PNC's president became aware of a copy of the two o'clock video at the company's headquarters, claiming it was 10th January when in fact it was 25th December, but the STA were informed of the actual date ahead of Nishimura's conference.[aa] Why he lied is unclear – it may have been a simple slip of the tongue – but apparently grief at the inevitable punishment his findings would inflict upon some of his life-long colleagues played a role. Nishimura tried to justify their actions in a suicide note, writing that they "were confident in their technical ability. But they may have found it difficult to explain their panic and confusion from the accident. It is most difficult for people to judge others and discover the truth."[ab] Explaining the decision to end his own life, Nishimura wrote, "I feel grave responsibility for [our][13] failure to restore trust and I feel very sorry for that." Bizarrely, Mr Nishimura's death drew media focus away from the hidden two o'clock video issue, and that specific matter disappeared from view (partly from a sense of shared responsibility among the media, who had hounded him for answers), but the entire episode undermined public confidence once again.[ac] "It's not an engineering but a social problem now," remarked the JAEC's Dr Yoichi Fujiie in 1996, in light of the Monju incident.[ad] In 2007, Japan's Supreme Court dismissed a lawsuit brought by Nishimura's widow against PNC, because her lawyers could not prove the company had ordered him to lie.[14, ae]

The original accident occurred when the protective sheath on a thermocouple electrical temperature probe broke off inside the pipe, which was contracted by Toshiba, along with all secondary piping construction work, to the same Ishikawajima-Harima Heavy Industries which built *Mutsu*. The probe bent, thus allowing molten sodium to force its way out of a gap at its base. A laser microscopic analysis revealed cracks in the probe sheath caused by flow-induced vibration fatigue that had not been accounted for in its design.[af] The probe itself had been required to match detailed specifications and undergo extensive technical review reflecting on lessons learned from Jōyō before approval, but the sheath covering it had fewer requirements. Nevertheless, a problem *was* spotted at the design stage and sent back for modification, but the probe's four designers were not specialists in fluid dynamics, according to the STA, and did not anticipate the kind of fluid vortex present around the sensor.[ag] As such, while they altered the method of attaching the sheath to the pipe to reduce thermal stress, they opted not to modify its shape. This confusion was perhaps compounded by a strange shuffle in roles between the four prime contractors from Jōyō to Monju.

---

13 The word "our" is my own assumption; the only existing translations of this quote all use a word that appears to be a typo, as it isn't English or Japanese. He left three notes in total, including one to his wife and another to his colleagues.

14 A lawyer named Kazuo Hisumi who worked on the case has an interesting blog post about it: https://blog.goo.ne.jp/tokyodo-2005/e/90e8b1da7efa1cbcc910ed184bfadec1.

With Jōyō, Hitachi handled design work for the reactor, Toshiba and Fuji made the fuel-handling equipment, Hitachi did the primary coolant system and Mitsubishi the secondary loop. With Monju, the responsibilities changed to Mitsubishi for the reactor, Fuji for the fuel-handling and some waste treatment facilities, Hitachi for the primary coolant system and Toshiba for the secondary loop.[ah] In the end, the missing sheath was not found until 28th March, almost four months later.

Japan's serious lack of crisis management and preparations became apparent in 1995. First, in January, the Kobe earthquake laid waste to a city that should have been prepared. Two months later, the Aum Shinrikyo cult murdered twelve people and injured almost a thousand more by releasing the sarin nerve agent in five coordinated attacks on the Tokyo subway during rush hour. Then Monju. The authorities and emergency services reacted sluggishly in all three cases, prompting changes to organisational structures, emergency procedures and medical surge capacity which produced a more effective response to the 2004 Chūetsu earthquake.[ai] It's not unreasonable to think a review of nuclear disaster preparedness was in order, but this did not happen.

* * *

PNC, still reeling from the Monju controversy, invited further bad press in 1997 when a fire broke out at their radioactive waste bituminisation facility at Tōkai-mura on 11th March. The site encased and solidified low-level liquid waste in molten asphalt (bitumen) for storage, and that day was trialling a new asphalt-waste mix, using 20% less asphalt than normal. A gradual chemical reaction inside one fresh barrel ignited the already-hot contents at 10:00 a.m. and quickly spread to several others nearby.[aj] Alarms alerted workers, who thought they had extinguished the fire after spraying a manually operated sprinkler system for just one minute – despite knowing it took around eight minutes from past experiments – but the room filled with smoke and they didn't approach the barrels for confirmation.[ak] The same staff turned off the sprinkler and fire alarms but rising radiation and temperature levels quickly forced a full evacuation. Several minutes later, a ventilation filter became clogged and the equipment for keeping the space at below-atmospheric pressure failed.[15] PNC reported the fire but not the subsequent radiation leak. At 8:00 p.m., just as workers were preparing to re-enter the building, built-up flammable gases ignited and rocked the facility in a large explosion, destroying a thick lead isolation door, which then allowed small quantities of radioactive materials to blow out into the environment.[al] The incident exposed 37 nearby personnel to trace amounts of radiation in what the STA declared the country's worst-yet nuclear accident, which was rated a 3 on the International Nuclear Event Scale.

Notwithstanding the lack of harm to personnel and the environment, PNC returned to the news cycle in a negative light once again. When they lied about workers extinguishing the fire before evacuating (only for the truth to be unveiled during an accident hearing), then destroyed photographic evidence, something clearly had to change. "The lessons from Monju were not learned," conceded one PNC spokesman to reporters. "My own mother was asking, 'What's going on with your company?'"[am] A leak of radioactive tritium at the Fugen reactor several months later (which PNC illegally neglected to report for well over 24 hours), followed by the revelation that 11 similar incidents had gone unreported, sealed the company's fate. "There is a difference between 'safety' in the technical sense and 'peace of mind' as a human perception," the STA admitted, "and the two at the moment do not match in the nuclear field." After six hearings, the agency decided on limited reforms and a name change to reduce nationwide distrust of the company and its operations; PNC became the Japan Nuclear Cycle Development Institute (JNC) in October 1998.[an] Some of PNC's functions were privatised, but the fast breeder research and most other responsibilities and staff carried over to JNC.

Katsuhiko Ishibashi, the outspoken Kobe University seismologist, published another daunting prediction in the October 1997 issue of Japan's *Science* journal, wherein he almost perfectly outlined the events that would transpire at Fukushima 14 years later.[16] He coined the term *genpatsu-shinsai* ("nuclear power plant earthquake disaster") to describe a scenario in which a massive temblor cripples a power

---

15 The area was a "red zone," where radioactive materials are handled, and thus was kept at a negative pressure to prevent any contaminants from escaping.

16 Professor Ishibashi expected this scenario would most likely occur at the Hamaoka Nuclear Power Station south-west of Tokyo, 400 kilometres from Fukushima Daiichi. Source: Ishibashi, Katsuhiko. "原発震災－破滅を避けるために" (in Japanese). *Science*, vol. 67, no. 10, October 1997. https://historical.seismology.jp/ishibashi/opinion/9710kagaku.pdf.

plant, severing its connection to the outside world and leaving it unaided as radiation leaks make recovery impossible. Ishibashi was not a natural critic but became worried after his Kobe Earthquake predictions about building vulnerabilities came true, despite expert assurances. Tokyo University professor of engineering Haruki Madarame, a highly respected academic who would play a prominent role in the Fukushima disaster, dismissed Ishibashi and his prediction as the work of an amateur, saying "in the field of nuclear engineering, Mr Ishibashi is a nobody."[ao] Madarame made no mention of a December 1990 severe accident assessment report by the US NRC, which concluded that an earthquake-induced loss of power was indeed "the most likely" external cause of reactor damage.[ap]

iii. Tōkai Tragedy

The year 1999 marked one of the most upsetting incidents in world nuclear history. On the morning of 30th September, three technicians arrived for work at the 20-by-20-metre, single-storey Fuel Conversion Building belonging to Japan Nuclear Fuel Conversion Co (JCO), a private nuclear fuel enrichment and reprocessing company established back in December 1980 by Sumitomo Metal Mining, one of the original big five *zaibatsu*. Sumitomo joined the nuclear fuel game in 1969, and JCO now ranked among the nation's largest private fuel companies by volume of fuel handled. The building, commissioned in 1988, formed part of one of fifteen nuclear facilities now located in Tōkai-mura.

JCO faced stiff competition for contracts during the 1990s. Turnover almost halved between 1991 and 1998, so, in a bid to save money, management laid off one-third of its employees, retaining around 150 but dropping the number of specialist technical staff from 34 to 20. Those who remained were encouraged to work faster with efficiency drives known as *kaizen*.[aq] The Japanese word for "improvement," *kaizen* is a customary work philosophy of making continual changes to boost speed, reduce costs and perfect quality control. Safety fell by the wayside as JCO streamlined their operation over and over to maintain its position as the only domestic company processing fuel for BWRs operated by TEPCO and five other utilities, plus the ever-changing custom requirements for their contract with JNC's Jōyō, Fugen and Monju reactors.[ar] Discarding their experts left JCO with nobody skilled in either process, according to a US NRC report, and there was no government or internal oversight because the facility was not considered dangerous.

In 1999, JCO received urgent orders for large batches of specialised 18.8% enriched uranium fuel for the Jōyō reactor, production of which is slow and yields low quantities.[17] To make matters worse, they had almost exclusively enriched fuel to 3–5% for ordinary light water reactors and possessed little comprehension of the extra precautions required when enriching to 18.8%. The STA had established a procedure that avoided the risk of spontaneous fission by setting a strict limit of 2.4 kilograms of uranium processed per batch. This method involved dissolving uranium oxide powder in nitric acid inside a special dissolution tank, then using a hose and pump to transfer the dissolved mix of uranyl nitrite into extraction columns[18] to remove impurities.[as] The columns were tall and narrow specifically to prevent fission via the formation of a critical mass of neutrons. From there, the liquid was transferred to a precipitation tank fitted with a water-cooling jacket for heat removal, where any solids were filtered out. This unusual need to create a uranyl nitrate solution was a direct result of US non-proliferation pressure in the 1970s, but JCO developed their own quicker method in 1996 which, the company later admitted, went deliberately unreported to the government.[at] Using this secret quicker process, technicians dissolved the powdered uranium oxide in a metal bucket heated by an ordinary portable electric cooking stove, then poured it straight into a precipitation tank of the wrong shape to prevent criticality. They skipped the extraction columns entirely.[au, av]

On that morning, 35-year-old technician Hisashi Ouchi (pronounced Oh-oo-chee) stood beside the shoulder-high precipitation tank, his right hand holding a funnel over a hole in the tank's lid. Slightly above him on a ladder, Masato Shinohara, aged 40, poured bucket after bucket into the tank through Ouchi's funnel. Their supervisor, 54-year-old Yutaka Yokokawa, sat at a nearby desk behind a thin dividing wall, while the other two worked off to his left. All three men had processed fuel enriched to 3–5% on numerous

---

17 Strictly speaking, Jōyō's fuel was not enriched in the normal sense. It went through a special chemical purification process that removes impurities and leaves it with 18.8% uranium-235.

18 These extraction columns have numerous different names, depending on the source, including "dissolution towers," "buffer tanks" and others. "Extraction column" appears to be the correct technical term, but I could easily be wrong.

occasions, but none – according to later court documents – were familiar with the process for enriching to 18.8%.[aw] Yokokawa alone had once received basic training on fission criticality and its possible hazards years earlier ("I did not understand well what it meant," he confessed) but was unaware that the shape and mass of a uranium solution could play a role.[ax] When no accidents occurred over time, he came to believe the dangers described in his training were exaggerated. Still, as this work for Jōyō was new to him, Yokokawa questioned a senior employee on the company's shortcut procedure and was assured of its safety, though some observers later speculated that this expert thought Yokokawa meant the usual lower enrichment.[ay] All three men therefore believed their work to be absolutely safe. At around 10:35 a.m., while Shinohara poured his seventh bucket, bringing the tank's total to 16 kilograms of enriched uranium – around seven times the legal limit – the mixture suddenly fissioned without warning, and each man saw a blue-white flash of Cherenkov radiation.[19] They stood beside an uncontained nuclear reactor.

Most radiation occurs during radioactive decay of an 'unstable' isotope, where an atom has an imbalanced ratio of protons to neutrons, hence breaking apart – decaying – to change into another isotope.[20] Three types of radiation are normally associated with nuclear accidents, each with their own dangers: alpha, beta and gamma.[21] Alpha particles can't travel more than a few centimetres through the air, nor penetrate human skin, but can still be fatal if ingested. Beta radiation is stopped by a ream of paper but can penetrate the skin if placed in close proximity. Gamma radiation is usually the most harmful type, as gamma rays travel long distances and penetrate all but the densest materials, like lead. There is another type, called neutron radiation, but it's extraordinarily rare for someone to be irradiated by this (only six cases prior to the Tōkai-mura accident), as it arises from direct exposure to fission.[az] Over 99% of fission neutrons are produced during fission, with the rest coming from radioactive decay. As such, neutron radiation does enormous damage to the human body, particularly to tissue enriched with water and fat.

---

19 Cherenkov radiation is a visible pulse of light, occurring when a charged particle such as an electron passes through a transparent medium like water or air at faster than that medium's speed of light. It is named after the Soviet physicist Pavel Alekseyevich Cherenkov (1904-1990).

20 Note, decay is not the only cause of radiation, but this concept is too complex for a concise explanation.

21 An excessive concentration of x-rays can happen in radiotherapy, but that isn't associated with criticality events.

Radiation alarms tore into the stunned silence as a massive concentration of neutron and gamma rays penetrated Ouchi and Shinohara. Yokokawa knew the alarm and ominous blue pulse meant a criticality; he yelled to his colleagues, who fled into a nearby locker room, where Ouchi immediately collapsed into a seizure.[22] Another man joined Ouchi's two colleagues, and together they tried to resuscitate and prevent him from choking "by inserting chopsticks between his teeth and turning his face to the side."[ba] Their quick thinking prevented a quick death: Ouchi began to vomit within minutes.[bb] Shinohara did too, less than an hour after exposure, but Yokokawa, being farther from the tank, took a lesser dose.[23] He remained in the vicinity for several minutes, checking on the tank and phoning to report a criticality to his superiors – who didn't believe him – before walking out of the Fuel Conversion Building unaided.[bc] The fourth man, meanwhile, rang the Tōkai-mura fire department at 10:43 a.m.

Three paramedics arrived in an ambulance within 15 minutes and, after an almost hour-long delay while they searched for a hospital willing to accept potentially contaminated patients, drove 24 kilometres to the National Mito Hospital, where doctors determined the men needed urgent specialist care.[24] A medical helicopter flew all three 75 kilometres south-west to the National Institute of Radiological Sciences (NIRS) in Chiba, Japan's only radiation research facility, where the nation's top specialists in radiation medicine awaited them. Throughout the journey, Ouchi suffered repeated "episodes of nausea, vomiting, and diarrhea."[bd] (As an aside, I find it shocking that Tōkai-mura – the country's largest concentration of nuclear facilities – did not have its own modest treatment centre for radiation injuries.)

Based on his visible symptoms and elevated body temperature, the NIRS staff initially estimated a dose of at least eight sieverts (Sv) for Ouchi. Shinohara's estimate was six to eight sieverts, while they thought Yokokawa, who displayed no strong symptoms, had received under four. The sievert is a modern alternative to the old unit of roentgens and is used to measure a dose of ionising radiation on the body. One sievert is 1,000 millisieverts (mSv), and one millisievert is

---

22 One source also describes this scene as occurring in a decontamination room in an adjoining building.

23 Credible medical journals written by people involved in their treatment vary in the precise details. Some say Shinohara's vomiting began half an hour after exposure, while others say "almost an hour" after.

24 In Japan, ambulances come from fire departments rather than being their own distinct emergency service.

1,000 microsieverts (μSv). Values are almost always given in either micro- or milli-sieverts, as it's extremely rare for someone to absorb as much as one sievert. You'd receive 10 microsieverts (0.00001 sieverts) from a chest x-ray, for example, while a full-body CT scan yields a dose of 10 millisieverts (0.01 sieverts). We are all exposed to background radiation at a rate of around 2 to 3 millisieverts per year at sea level, while 100 millisieverts, spread over a year, is the lowest dose with a clear increase in the likelihood of developing cancer.

A single sievert, taken in one hit, will cause acute radiation syndrome (ARS), with symptoms of nausea, vomiting, fainting and massive bleeding, but won't kill a healthy adult. In 1945, half of those who received 4.5 sieverts from the bombs at Hiroshima and Nagasaki died.[be] Everyone exposed to 6 sieverts died, matching the approximate average dose received by the Chernobyl first responders killed in 1986. With very few exceptions, nobody had ever taken Ouchi's estimated 8 sieverts in one hit and survived, but this was 1999: decades after most criticality accidents and 13 years after Chernobyl. Much was learned and shared with the international medical community from that accident. Though there were no guaranteed treatments for such massive doses, the NIRS team were optimistic that, with modern medicine and experimental procedures, Ouchi stood a fighting chance.

* * *

The amount of radiation a person can absorb and survive depends on the duration of exposure. It is possible to survive much more than 8 sieverts over a long period, though it's rare. The person thought to have survived the highest-ever dose is Albert Stevens, who unknowingly accumulated an astonishing 64 sieverts over two decades, a fact only revealed after horrifying human experiments were exposed by the Pulitzer-winning investigative journalism of Eileen Welsome in the early 1990s.

A series of accidents among scientists handling plutonium on the Manhattan Project raised concerns about the dangers of inhaling plutonium dust.[25] To help develop a diagnostic tool for analysing the amount of plutonium inside someone, researchers began a

---

25 Plutonium, unlike other well-established elements at the time such as radium and polonium, is difficult to detect while inside the human body.

series of secret experiments on terminal patients at the University of California Hospital in San Francisco, injecting 18 of them with a plutonium solution.[bf] "These people," said the hospital's chief radiologist, Earl Miller, who oversaw the plutonium injections, "the people that were chosen usually in these studies, they were doomed. They were ready to die."[bg]

Stevens, a house painter in his late fifties, was admitted to the UoC hospital with terminal cancer and injected with plutonium on 14th May 1945 without his knowledge. But after doctors removed most of what they thought was his cancer during surgery, subsequent tests showed it had in fact been a large malignant ulcer. Stevens became a human guinea pig for the researchers who injected him, dutifully providing urine and stool samples each week for the next two decades to doctors who claimed to be studying his miraculous cancer recovery. He died aged 79 of heart disease, having survived many times a fatal dose of radiation. Nobody ever told him about his continual irradiation.

* * *

Ouchi's symptoms had subsided by the time he reached hospital, and he appeared to be in almost perfect health. The nurses later said that he was cheerful, spoke clearly and remembered what had happened, and that his only visual signs of ill-health were a slight red swelling to his face and right hand. This brought hope to the attending physicians, but a tissue analysis soon revealed his actual dose was a catastrophic 17–20 sieverts.[bh] (A posthumous study showed that Ouchi's upper-right abdomen received the highest absorbed dose, at 61.8 sieverts.)[26] His lymphocyte count (the amount of a type of white blood cell that is particularly sensitive to radiation), which is normally between 25% and 48% of all white blood cells, was down to 1.9%. This was somewhat expected, since intense radiation always damages the body's ability to regenerate

---

26 The proper dose was 61.8 "grays," but I've written sieverts for simplicity with topics discussed later. A gray is defined as the absorption of one joule of radiation energy per kilogram of matter. The dose broke down as 27 grays of neutron radiation plus 34.8 grays of gamma radiation, which doesn't necessarily translate properly to sieverts, since gamma and neutron rays have different gray-to-sievert dose equivalents (see why I've kept it simple?). The information is available here: "Brief note and evaluation of acute-radiation syndrome and treatment of a Tokai-mura criticality accident patient. *Journal of Radiation Research*, September 2001: http://paperity.org/p/34453684/brief-note-and-evaluation-of-acute-radiation-syndrome-and-treatment-of-a-tokai-mura.

cells first, but the drop signalled an inability to fight infection. Some in the team saw this as a lost cause from the outset and thought nothing could save him, but others were determined to try.

With his immune system crippled, Ouchi needed an urgent cell transplant. More specifically, a transplant of white blood cell-re-generating peripheral hematopoietic stem cells, first used in 1958 at a reactor criticality accident in Vinca, Yugoslavia.[bi] The procedure usually involves extracting bone marrow from a compatible donor under general anaesthetic, as was the case with Chernobyl victims, but a new technique pioneered on cancer patients used cells from blood. Most stem cells grow in bone marrow, but a special drug boosts the small number present in a donor's bloodstream. Doctors draw the cells from a vein, then painlessly inject them into a patient – ideal with radiation victims, whose bodies are delicate. NIRS was not experienced in this type of surgery, so after three days, both Ouchi and Masato Shinohara, whose confirmed dose was around 10 sieverts, were transferred to the University of Tokyo Hospital (UTH) at the suggestion of Professor Kazuhiko Maekawa.

Maekawa was a specialist in emergency medicine at UTH who knew the pair stood the greatest chance of survival there. He had recently attended an information exchange conference where he was "astonished by the inadequate education in radiation emergency medicine for hospital physicians and medical staff working near nuclear facilities."[bj] He knew Japan could do better but was surprised when the men arrived. "The amount of radiation that Mr Ouchi had presumably received was fatal," he said a year later; "that was obvious to everyone in the team. However, Mr Ouchi looked very well at that point and we didn't have the impression that he had been exposed to such a high level of radiation. Therefore, we were determined to do everything we could."[bk]

* * *

The uncontained criticality at Tōkai-mura continued on and off during much of this time, with JCO staff often unable to approach. As the uranium solution boiled, pockets of steam formed in the fuel and stopped the fission reaction. Unfortunately, the jacket continued to circulate cooling water around the outside of the precipitation tank, which caused those voids to collapse as the uranium mixture cooled, triggering further reactions. Police cordoned off the area, but inadequate contingency planning (the facility's licence review concluded there was "no possibility of criticality accident") meant intense confusion arose over what had happened, whether it was still happening, and then how to fix it.[bl, bm] Compounding this inaction was the now-traditional lethargic response, with the STA taking over three hours to dispatch a radiation-measuring team to the site. Neutron-detection equipment did not arrive for over six hours. The wider public did not learn of the danger for almost five, and, when they did, in some instances it was local town officials acting of their own volition.

Before 2:40 p.m., JCO twice asked Tōkai-mura mayor Tatsuya Murakami to evacuate everyone living within 500 metres (1,640 feet) of the site, pointing out that high radiation readings had forced JCO's own employees to retreat to a position 280 metres from the Fuel Conversion Building yet there were houses just 150 metres from it. The mayor's office could not reach the STA for advice, but the Ibaraki Prefectural government cautioned against moving anyone because the readings were below the (bizarrely high) regulation 50-millisieverts-per-hour threshold for evacuations. With nearby roads and rail lines closed, Murakami nevertheless decided at 3:00 p.m. to evacuate 161 people living within 350 metres of the facility in buses to a community centre two kilometres (1.2 miles) away. A further 310,000 people in a ten-kilometre radius – including the original JAERI campus, five kilometres to the west – received precautionary instructions to shelter indoors[27] seven hours later, though there was no risk of contamination at this distance. Virtually no fission products left the building and almost all radioactivity off-site consisted of neutron or gamma rays, which don't stick to dust, clothes or hair. While the accident continued, round-the-clock media reports seared a new phrase into the public lexicon – *rinkai jiko* ("criticality accident") – and described the situation as out of control, spreading fear across Japan and terrifying the local community. In the chaos, 46 others were exposed to low levels of radiation, including the staff of a nearby golf course.[bn]

---

27 The prefectural government offered the following bullet-list instructions: "Stay indoors. Shut all the windows and switch ventilation off / In case of travelling in cars for unavoidable circumstances, keep all the windows of the car shut and avoid using the ventilation fan / Tap water is safe, because the source of water supply has been changed / Do not drink well water or rain water / People who voluntarily took refuge [at] any downwind refuge point are advised to take further refuge, leaving the downwind area." As you can imagine, these instructions did not put people's minds at ease. The order to stay indoors was rescinded at 4:30 p.m. on 1st October.

The government eventually held its first press conference at 9:00 p.m., in which Chief Cabinet Secretary Hiromu Nonaka admitted to the "strong possibility that abnormal reactions are continuing within the facility.... There are concerns about radiation in the surrounding areas."[bo] The conference owed its delay, in part, to unhelpful technical advice given to the Prime Minister's Office, whereby "the experts briefing the prime minister's team infuriated everyone by merely insisting that the accident should never have happened."[bp] To everyone's relief, rainfall helped prevent what little radioactivity had escaped from drifting over nearby communities.[28] Two-year-old photographs of Tōkai-mura's damaged bituminisation facility were immediately mistaken on internet message boards as being from this accident, sparking angry speculation about more cover-ups. Media reports throughout the night aggravated people's concerns, including *Reuters* and others quoting an Ibaraki prefectural official saying "As of late Thursday night [30th September], 3.1 millisieverts of neutrons per hour, or about 15,000 times the normal level of radiation, was detected two kilometres from the accident site."[bq] This was unusually high but not enough to harm anyone without prolonged exposure. Yet, when people hear "radiation is 15,000 times higher than normal" without an adequate explanation of what that actually means for their health, they understandably panic. A JCO official visited the emergency evacuation centre and begged the evacuees' forgiveness, saying, "We have no words to express our apologies. We cannot escape our responsibility."

JAERI developed a plan to halt the runaway reaction. At 2:30 a.m., sixteen hours after the accident, employees of JCO, JAERI and JNC began a four-hour operation to stop water flowing to the precipitation tank's jacket – water which was still acting as a moderator and reflector, enabling the criticality to continue. First, they closed the water tower valve and opened the drain valve, then dismantled the pipe and forced argon gas into the system to flush everything out. Next, a fire engine pumped in aqueous boric acid to neutralise the fission reaction.[br, bs] They could only approach contaminated areas for a few minutes at a time, as preliminary estimates suggested radiation could be in the sieverts-per-hour range.

---

28 Radioactive particles are captured by the falling rain, bringing them down to earth, where they con-taminate the surrounding environment. That can be bad if the rain falls onto a city, but the rain was beneficial in this case because radioactivity was prevented from spreading far out over populated areas.

Fifty-six employees were exposed to radiation during the operation.[bt] JCO initially reported that one man sent to photograph the piping in the operation's planning stage received a dose of 20 millisieverts. In fact, his radiation monitor turned out to be a two-digit device that reset its counter whenever it reached 99, and he had actually absorbed a very high 120 millisieverts.[bu] Among everyone involved, doses ranged from an insignificant 0.6 up to that high of 120 millisieverts, according to the Nuclear Safety Commission.[bv] Most were well below 50 millisieverts, but six people exceeded the legal safe limit. The three paramedics who collected Ouchi, Shinohara and Yokokawa received around 13 millisieverts each – an avoidable outcome, since JCO didn't warn them the accident was nuclear in nature. Scientists estimated that nearby residents received, at most, an average 21 millisieverts: not good, but thankfully not life-threatening either.[bw] Fission finally stopped at 6:00 a.m., almost 20 hours after it began. Over 10,000 residents requested medical check-ups over the next ten days. One of the government's top spokesmen apologised, commenting that "as a modern nation, it is shameful that this kind of accident occurred."[bx]

JCO first placed blame at the feet of Ouchi, Shinohara and Yokokawa, painting them as incompetent, corner-cutting fools, but within days it became clear this was a twisted version of the truth. In response, the STA revoked JCO's operating license on 6th October, and police raided their headquarters and Tōkai-mura offices – the same day that Prime Minister Keizo Obuchi visited the town of Tōkai-mura, where he made a show of eating local produce.[by] Afterwards, he announced his intention to tighten restrictions to ensure such an incident never happened again, saying "The accident has become a concern not only to Japan but to the whole world.... To come up with preventive measures is the way to recover the people's trust toward nuclear energy." The president of Sumitomo Metal Mining resigned two days later.[bz] The IAEA, meanwhile, acknowledged that Ouchi's and Shinohara's chances of survival were low.

* * *

Twice a day, nurses drew samples of Ouchi's blood for a broad spectrum of cutting-edge analyses. Preventing infectious bacteria from entering his system was vital. As such, the hospital converted his room into a special sterile environment, with its own air purification systems and an adjoining clean-room for incoming medical staff. They received disturbing news after a week. Microscopic images of Ouchi's bone marrow cells revealed a bent, tangled and incomprehensible mess of DNA instead of the ordered structure one would expect. His blood's platelet count had also dropped to around 10% of a healthy person, meaning it would not clot if he bled.

Doctors began to search national blood banks for a compatible hematopoietic stem cell donor on the day of the accident. The odds of finding a match were tens of thousands to one, but Ouchi was in luck: he had a compatible sister. For two days, she donated blood that doctors mined of its stem cells and drip-fed back into her brother's bloodstream. It was the first time ever that this specific type of transplant had been performed on a radiation accident victim, and the process was smooth, but it would take over a week to learn the results. Shinohara also received a transplant three days later, but his came from the umbilical cord of a newborn baby because they couldn't find a matching donor.

As Ouchi's symptoms became more severe with each passing day, he felt an unceasing thirst and began struggling to eat and sleep.[ca] One day, when the nurses peeled back the surgical tape securing his drip and monitoring sensors to replace them with fresh dressing, his skin tore off with it, like duct tape ripping up tissue paper. Radiation had destroyed his skin's ability to regenerate, so the delicate base layers became exposed as the top layers fell off. They abandoned the tape after a week and a half, but soon even the friction of washing him disintegrated his skin. The hospital department heads secured special licences to import drugs not available in Japan and held twice-daily meetings to ensure Ouchi received the best possible care. Still, there had never been a criticality accident like this in Japan, so they turned to an American specialist in hematopoietic stem cell transplants named Dr Robert Gale.[cb] Dr Gale pioneered this method of treatment for radiation victims and assisted Soviet doctors in Moscow shortly after the Chernobyl disaster. He participated in all the daily briefings at the Tokyo hospital and offered advice where he could.

Around this time, Ouchi started to suffer from dyspnoea (shortness of breath) caused by liquid in his right lung. Under ordinary circumstances, a surgeon would puncture the lung to remove any liquid obstructing his breathing, but radiation damage made healing from such an incision difficult. Nevertheless, doctors felt the procedure was their only option. Ouchi had so far mustered a brave and cheerful facade, especially during the daily visits from his family, but cried out in agony during the operation, begging them to stop. By the eleventh day, doctors had no choice but to insert an artificial respirator tube down his throat, forever ending his ability to speak. From then on, nurses administered all food and water via an intravenous drip in his neck.

Many people characterise Ouchi's treatment from this point as little more than cold exploitation to advance medical understanding of radiation damage, and with the benefit of hindsight this is an understandable position. But, from reading and watching interviews with the men and women who cared for him, it's clear this could not be further from the truth. True, a great deal was learned from this accident, just as with Chernobyl, but that in itself is not an argument against his continued treatment. On the contrary, the medical team tried in earnest to lift Ouchi's spirits by playing his favourite music and telling him stories, even if he gave no indication of having heard. Many knew his likelihood of survival was slim – however hard they forced it to the back of their minds – but each was determined to try. Their efforts were rewarded with a glimmer of hope. The transplants were a success: after two weeks, Ouchi's body accepted his sister's cells and by day 18 his count was well within normal limits.

His condition stabilised for a time, but on day 27 his intestines' mucus layer died, meaning he could no longer digest or absorb nutrients. His body now flushed everything out as green, watery diarrhoea, and from here, nobody knew what to do. Specialists from the US, Russia, France and Germany flew in to lend their knowledge to Ouchi's fight for survival, and all agreed they should wait. The skin from his right arm had fallen off by this time, leaving behind raw red and yellow tissue. Over the next few weeks, the damage spread, until the entire front of his body was a sticky, skinless mass that leaked a litre of fluid each day. Even his eyelids fell apart, often causing his eyes to bleed, despite a special ointment protecting them. A ten-strong team, including Professor Maekawa, who slept at the hospital, swaddled him in gauze and antiseptic each day – a tender but gruelling and sweaty task in a room heated to 30°C to maintain Ouchi's core body temperature.

A small patch of new skin and some intestinal mucus regeneration bolstered their resolve. A skin sample from Ouchi's sister was hastily replicated in a laboratory and applied to Ouchi's body, but the grafts didn't hold. Yet, even now, a month and a half since exposure and with a steady intake of painkillers and sedatives, his body discharging almost 10 litres of fluids and blood each day, he continued communicating through small gestures during brief bouts of consciousness. More and more, however, the increasingly demoralised medical team questioned their efforts – even Maekawa. He shared and acknowledged his subordinates' doubts but encouraged them to keep going for Ouchi's sake.

His heart stopped three times on the morning of 27th November, 59 days after exposure. He survived, but the conflict returned. Some, including Ouchi's family, believed that his beating heart indicated a fight for survival and felt a renewed determination to help him. Others, including Maekawa, were despondent. He knew the incident had caused lasting harm. The medical team thought it likely that vital oxygen had reached Ouchi's brain via artificial means during his cardiac arrest, but his liver and kidneys ceased to function and he all but stopped responding to even reflex stimuli.

On the 81st day, Maekawa reluctantly acknowledged that their efforts had been in vain. They had pulled miracle after miracle out of the bag during the last two and a half months, but Ouchi was now so far gone that recovery was impossible. Mould had begun to spread across his body, he had developed gastrointestinal bleeding, respiratory and kidney failure, and he was completely reliant on machinery.[cc] "We tried every possible treatment and medicine," Maekawa said a year later, before pausing in thought. "I think, even in hindsight, we could hardly have taken further measures against the bleeding in the digestive tract."[cd] Ouchi's wife, young son, mother and father were informed that next time his heart stopped, the doctors would not resuscitate him.

Yokokawa, the third irradiated JCO employee, had remained at NIRS and made a steady recovery with the help of platelet transfusions. He was discharged on 20th December.[ce] The very next day, at 11:21 p.m., Hasahi Ouchi's blood pressure entered a sharp decline and the 35-year-old passed away. Grief overwhelmed both his family and the hospital staff. "I don't want him to thank us," said one of the nurses in an NHK documentary in 2001. She, like the other doctors and nurses interviewed, struggled to talk throughout, her expressions racked with indecision and sorrow. "I would accept his resentment. All I want is to know whether we did the right thing.... I couldn't have endured doing the treatment if I hadn't had the conviction that I was doing it to protect his life."[cf]

The entire nation shared in the mourning. National and international media published occasional updates on Ouchi's condition throughout his ordeal, and it's perhaps because of this that the public rained fury down on JCO and the government. How could they allow such an accident to claim an innocent life?[29] It was a tipping point for many – distrust of nuclear power and the companies and institutions promoting it reached new highs, particularly in the Tōkai area. Tōkai-mura mayor Tatsuya Murakami, who had previously supported nuclear power, said that he "lost confidence in the government, and ... became convinced that [Japan] lacks adequate capabilities to maintain nuclear power plants."[cg] Days after the actual criticality, a record 74% of respondents to a *Mainichi* poll considered nuclear power to be unsafe, while only one in five supported further development, though even this spike flattened in time.[ch]

Ouchi was the first official victim of nuclear power in Japan. Radiation exposure had never been acknowledged as a cause of failing health – bar the atomic bombings – and had been ignored in the mainstream media for decades, but his death brought the conditions and health of subcontractors into the limelight once more, albeit briefly. According to the BBC:

> At least 700 people working in the nuclear industry in Japan may have died from exposure to dangerous levels of radioactivity. The incident at the Tōkai-mura plant one month ago has revealed dangerous practices likened by some critics to "modern slavery" within the industry, putting the lives of untrained temporary workers at risk. Employment brokers are recruiting temporary workers from Japan's growing number of homeless people to do jobs like cleaning nuclear reactors. The recent poor safety record of the industry has made it hard for them to recruit staff, but the homeless are tempted by promises of much higher wages than other jobs can offer.[ci]

Masato Shinohara received another skin graft (to his lower legs) on the day Ouchi died, and a further graft to his face also succeeded,

---

29 They sometimes got little details wrong, like saying Ouchi's *brother* donated bone marrow. See: Normile, Dennish. "Experimental treatment for Japanese radiation victim." *Science* magazine, 4th October 1999. https://www.sciencemag.org/news/1999/10/experimental-treatment-japanese-radiation-victim.

although the connective tissue began to fail during the third month after exposure. Still, despite exposure to a fatal radiation dose, his bone marrow function returned and he appeared to recover thanks to the hard work of UTH staff – even enjoying escorted trips around the hospital gardens in a wheelchair on New Year's Day – but in late February 2000 he contracted aspiration pneumonia. Shinohara spent the following two months connected to a respirator as Maekawa and his team battled against inevitability, but on the morning of 27th April, after 211 days, he died of multiple organ failure. STA head Hirofumi Nakasone (son of Yasuhiro Nakasone) visited the hospital to express his sympathies to Shinohara's family. Grief-stricken once more, Maekawa announced to the media that treating Shinohara had forced him to confront the limitations of medical science and that he was "overwhelmed by a sense of helplessness in the face of something caused by human arrogance."[cj]

* * *

Some good did come from the tragedy. Unlike with previous accidents, this time the government took real notice and revised several laws related to supervision, the handling of nuclear fuel and disaster preparations. The wider industry, too, acknowledged a problem. "There is an opinion that it is not appropriate to generalize the whole of nuclear energy," wrote Kazuo Furuta, the man leading a working group investigation for the Atomic Energy Society of Japan. "However, given the fact that an accident has occurred that was not expected to happen in Japan, it is unreasonable to assert that there are no other nuclear safety blind spots.... Therefore, all nuclear officials must humbly accept the lesson of the accident and try to make the most of it."[ck]

And, to a certain extent, they did. From this point on, nuclear facilities of *all* kinds were required to undergo periodic government inspections, and all nuclear accidents were to be immediately reported to the Prime Minister's Office, not just to the local government.[cl] As was customary, the Nuclear Safety Commission wrote an accident report detailing the various causes and consequences. In addition to the elements mentioned above, they found a front-heavy bias on safety checks and that, while facilities and equipment *were* given thorough inspections initially, ongoing operating procedures went unmonitored. The possibility that companies or personnel might deviate from established guidelines was not considered at all, so the commission recommended implementing "specific safety designs against wrongful operation while taking into consideration of potential dangers." The report even acknowledged Japan's ingrained problem with the nuclear village's safety myth, writing "The present circumstances now call for a shift in consciousness from the 'safety myth' in nuclear power and the 'absolute safety' idea to a 'risk-based safety assessment.' The risk assessment concept is being established in the United States and European countries; however, efforts must be made to promote an understanding of this concept in Japan."[cm] The solution called for a "safety culture," where safety takes absolute precedence over all else. Had this attitude promulgated throughout the industry, Fukushima may well not have happened.

The NSC report also highlighted the extreme stress experienced by nearby residents, including a social stigma similar to that experienced by the *hibakusha* radiation victims of WWII. It stated that, in addition to the importance of public education on the true dangers (or lack thereof) of radiation exposure, "It is necessary in implementing measures to note that even if an accident comes to an end, victims must receive health and mental care for a long period of time. For this purpose, professional advice on health and mental care for victims should be taken from specialists in psychiatry, psychology, social psychology, and sociology."[cn]

Residents and businesses near the accident site filed over 7,000 compensation claims within a year. Most were for economic losses due to frightened gossip about radiation, forcing farmers to remove the Tōkai region label from their produce, and almost all were settled for a total of ¥12.73 billion (around $120 million in 2020). The Japan Atomic Energy Insurance Pool was used for the first time ever to cover ¥1 billion of this amount, while Sumitomo paid most of the rest because JCO's insurance covered less than 10% of the total.[co] Despite radiation anxiety being a clear cause of lost income, the updated nuclear accident compensation law stated that "mental anguish alone, without any personal injury, is not recognised as damage unless the claimants can irrefutably prove a causal relationship and the proportionality of the amount of compensation sought."[cp] In other words, victims of social alienation would not be compensated.

It transpired that the government-run System for Prediction of Environmental Emergency Dose Information (SPEEDI), a computer program using data from a network of radiation monitoring sites spread across Japan, was not used during the Tōkai-mura incident because "for budgetary reasons, the facility was not included among the facilities being monitored." The *Yomiuri* reported that "according to the Science and Technology Agency, under the current system, SPEEDI only has topographical data on nuclear plants and nuclear fuel reprocessing plants, which are required to have disaster-prevention measures. In any event, the agency did not ask for SPEEDI data in order to predict dispersal patterns after the nuclear accident."[cq] It's unclear if they didn't ask for it because they knew it wasn't available, because of the low volume of fission products, or because the Tōkai area is plastered with an array of radiation monitors belonging to various parties. Perhaps the data they received indicated the system wasn't necessary in this instance. Regardless, SPEEDI would become embroiled in further controversy during the Fukushima disaster.

The accident seriously dented public and political enthusiasm for nuclear power, prompting the power companies to reduce the number of planned reactors from around 16–20 down to 13 over the next decade. Even residents of Tōkai-mura – 10,000 (one-third) of whom worked in the industry, with almost all the rest connected to it in some way – felt a lingering trepidation. I always consider quoting accident victims a little redundant because *of course* they're upset, but after reading through comments made over the subsequent 12 months, the following stuck out to me: "The villagers are very careful about what they say and do. If they speak out, people would just criticize them as opposing [nuclear energy]."[cr]

It didn't help that everyone involved appeared to go out of their way to breed mistrust. Among a multitude of health check-ups conducted on all nearby residents, for instance, the local government ordered tests for a substance known to indicate DNA damage on everyone within 350 metres of the criticality. When the results revealed eight people beyond the normal limit, including one almost 40% beyond, rather than reassure them, "the prefectural government officials refused to conduct further tests because they were not sure whether they would be trustworthy. Furthermore, they did not give the test results to the people, saying it may cause anxiety among the residents," according to the *Associated Press*.[cs] I'll never understand how authorities, without malice, unintentionally make problems worse in these situations.

On 11th October 2000, police arrested six former and current senior managers of JCO and charged them with professional negligence, including the facility's manager and chief of production, along with Yutaka Yokokawa.[ct] Critics (and the defendants' lawyers) decried the move as another example of sidestepping the serious lack of industry oversight, especially when JCO president Hiroharu Kitani was not among those charged.[30] Nor were officials at JCO's parent company, Sumitomo, despite early police suspicions that they knew about the illegal processes.[cu] The presiding judge frowned upon speculation that government agencies allowed accidents to happen through inadequate regulation, reversing the argument while declaring in the March 2003 trial, "Allegations that the administrative authorities' supervision was insufficient are just a shift of blame."[cv] JCO was fined ¥1 million ($11,700 in 2020), while Tōkai reprocessing facility manager Kenzo Koshijima received a three-year sentence, suspended for five years, and a ¥500,000 ($5,850) fine. The other five, including Yokokawa, were handed suspended prison sentences of between two and three years. In other words, nobody was really held accountable.

The defendants' sentences were lighter than expected because it became clear at trial that they held only partial responsibility.[31] PNC, the state-run operator of the Jōyō reactor for which the fuel was being prepared, had asked JCO to modify their manufacturing process to ensure a consistent density while knowing that JCO's licence did not cover such a process.[cw] Nobody from the NSC which approved the licence, nor the STA which administered the company, nor PNC which requested the change, were asked to testify at trial. Neither prosecution nor defence appealed the ruling, and JCO announced it had abandoned plans to resume uranium processing the following month.[cx]

* * *

---

30 Kitani only took up the post three months prior to the accident and was unaware of the shortcut procedures. His predecessor, who, reports suggest, likely did know, was not charged either.

31 On a final distasteful note on this accident, JCO briefly opened the building where the accident occurred to tourists in December 2004. I don't know what they were thinking.

Speaking of fuel, the 20th century dealt one final blow to Japan's nuclear aspirations, right on the cusp of completing the cycle. Two weeks before the tank of uranium at JCO fissioned in Ouchi's face, Britain's *Independent* newspaper revealed that British Nuclear Fuels Limited (BNFL) – the UK-government-owned company reprocessing spent fuel bound for Kansai Electric's (KEPCO's) Takahama plant – had admitted to "irregularities" in its MOX fuel fabrication records.[cy] BNFL's MOX Demonstration Facility resided at the former Windscale site, which was renamed Sellafield in 1981 after the Three Mile Island accident drew unwanted attention back to the UK's own nuclear embarrassment. A complex assembly line at the facility pressed plutonium/uranium-oxide granules into pellets precisely 8.2 millimetres wide by 4 millimetres tall in a process that took months. Any deviation from this exact size – the maximum variance was ±0.0125 millimetres – would deform the fuel rods in a reactor, so all completed pellets underwent a stringent multi-stage quality check.[32]

The first stage involved automated scanning at three points by a laser micrometer, and those which passed were visually inspected for deformations. Workers then triple-checked a 5% sample with the micrometer, manually typing each measurement into a spreadsheet.[cz] At some point, however, they stopped performing this final check and instead duplicated existing data to save time. Five lost their jobs.

KEPCO flew a team of investigators to Sellafield, where they soon discovered a second batch of fuel with falsified data. Then a third. A government investigation report in February the following year revealed the practice had been routine at four of the facility's five shifts, that "the occurrence of non-compliant behaviour is not at all surprising" because of the work's tedious and repetitive nature, and that this stretched back to at least 1996.[da] The fuel, the report argued, remained safe because of the first fully automated check. Nevertheless, amid massive international fallout, the British government shut down BFNL's MOX facility and, in December 1999, MITI banned KEPCO indefinitely from loading the fuel at Takahama, where it was intended for the nation's first large-scale use after years of testing and political obstacles.[db, 33] "With these

---

32 There's 0.17 mm of helium-filled space between each pellet and the fuel rod wall. Under load, the pellets change shape depending on output and must not touch this wall lest the rods themselves crack or bend.

33 Tsuruga Unit 1 was the first commercial reactor to have a small test load of MOX fuel, in 1986.

new reports of dishonesty," remarked minister for MITI Takashi Fukaya, "we have to say that trust in BNFL has collapsed. Until BNFL regains that trust, we cannot import its fuel."[dc] The utility demanded Britain compensate them and take back the two batches[34] delivered in September (days before the JCO accident) and October, which they did at a cost of £200 million (in 2020).[35] In return, Japan agreed in July 2000 to lift their moratorium on further MOX deliveries from BFNL, but by this time replacement contracts had been signed with French company COGEMA ("Compagnie générale des matières nucléaires," which became Areva and then Orano), the same company used by TEPCO for its reprocessing.

As for Sellafield, it developed a horrendous reputation over the years. In 2007, for example, it came to light that the site's medical officer had ordered the removal of organs from 65 deceased employees (plus 12 more at other British nuclear sites) for study between 1962 and 1992, usually without the knowledge or consent of their families.[dd] The perpetrators took some extreme precautions. "The removal of the femur would leave an obviously limp leg, requiring reconstruction," notes the government enquiry report. "The mortuary technicians would use a broom handle to replace the missing bone and give the appearance of normality. The mortuary technicians' skill – and the way in which undertakers presented the body, often under a shroud stapled to the sides of the coffin – meant that most families did not notice that the body had been stripped of organs."[de] Nobody was prosecuted. Research on the radiological effects on human tissue was and remains important, and Ouchi's organs were preserved for medical research with his family's permission, but the subterfuge employed at Sellafield was barbaric.

---

34 This request was somewhat complicated because the fuel technically still belonged to the United States, from whence it had originally come.

35 £40 million compensation and £73 million to retrieve the fuel in 1999 money.

CHAPTER 6

Harbinger: The 2000s

i. Shakeup

The new millennium brought boundless optimism to the world, as if everyone could enjoy a fresh start. Japan shared in the festivities, but its nuclear industry now struggled to shake an image of recklessness. With confidence in Japan's image of unrivalled economic and political superiority shattered by the "lost decade," ever-expanding adoption of the internet drove engagement and recruitment among anti-nuclear activists. When combined with an increasing distrust in utility and governmental reassurances about nuclear safety – despite their ever-increasing funds for promotion and compensation – lead times for new reactors grew and grew.

As the calendar opened to the 21st century, nuclear energy still accounted for around one-third of the national total, generated by an armada of 17 power plants in 14 prefectures: from Tomari NPP[1] on the far-north island of Hokkaido to Sendai NPP[2] on Kyūshū, the southernmost of Japan's main four islands. Five new reactors came online during this decade: four at existing plants (Onagawa, Hamaoka, Tomari and Shika), but one – in Aomori Prefecture, just 20 kilometres from the harbour where *Mutsu* first launched – was brand new. The Higashidōri Nuclear Power Plant was the last built before 2011 and remains the most recent as of 2020. Tōhoku Electric and TEPCO share the unique site, although only Tōhoku

---

1   Tomari has three PWRs, with a combined output of 2,070 MW. A worker died there in August 2000 after he fell into a sump tank at the plant's waste treatment facility. He was taken to hospital but succumbed to his injuries. The incident was criticised for its similarities to the Tokai-mura accident a year earlier, in the sense that ambulance drivers were not informed about nor protected from the radiation on the man's contaminated clothes, and the hospital he was taken to was also not equipped to deal with radiation.

2   Sendai was built in the 1980s and is run by Kyūshū Electric Power Company. It has two 890-megawatt PWR reactors and sits 100 kilometres from the old port city of Nagasaki.

operates a working reactor, since construction work on Tōhoku's second reactor and TEPCO's first were postponed in 2011.

Fallout from the previous year's accident reverberated throughout the industry. As with each previous occasion studied by the NSC, they determined that blame lay with a combination of poor or non-existent training, disregarding safety procedures and a lack of government oversight to uncover any violations, thus allowing the problems to begin and gradually worsen until inevitability struck. This was once again followed by lies and attempted obfuscation of events. The same old pattern repeated itself – "the government was widely criticised for lax supervision and a slow response to the country's worst nuclear accident in September," read one *BBC* article – but, on this occasion, two people suffered horrendous deaths on the world stage.[a]

Given that this latest accident was clearly preventable, MITI and the NSC recommended a shakeup to usher in the third phase of nuclear regulation. The NSC itself was split off from the Science and Technology Agency and attached directly to the Prime Minister's Office in April 2000, increasing in size from 20 to 92 members, before moving again with the JAEC to the Cabinet Office in 2001. Here they enjoyed greater independence, as the Cabinet Office sits above the PM's Office and all other organs in the political hierarchy.[b] Unbelievably, however, any remaining language in the law that specifically ordered government ministries and private businesses to respect the NSC's decisions (remember, they had no legal authority) was *removed*.[c]

Legislators did alter laws relating to nuclear fuel to "strengthen the nuclear safety requirements for the management, operation and inspection of nuclear processing plants and nuclear energy facilities."[d] And, on 17th December, the Diet approved the creation of the Act on Special Measures Concerning Nuclear Emergency Preparedness, "to strengthen nuclear disaster control measures ... by providing special measures for the obligations of nuclear operators concerning nuclear disaster prevention, the issuance of a declaration of a nuclear emergency situation and the establishment of nuclear emergency response headquarters."[e] The Act attempted to clarify areas of uncertainty about what was and wasn't the responsibility of nuclear utilities and the national and local governments, as well as the timeliness and route of communications from a given company. The previous emergency law dealt primarily with natural disasters and had caused some confusion over who should do what when implemented during the JCO criticality.

In a broader move already planned from 1998, the government enacted sweeping agency restructurings in 2001 to reduce costs, clarify inter-agency responsibilities and generally provide more effective political leadership.[3] The Ministry of International Trade and Industry (MITI) was renamed the Ministry of Economy, Trade and Industry (METI) but remained responsible for regulating nuclear utilities and creating and promoting nuclear energy policy. Within METI, a new safety regulating body was established: the Nuclear and Industrial Safety Agency (NISA). The government didn't forget the Science and Technology Agency; PNC's (now JNC's) problems and the lack of viable progress with Fugen and other projects meant bureaucrats saw the STA as a disappointment. The agency merged with the Ministry of Education to form the Ministry of Education, Culture, Sports, Science and Technology (MEXT), ending the STA's three decades in control of nuclear projects. MEXT was made responsible for, among other things, "the development of nuclear technologies; safety regulations governing research reactors; protection against radiation hazards; the use and transportation of nuclear materials; [and] the use, storage and transportation of radioisotopes and peaceful uses of nuclear energy (safeguards)."[f] In any future severe nuclear accident, the prime minister would set up a crisis centre to monitor events as they unfolded, potentially ordering evacuations or for the Japan Self-Defense Forces (JSDF, the country's unified military, established in 1954) to intervene.

ii. When It Rains, It Pours

In July 2000, Kei Sugaoka, a Japanese-American former General Electric International "physics engineering aide," wrote a letter to the Japanese government blowing the whistle on two instances he had witnessed of TEPCO falsifying inspection reports at Fukushima Daiichi. Sugaoka had worked for GE inspecting reactors in the United States, Italy and Japan between 1976 and 1998, when he was laid off during a round of cost-cutting measures. In August

---

3   This restructuring was a pretty big deal and made changes in all sorts of areas. It's too vast (and boring) to go into here, but plenty of information is available about it online.

1989, his team discovered cracks in the steam dryer at Daiichi Unit 1. Such cracks can occur from moisture condensation or the stresses of heat and neutron bombardment degrading the metal over time. This is not necessarily a problem, according to the NRC, provided the dryer "does not continue to experience significant crack propagation and the safety consequences of any loose parts that may be generated have been previously analyzed to be acceptable," but, depending on their severity, cracks can require repairs or an entire replacement dryer assembly.[g]

"All I know is, I've never seen a dryer cracked like that," Sugaoka said with a laugh as he described his findings during a 2003 television interview.[h] "I've inspected many dryers – I'm very familiar with dryers and cracking in dryers." Upon further examination, he realised the dryer had been installed the wrong way round, a fact which – along with the large cracks – was omitted from the inspection report. Sugaoka's crew recorded a video of the damaged components, but TEPCO managers then ordered them to edit the footage, removing sequences showing cracks before submitting it to the government. Back in the United States, his superiors dismissed his concerns about the troubling practice of concealing defects in Japan. He returned to work, despite his misgivings, but felt compelled to speak up in dissatisfaction after being laid off.[4] NISA notified TEPCO of the accusations and began to investigate, while TEPCO launched an internal investigation of their own.

The following summer, four geologists supported by MEXT presented new research on the Jōgan tsunami from AD 869 in the *Journal of Natural Disaster Science*. They demonstrated, through sediment analysis and other methods, that waves during that event peaked at 8 metres (26 feet) high and ran at least four kilometres inland on the Sendai Plain, a large area of the east coast just north of Fukushima, indicating the earthquake had a magnitude of between 8.6 and 9.0.[5] The team further showed how earlier tectonic slips created other mega-tsunami prior to even the 869 event, writing:

4   Sugaoka first filed a discrimination complaint over his layoff before trying and failing to sue General Electric before eventually writing to MITI. This makes me somewhat question his motives for writing to them, but the outcome remains the same. Source: Upton, John. "Inside Japan's failing nuclear reactors." *Bay Citizen*, 16th March 2011: https://web.archive.org/web/20110324061245/http://www.baycitizen.org:80/disasters/story/inside-japans-failing-nuclear-reactors/2/.

5   Surveys have found deposits four kilometres inland but speculate they may have gone farther. I've seen claims of up to 7 km, although this figure came from spurious news articles and I can't find anything to back it up. Sediment 3 km inland was found while conducting research ahead of construction on Onagawa Unit 2.

The depositional ages inferred from [carbon] dating suggest that gigantic tsunamis occurred three times during the last 3,000 years. The respective calendar age ranges of the lower two layers are BC 140–AD 150 and ca. BC 670–910. The recurrence interval for a large-scale tsunami is 800 to 1,100 years. More than 1,100 years have passed since the Jōgan tsunami and, given the reoccurrence interval, the possibility of a large tsunami striking the Sendai plain is high.[i]

Their warning was overlooked because the Philippine Sea Plate fault line farther south seemed more likely to trigger the fabled "Big One," a long-awaited quake of astounding savagery.

A few months later, on 19th December 2001, TEPCO submitted a single-page memo to NISA outlining their tsunami expectations at Fukushima Daiichi in response to a request sent to each utility. Using guidelines written by the Japan Society of Civil Engineers, the memo outlined TEPCO's belief that regional geology could not produce an earthquake greater than magnitude 8.6 and no nearby tsunami could thus exceed a height of 5.7 metres (18.7 feet).[6] The document's minuscule length did not raise eyebrows at NISA. "This is all we saw," said the now-head of NISA's earthquake-safety division, Masaru Kobayashi, in 2011. "We did not look into the validity of the content."[j]

By 29th August 2002, NISA had unearthed enough evidence of wrongdoing from Kei Sugaoka's tip-off to announce their ongoing findings at a press conference. Investigators reported 29 falsified reactor defect inspection reports at 13 of TEPCO's 17 reactors at Fukushima Daiichi, Daini and Kashiwazaki-Kariwa, 11 of which were never fixed during the intervening time.[k] The next day, forewarning of events to come, the government's Central Disaster Management Council announced that "a long-feared massive earthquake hitting the Tōkai region in central Japan could result in the deaths of 8,100 people, destruction of 230,000 houses and buildings and daily economic losses of 345.1 billion yen."[l] In a second instance of unfortunate timing, TEPCO were forced to shut down Unit 2 at Fukushima Daiichi days later – unrelated to the ongoing scandal – due to a radiation leak.[m]

---

6   The JSCE committee responsible for the guidelines was criticised because 22 of its 35 members had strong ties to the nuclear power industry, though given how pervasive that industry is in Japan, I'm not sure the criticism is entirely justified. The society stood by the guidelines after Fukushima but said they did not check how TEPCO had applied them.

Sugaoka received death threats after METI provided his name to TEPCO within months of his original letter, in a direct violation of whistle-blower protections. There were no serious repercussions.[7] He later admitted, "I never thought it was that big – that many cases of cover-ups." It transpired that a number of TEPCO's own employees had witnessed poor working practices and informed the government of safety concerns. Their identities were also revealed to TEPCO, but again NISA took no action. After this, long-term Fukushima governor Eisaku Satō claimed whistle-blowers began to approach him directly, bypassing TEPCO and NISA for fear of reprisal.[n]

Prime Minister Junichiro Koizumi scolded TEPCO, telling reporters that "a power firm should give its highest priority to ensuring the safety of nuclear plants." Rural host communities and their local governments agreed, expressing alarm that, as Satō put it, "the tip came two years ago, but the reactors are still in operation with cracks in them.... I do not think [METI] is qualified to comment on maintaining and improving safety."[o] Satō was a strong supporter of the nuclear industry because of the jobs and subsidies it brought, along with the pride of hosting such an important facility, but gradually changed his mind as the number of incidents mounted. TEPCO had postponed plans to use MOX fuel at Fukushima Daiichi back in February 2000 in the wake of the BFNL scandal, until it could verify the safety of its own Belgian-made MOX. Nevertheless, with nuclear companies appearing to fake safety checks all around him, Satō rescinded permission for TEPCO to use it in September 2002 – a decision he would soon attribute to ending his career.[p] This followed a similar decision in Niigata Prefecture blocking TEPCO from using MOX at Kashiwazaki-Kariwa. That fuel still sits there unused today, almost twenty years later.

NISA justified the delay in investigating Sugaoka's allegations by saying that METI "repeatedly requested TEPCO to confirm the alleged matter on the basis of a newly received allegation.... However, the investigation did not smoothly progress because of little cooperation of TEPCO." METI instead went around TEPCO and straight to General Electric to produce records in November 2001, but TEPCO's refusal to admit to the cover-ups until August 2002 prevented METI and NISA from revealing them to the public. (Why is unclear: in most other situations, the government could

---

7   NISA's chief and a couple of other officials were reprimanded and sacrificed a small portion of their pay.

simply compel them to share documents.) Minister for METI Takeo Hiranuma apologised, saying that "taking two years [to investigate] is too long in light of common sense. It should have been done more swiftly."[q] NISA established subcommittees to review the investigation's slow pace and the adequacy of voluntary inspections that clearly weren't being done properly.

More falsifications emerged, spanning the 1980s and 1990s and involving dozens of employees. One case worth highlighting involved faked air-leak tests at Fukushima Daiichi's Unit 1 containment building in 1991 and 1992. BWR Mark I and II containments are known to perform badly in leak tests.[r] TEPCO knew the daily leak rate surpassed the regulation maximum of 0.348%, so instructed Hitachi engineers to inject compressed air into the space to ensure that pressure tests would indicate an uncompromised integrity. [8, s] Eight employees received reprimands, and the plant's Operation and Maintenance Division chief was fired. Beginning in April 1997, TEPCO became so concerned by the revenue lost during inspection downtime that they explicitly began paying maintenance subcontractors based on the number of days saved by working as fast as possible.[t] It was during this period that, while repairing the core shroud of Daiichi Unit 3, a TEPCO subcontractor recruited foreign welders to do the work because the regulations governing their radiation exposure were less stringent than domestic workers, as one of the project's managers revealed in 2011.[u]

TEPCO initially claimed their internal investigation indicated that nobody above an individual plant's own division chief was involved in the sweeping corruption, but later discovered that at least one current unnamed board member and up to 30 others at the company's Tokyo headquarters had participated.[v, w] Upon being informed of cracks at Unit 3 of Fukushima Daiichi and that an ultrasound test was recommended, "the senior official at the headquarters gave specific instructions not to do any ultrasound examination."[x] In early September 2002, TEPCO announced the resignations of President Nobuya Minami, Chairman Hiroshi Araki, a vice president and two senior advisors. Minami also

---

8   TEPCO employees did not pump in the air themselves. Hitachi workers were performing the tests, but TEPCO employees ordered them to inject the compressed air. Hitachi punished their responsible employees when all this came out, and company president Etsuhiko Shoyama and other executives took pay cuts to apologise. NISA's brief report on what happened can be found here: "Falsification of leak rate inspection of the reactor containment at the Fukushima-Daiichi nuclear power station." Nuclear and Industrial Safety Agency, 11th December 2002: https://web.archive.org/web/20030322172216/http://www.nisa.meti.go.jp:80/text/kokusai/TEPCO021211.pdf.

resigned his position as president of the influential Federation of Electric Power Companies (FEPC).[y] He recommended 62-year-old vice president Tsunehisa Katsumata, the "razor-sharp" man conducting TEPCO's own internal investigation, as his successor. Katsumata had joined the company in 1963 after graduating from Tokyo University's Faculty of Economics and, like his predecessor, hailed from the utility's Planning Division. The fourth of five brothers, all of whom have had distinguished careers, Katsumata became famous internally for his surgical cost-cutting.[9] Two weeks after the scandal broke, NISA decided not to pursue criminal charges against TEPCO, citing insufficient evidence of legal violations. They eventually judged 11 of the 29 original cases to have had "no safety significance," but "cautioned TEPCO severely" for the incident and ordered Daiichi Unit 1 shut down for a year as punishment for the laws they *did* violate. [z, 10]

At the scandal's zenith, industry magazine *Nucleonics Week* revealed old reports by the IAEA, whose plant inspectors had not been invited back to Japan since 1995 after finding serious problems with the country's attitude towards safety.[11] The reports had highlighted dozens of failings and didn't hold back on scathing comments, saying plants had "insufficient event analysis on near-misses," "weakness in emergency plan procedures" and a "lack of training for plant personnel on severe accident management," along with inadequate evacuation plans.[aa] The IAEA being this blunt about safety problems was quite unusual, as they often downplay these things.

Still more utilities and subcontractors became embroiled in the scandal as they reviewed old inspection reports. These included Japan Atomic Power Co, which repeatedly hid cracks up to 47 centimetres long in the reactor core shroud at Tsuruga during routine inspections in 1994, 1996 and 1998; Chūbu Electric, the nation's third-largest utility, which temporarily shut down all of its reactors after they admitted failing to report cracked water pipes at Hamaoka;

---

9 The Katsumata brothers are all Directors, Presidents and Professors of various organisations and institutions.

10 NISA gave a report submission order for this case according to the laws and investigated the falsification based on the submitted report. NISA determined that the act of TEPCO constituted a violation of the "Law for the Regulations of Nuclear Source Material, Nuclear Fuel Material and Reactors" and the "Electric Utilities Industry Law." Source: "Status report – Countermeasures against falsification related to inspections at nuclear power stations." Nuclear and Industrial Safety Agency, 10th December 2002: https://web.archive.org/web/20030826123118/http://www.nisa.meti.go.jp:80/text/kokusai/TEPCO021210.pdf.

11 The old reports were from inspections of Fukushima in 1992 and Hamaoka in 1995.

Tōhoku Electric, whose Onagawa plant also had a cracked reactor shroud and water pipes; Kansai Electric, with a far less significant loss of inspection reports; and Chugoku Electric, which again did not report reactor shroud cracks and failed to inspect hundreds of components at Shimane NPP.[12, ab, ac] In many instances, employees at vendors General Electric, Hitachi and Toshiba participated in the deception at the utilities' behest, when they weren't a victim of it themselves.

TEPCO confirmed Tsunehisa Katsumata as its new president on 16th October 2002 (he became Chairman of the FEPC in 2005). "I was really shocked to find out about the cover-ups," he admitted during a press conference, "because I had thought the nuclear division, where the manuals stated what to do in great detail, was the most strict division in terms of following regulations."[ad] He promised change, offering profuse apologies and an uncharacteristic criticism of the company he now led in a statement the following March. "Although it goes without saying that no circumstances can excuse the misconduct committed by our workers," he said, "I believe that I have to explain these circumstances in order for you to understand why engineers in our nuclear power division have acted inappropriately."[ae]

He first outlined how the specific reactor components where falsifications occurred were not considered to be safety-critical, and thus were discovered during voluntary inspections not overseen by a visiting government inspector. As such, he argued, those maintenance workers who found the cracks had no clear guidelines as to what constituted a safety risk, and they worried that reporting every problem to NISA would entail excessive reactor downtime. He further explained:

> this fear resulted in a conservative mentality that led them
> to avoid reporting problems to the national government as
> long as they believed that safety was secured.... In addition,
> the engineers were so confident of their knowledge of
> nuclear power that they came to hold the erroneous belief

---

12 A core shroud is an open-ended cylinder that fits around the other reactor components inside the core, used to hold fuel assemblies in place and direct the flow of water. Core shroud cracking became a big deal after first being discovered in BWRs in 1990. Such cracking was soon recognised as a widespread problem, and regular inspections were required all over the world. Here is an interesting NRC document about what they are and how they were repaired: "BWR vessel and internals project BWR core shroud inspection and flaw evaluation guidelines (BWRVIP-76NP)": https://www.nrc.gov/docs/ML0036/ML003688790.pdf.

that they would not have to report problems [...] as long as safety was maintained.... [They] eventually came to believe that they would be allowed not to report faults if the faults did not pose an immediate threat to safety and, as a result, they went as far as to delete factual data and falsify inspection and repair records.... Nuclear Division Members tended to regard a stable supply of electricity as the ultimate objective, and they repeatedly made personal decisions based on their own idea of safety.

Katsumata appeared humble, acknowledging "This way of thinking shows that our safety culture was inadequate." He reaffirmed his commitment to change.

Even after conducting thorough safety checks, TEPCO faced substantial unanticipated opposition from local governments, which refused to permit reactor restarts until the company took substantial steps to restore public trust. As such, only five of the seventeen reactors had resumed operation a year after NISA's August 2002 press conference, forcing TEPCO to reactivate moth-balled LNG plants and purchase electricity from other sources to cover the shortfall. They adopted the mealy-mouthed slogan "Create a mechanism that does not permit people to perform any dishonest act, and create a culture that encourages people to refrain from performing any dishonest act" to help reform the company's tarnished image, while revising a number of operating principles and practices.[af] METI updated the law on voluntary inspections to ensure the episode did not repeat itself and began conducting unan-nounced inspections. In reality, however, the plants were informed ahead of time.[ag]

The falsified inspection reports did not conceal catastrophic safety problems and NISA took no action, but the entire affair arrived on the heels of the 1990s accidents and reinforced a belief among some Japanese that nuclear power companies – TEPCO, in partic-ular – could not be trusted, nor could the government be trusted to regulate them. When a string of similar falsification scandals unconnected to the power industry appeared one after the other at around the same time, a new popular term arose: *Tōden mondai*, "the TEPCO problem." And yet, by the mid-2000s, an *Asahi* poll indicated that support for nuclear power surpassed opposition for the first time since Chernobyl.[ah] Indeed, compared to the 1999 criticality accident, the cover-ups in the early 2000s received little news coverage, resulting in a more limited public awareness outside those paying close attention.[ai]

Then, in 2006, the scandal spilled out into other power sectors, with falsifications at thermal and hydro plants. It was huge: over 1,000 instances by the end of the year, rocketing to a downright ridiculous 10,000 cases of "concealed problems or altered data at power plants" reported by the FEPC in March 2007.[aj] The list ran the gamut from the potentially dangerous[13] to the plain lazy,[14] with hydroelectric plants accounting for most, but it also included hundreds of new cases at nuclear plants.

TEPCO disclosed 199 instances between 1977 and 2002, some of which were potentially far more serious than previously reported.[ak] At Kashiwazaki-Kariwa Unit 1 in May 1992, the residual heat removal pump (the main cooling system during and after shutdown) failed the day before a government inspection. Technicians knew the inspector would flag this system failure, so they switched the pump over to a test-run mode. When operators then demonstrated the pump by pressing the control room button, the relevant indicator light blinked on – it passed the test, but nothing had actually happened, and the plant ran for two days without a fully functioning cooling system before it was fixed. Further cases appeared well into 2007, including a few accidental criticalities caused by control rods slipping and multiple occasions where TEPCO and Tōhoku Electric had failed to report emergency reactor shutdowns in 1984, 1985, 1992 and 1998 to the government as required, plus one instance of Unit 1 at Hokuriko Electric's Shiga plant going critical by accident after workers mishandled fuel rods.[al, am, an] Three nuclear research facilities in Tōkai also closed for a while after another whistle-blower revealed problems there.[ao]

I'm not going to list every single instance; you get the idea at this point. Sporadic reports continued until the 2011 earthquake,

---

13 The *Yomiuri* newspaper reported in late 2006 that "the results were disastrous. At 68 dams throughout the nation, deformation and the fabrication of survey data, including water levels as well as failures to make necessary reports to authorities, were discovered." Source: "Power firms struggling to generate public trust." *Yomiuri Shimbun*, 30th December 2006: http://www.yomiuri.co.jp/dy/editorial/20061230TDY04005.htm Archived link: https://web.archive.org/web/20070106225931/http://www.yomiuri.co.jp/dy/editorial/20061230TDY04005.htm.

14 KEPCO's Ōi Nuclear Power Plant revealed one instance where the ocean outlet water temperature was higher than the temperature of water leaving the turbines. Nature and logic dictate that water still in the system should always be hotter than water flowing back into the ocean, but on this occasion it wasn't. Instead of investigating the discrepancy, perhaps by checking for damage to temperature probes, KEPCO just falsified the data. Obviously, the local aquatic ecosystem wasn't considered either.

but most were insignificant. The point is that all these companies ignored the rules for their own convenience.[15] Despite everything, one of NISA's final reports on the initial TEPCO falsifications highlighted many instances of good practice with plant maintenance, and no doubt this was true most of the time, but they were lost in the noise.[ap]

The 2000s brought shame to the power industry. News channels repeatedly aired footage of company executives bowing in ritualised *owabi kaiken* ("apology with an audience"), yet the cover-ups did little to move the needle of public opinion. The government responded to the barrage of deceit by ordering TEPCO, JAPC, Hokuriku Electric and Chugoku Electric – those responsible for the most serious infractions – to conduct additional, more stringent checks at their plants, and ... that was all.[aq, ar] METI knew that to shut down all those power stations would create a crippling energy shortage, so they did nothing. No fines, no prosecutions, nothing – for which the ministry took a battering in the press over its soft approach. In February 2007, a magazine quoted Masao Takuma, Kashiwazaki-Kariwa's former manager during the 1992 ECCS pump failure incident and then-current TEPCO advisor, expressing his opinion: "People at the site have great pride in their technology," he said. "However, the regulations covering nuclear power are very strict. It seems that this had the opposite effect to that which [was] intended. People ended up thinking that all that was necessary was to pass the inspections."[as] Many others, including numerous power company executives, proposed similar versions of Takuma's argument: nuclear rules were too tight and every tiny problem had to be reported, leading to downtime, lost profits and lost electricity. The argument has merit up to a point but disregards clear signs of systemic problems, as evidenced by the huge number of falsified inspections at hydroelectric plants, and does not excuse the most severe cases, especially when this lax attitude led to fatalities.

---

15 In April 2009, for example, Hitachi announced two instances of falsified data at the Hamaoka and Shimane NPPs, one in 2001 and the other in December 2008. As noted in a *Yomiuri Shimbun* article, "According to the companies, the falsified data related to the final processing of welding that connected pipes to moisture separator reheaters. The reheaters remove moisture from steam generated by turbines and heats it up again. To strengthen the welded area, the pipes must be heated to about 700 C after welding and then cooled down slowly. But the connected pipes were cooled down at twice the stipulated speed because the person in charge turned off the device heating the pipes to step away from the work area. The person reportedly erased the recorded data and entered data in its place that conformed with regulations." Source: "N-plant pipe data falsified." *Yomiuri Shimbun*, 14th April 2009: http://www.yomiuri.co.jp/dy/national/20090414TDY02310.htm Archived link: https://web.archive.org/web/20090415155203/http://www.yomiuri.co.jp/dy/national/20090414TDY02310.htm.

iii. Deaths at Mihama

Monday, 9th August 2004 marked 59 years since the Hiroshima bombing, a day of national mourning. It was also the date Japan suffered its worst-ever accident at a nuclear plant in terms of direct fatalities, and the first fatal accident during operation. Hundreds of subcontractors and staff at KEPCO's Mihama plant were busy preparing for Unit 3's periodic inspection and maintenance, which would take it offline for an extended period. KEPCO depended on nuclear power for 60% of its electricity output – the highest among its peers – and a speedy downtime hinged on comprehensive preparation. Missing anything could result in expensive delays. Back in February 1988, for example, both main water pumps at Hamaoka Unit 1 failed simultaneously – thought to be impossible – three months after an inspection because, even though the pumps ran on separate electrical circuits for redundancy, the switches protecting them from overheating did not. The electromagnetic relay connecting them burned out seven years earlier than the manufacturer's guarantee period, cutting power and forcing operators to shut down the reactor, thus emphasising the wisdom of a thorough inspection.[at]

At 3:22 p.m., fire alarms tore through Mihama's control room and turbine hall, indicating a fire at one end of the vast hall's second floor. Eleven people stood near the 'A' loop condensate pipe when it ruptured up near the ceiling.[au] Steam "spewed" out, according to a KEPCO spokesman, at 1,700 cubic metres per hour, instantly boiling alive four men and scalding seven others.[av] Operators sent to investigate found the space filled with steam and, unable to see the affected area, ran back to the control room and began an "emergency load reduction."[aw] Two minutes later, a water/steam imbalance alarm tripped,[16] causing the reactor to automatically shut down.

A fifth man with burns to 80% of his body passed away two weeks later, but the other six survived their injuries. All eleven were subcontractors from Kiuchi Instruments, a company hired to maintain the many gauges, dials and thermometers throughout the plant. KEPCO's CEO, Yosaku Fuji, offered a 50% reduction

---

16 According to NISA, "The trip signal is activated when feedwater flow rate to the steam generator is less than steam flow rate under steam generator water-level-low condition." Source: http://www. meti.go.jp:80/english/newtopics/data/mihama3e.html. Archived link: https://web.archive.org/ web/20041211234323/http://www.meti.go.jp:80/english/newtopics/data/mihama3e.html.

in his pay for three months by way of apology. Luckily, Mihama operates PWRs, so there was no radiation risk from the secondary loop rupture.

The accident was caused by flow-accelerated corrosion (FAC), a slow thinning of the pipe wall over almost 30 years. A raised internal aperture, part of a system to help monitor flow rate, pushed steam against the 10-mm-thick pipe wall two metres down-stream of the aperture. There, the steam ate away at the pipe until it was just 0.4 mm thick: little more than a sheet of paper. "To put it bluntly, it was extremely thin," said minister of METI Shoichi Nakagawa after visiting the plant. "It looked terrible, even in the layman's view."[ax] Two almost identical accidents at American PWRs had occurred in 1985 and 1986 at the Trojan and Surry nuclear plants, the latter of which resulted in four deaths.[ay, az] Westinghouse then issued recommendations to all its partners for extra vigilance regarding pipe thinning, after which Japanese regulations required utilities to check 25% of FAC-vulnerable piping within each 10-year period. Areas within twice the pipe diameter of a probable source of corrosion were considered at risk, but Mihama's pipe burst one metre beyond that limit – an oversight that cost five men their lives.

Workers had only ever inspected the pipe's exterior, rather than using ultrasound to "see" any internal degradation.[17] The utility subcontracted Mitsubishi Heavy Industries to inspect and maintain Mihama's pipes between 1989 and 1996, but when Osaka-based Nihon Arm took over in 1996, instead of the anticipated computer database of past records, Mitsubishi gave them mostly hand-writ-ten notebooks.[ba, bb] Three years later, Mitsubishi became aware of a greater risk of corrosion than previously thought and made repeated unsuccessful attempts to contact Nihon Arm. Luckily, after spending years computerising Mitsubishi's notebooks, a Nihon Arm engineer noticed the fateful pipe was missing; he added it to the list of items requiring inspection and notified KEPCO in April 2003. KEPCO claimed to receive "no information that the portion of the pipe was posing an immediate danger" from either company, according to one company executive.[bc] This meant, in a tragic twist, that technicians

---

17 This appears to mean that they never actually checked the inside of the pipe. It was widely reported (including by NISA) that KEPCO never once checked the pipe, but exactly what this means is unclear. "We conducted visual inspections but never made ultrasonic tests, which can measure the thickness of a steel pipe," said Haruo Nakano, a Kansai Electric spokesman. Source: *Japan nuclear firm investigated.* BBC, 10th August, 2004. http://news.bbc.co.uk/1/hi/world/asia-pacific/3550604.stm

had planned to ultrasound the pipe five days after it burst. Mihama's deputy manager echoed his superior's remarks, telling reporters "we thought we could postpone the checks until this month. We had never expected such rapid corrosion."[bd]

A sweeping inspection of KEPCO's nuclear plants revealed dangerously thin pipe walls at ten of its eleven reactors and multiple further instances of components omitted from inspection reports.[be] "Nonconformity with technical standards had become a norm from around 1995 to the present," remarked one report, "by interpreting technical standards independently and postponing repairs."[bf] KEPCO promised to strengthen the organisation of its Nuclear Power Division and improve maintenance practices, while inspections at all 23 Japanese PWRs followed.

National opinion polls cited by the *BBC* indicated that half the population thought the number of nuclear facilities should be lowered as Japanese media outlets fretted over ageing facilities such as Mihama undermining public support for nuclear power.[bg, bh] Matsutaro Shōriki's old paper the *Yomiuri* cautioned that "care must be taken not to overemphasize the dangers involved in the operation of nuclear power stations, which could lead to an overreaction."[bi] The *Asahi* disagreed, claiming in an editorial that "it is essential that power companies are prepared to judge where there are weaknesses in the system and inspect them in time. If only for that purpose, all the power companies should get together to exchange information on even the most minor problems." The paper also pointed out that "ample information in this regard is available through numerous databases in the United States."[bj]

Though the Mihama incident remains the deadliest accident at a Japanese nuclear power plant (even including the Fukushima disaster) and is thus worth discussing, the fact that it occurred at a *nuclear* plant is irrelevant in most respects. Any facility using steam under pressure is susceptible to such an accident, and on the scale of fatal industrial accidents it was relatively minor, but it highlighted even more negligence in the eyes of the public. The incident was all over the news for a couple of weeks, including news of police raids on KEPCO offices. Mihama's reactor did not restart again for two and a half years. Prosecutors charged five people from the plant with professional negligence shortly thereafter, all of whom settled out of court and received fines.[bk]

iv. The Balance of Power

The government merged JNC and JAERI to form the Japan Atomic Energy Agency in October 2005. While JNC largely avoided PNC's accidents and bad press, the JAEA would not. Earlier that year, after a decade of review, extensive upgrade work began at the fast breeder Monju. Engineers expanded existing leak detection systems to aid quick understanding of the location and severity of any compromised pipes. These new systems were also linked to the plant's ventilation ducts, which would now automatically shut to prevent aerosols from spreading, further assisted by the secondary loop being split into four isolated areas. The thermocouple that had caused the original problem was redesigned, and a nitrogen gas injection system was installed around the plant to starve any fires of oxygen.[61]

Global support among lay people and experts for breeder reactors waned over time (though not always at the highest levels of government). They were just too expensive to build and maintain, and dealing with sodium exacerbated the reliability problems experienced at ordinary plants. In December 1996, France halted what was, at the time, the last electricity-generating breeder reactor in Europe: the prototype Superphénix. Prime Minister Lionel Jospin announced its permanent closure the following June after a successful public legal challenge against its operation. The move sent shockwaves throughout the Japanese nuclear community, as France was the only other nation devoting significant resources towards nuclear reprocessing, but environmentalists hailed it as a great victory.

* * *

After the carnage wrought by the 1995 Kobe earthquake, the NSC spent five years examining data from academic papers and the latest worldwide safety practices in preparation for a full review of Japanese earthquake regulations. The Seismic Review Committee – comprised of thirteen academics, two power company managers and one person from NUPEC,[18] all chosen by the NSC and

---

18 NUPEC, the Nuclear Power Engineering Corporation, was a private company formed in 1976 "as a result of cooperation from such private industries as power generation, electric and heavy machinery manufacturing, and general construction along with advice and assistance from the government of

NISA – met multiple times from approximately late 1999 (sources vary) through August 2006, ostensibly to integrate these global standards.[bm] The seismologist Katsuhiko Ishibashi, as one of the few non-industry committee members, was unimpressed, later lamenting, "We went around and around for five years, but the outcome was predetermined."[bn]

The prolonged committee review period included feedback from the FEPC, essentially indicating which changes it would permit and which it would not. For example, the FEPC reluctantly agreed to implement back*checks* on existing plants using the updated guidelines, but "with respect to the reference which says ... 'it is important to apply [back*fits*, i.e. modifications]' ... we are concerned that such a statement can reinforce the claim that the seismic safety of the currents plants is insufficient, and can influence lawsuits aimed at stopping the construction or operation of nuclear plants." The same old argument justified sedate security upgrades at Kashiwazaki-Kariwa in the wake of 9/11. "We are in the process of making those changes [to prevent terrorist attacks]," one TEPCO official told George W. Bush's chief of staff Paul Dickman, "but we don't want to do them all at once because we don't want people to think that we have been operating them unsafely in the past."[bo] The 500-page 2012 report by the National Diet of Japan Fukushima Nuclear Accident Independent Investigation Commission (NAIIC) is damning in its assessment of this back-and-forth process:

> The [utilities, via the FEPC] stubbornly refused any moves toward backfits for the assessment of seismic safety or strengthened regulations, including the regulation of severe accident countermeasures. As a result, no progress was made in Japan toward introducing regulations necessary to reduce accident risk, and the country failed to keep pace with world standards by not fulfilling the concept of the five-layered defence-in-depth. The approach taken in reviewing regulations and guidelines did not follow a sound process of establishing regulations necessary to ensure safety, and the regulators and the operators together looked for points of compromise in the regulations in order to maintain appearances [of] regulation and satisfy the condi-

---

Japan and cooperation from learned experts." NUPEC provided research, development and testing of nuclear power technologies.

tions for one of their major premises: that "existing reactors should not be stopped."[bp]

This phenomenon of the utilities dictating to the regulators became known as "regulatory capture."

On 16th August 2005, a magnitude 7.2 earthquake struck the east coast just thirty kilometres (18 miles) from where the devastating 2011 quake would hit. At 251 Gal, the shaking exceeded the nearby Onagawa plant's seismic limits in some non-safety-critical areas, causing minor damage. It made NISA nervous. A magnitude 9.1 had rocked Indonesia eight months earlier, causing a massive tsunami – the deadliest in history, claiming 225,000 lives – that swept across the Bay of Bengal, where it reached India's Madras Atomic Power Station. Tsunami don't typically happen around India and the plant was not designed to withstand one, but it featured a clever safety system built for cyclones. "Before a cyclone hits, a storm surge comes," explained Dr L. V. Krishnan, former director of Safety Research and Health Physics at the Indira Gandhi Centre for Atomic Research. He continued:

> Half a kilometre into the sea there is a huge well that … is connected to another well on the shore on land … by an under-seabed tunnel. As the storm surge comes in and the water level rises in the well in the sea, the level of water also rises in the well on the shore. The moment the water level rises beyond prescribed limits in the well on the shore, the seawater pump trips.[bq]

This trip alerts operators, who then shut down the reactor, which is exactly what happened at Madras.

Was Japan similarly prepared for unexpected natural events? NISA ordered TEPCO to investigate possible weak points at its plants on 14th December 2005, specifically asking what would happen if the seawater inlet pumps were submerged.[br] TEPCO reported back on 11th May 2006. If a 10-metre tsunami hit Fukushima Daiichi, it would breach the plant wall, swallow the pumps and, with intakes for the water-cooled backup diesel generators on low ground, they would stop and the reactors would melt.[bs] All internal and external power systems would fail if a 14-metre tsunami hit, ensuring the destruction of every reactor. At a hearing that October, NISA stressed:

at some sites, the difference between the projected tsunami height and the height of the ground level of the premises is small – around tens of centimetres. Although this is permissible for the evaluation, tsunamis are a natural phenomenon and some that are higher than the design assumption may strike. If a tsunami exceeds assumptions, the safety margin will be zero due to the malfunctioning of seawater pumps for emergencies and damage to the reactor core.[bt]

The agency made it clear that this scenario could lead to a station blackout, which itself could result in a serious accident. TEPCO, in response, increased their tsunami estimate by a mere 40 centimetres to 6.1 metres and had elevated some equipment such as the seawater pump motor by November 2009.

Katsuhiko Ishibashi quit the Seismic Review Committee during its final meeting over frustration that discussions had been "unscientific" and that fellow members refused to implement an absolute worst-case design standard or re-evaluate the survey standards for active fault lines, saying "the committee's misunderstanding leads to a strong underestimation of powerful earthquakes."[19, bu, bv] An engineering professor and fellow committee member later commented, "I understood what Ishibashi was saying, but if we engineered factoring in every possible worst-case scenario, nothing would get built. What engineers look for is consensus from the seismologists and we don't get that."[bw] It's hard to disagree with his logic on the face of it (you don't build a library to withstand a meteor strike), but seismology involves plenty of guesswork: scientific, researched, computer-modelled, educated guesswork, but nobody can predict with 100% accuracy what the planet will do next, nor when. Combined with the potential hazards of a compromised nuclear reactor, disagreeing gets easier.

The May 2006 JAEC long-term plan called for further increased reliance on nuclear energy, from 30% to 40% by 2030, as part of

---

19 Ishibashi was not the only academic to be frustrated by government committees. A former Tokyo University professor of seismology named Kunihiko Shimazaki also tried to get the government to take earthquake and tsunami warnings at nuclear plants seriously but was repeatedly stonewalled. In an interesting interview with the *New York Times*, he claims the committee did not want to force TEPCO to spend money. Fackler, Martin. "Nuclear disaster in Japan was avoidable, critics contend." *New York Times*, 9th March 2012: https://www.nytimes.com/2012/03/10/world/asia/critics-say-japan-ignored-warnings-of-nuclear-disaster.html.

international obligations to reduce greenhouse gas emissions.[20] Despite this advancing of the nuclear agenda, the domestic industry experienced global drops in new reactor orders despite fresh fears of Middle East oil dependence after a spike in oil prices. The lull pushed General Electric and Hitachi to join forces in November 2006 to expand the portfolio of each and to provide aid to Hitachi, after fixing a batch of faulty turbines cost it over $100 million. This was the latest in a string of industry-shifting alliances during late 2006, which also saw Toshiba buy a 77% stake in Westinghouse for $4.2 billion and Mitsubishi Heavy Industries join with French nuclear group Areva (known today as Orano) to co-design a new generation of PWR plants: the first time European and Japanese nuclear firms had formed a major alliance.[bx, by]

Japan released its updated seismic guidelines in September 2006 following a year of public comment and review. Improvements included extending the age of active faults from 50,000 to 130,000 years, recommending new geomorphological surveys to provide empirical and theoretical estimations of ground motion, and introducing the concept of "residual risk" to consider unexpected events.[bz] The NSC declared the revisions to be "based on the latest findings in seismology, earthquake engineering, etc. and the accumulated experience of safety examinations to date."[ca] Indeed, they did boost overall plant strength by around 20%.[cb] The guidelines, however, once again overlooked accompanying phenomena like tsunami, which received only vague mentions such as "safety functions of the Facilities shall not be significantly impaired by tsunami which could be reasonably postulated to hit in a very low probability."[cc]

Strangely, the committee also *lowered* some existing standards, such as removing the blanket minimum earthquake resistance of a magnitude 6.5 within ten kilometres of a given plant. Staff now reviewed each plant on a case-by-case basis, as opposed to case-by-case with a minimum of 6.5, despite the enormous threat of earthquakes. This does not appear to have had much undue negative influence, however, as the new design-basis earthquake criterion – which every plant must survive – became higher than before. The problem was, though the new guidelines required power companies to conduct backchecks on their facilities, the NSC decided that

---

20 As you may have noticed, these percentages went up and down (or stayed the same) over time when things did not pan out as expected.

"confirmation of the seismic safety of existing facilities is legally outside regulatory actions."[cd] NISA and the NSC first scoffed at the FEPC's insistence on a three-year grace period to complete the backchecks but soon – once again – caved under pressure. By the time of the Fukushima disaster in 2011, four and a half years later, TEPCO had still not submitted their report.[21] Internally, the work was scheduled for completion in January 2016, over two decades after the Kobe earthquake that had started the process and one decade after the backchecks began. NISA did nothing meaningful to expedite the progress nor combat utilities' blatant disregard for government guidelines.

To summarise: the government, impotent in the face of private company domination, belatedly released improved earthquake regulations which remained far below the standard set by other nations and unfit for an area with such frequent and violent seismic activity. The reason is simple: the FEPC and the wider nuclear village viewed external threats as unrealistic. Rather than account for potential tsunami discovered by new technologies and methods, the NAIIC concluded, "TEPCO's understanding was that it was the impact of the risk on its business that had increased, not the likelihood of the risk. This meant … they were only aware of risks of taking counter-measures, shutting down existing reactors and facing lawsuits."[ce] This contrasted sharply with nuclear plants in the United States, which had been prepared for multiple external triggers, including earthquakes, floods and tornadoes, for over 15 years by this point.[22]

* * *

On 23rd October 2006, police arrested 67-year-old Eisaku Satō – the outspoken TEPCO critic and governor of Fukushima since 1988 – after he was forced to resign during his fifth term in office.[cf] He stood accused, along with his 64-year-old brother, Yuji Satō, of accepting bribes for illegally granting a large construction contract to the Maeda Corporation via bid rigging in the form of *ten no koe*,

---

21 TEPCO did submit interim reports for Fukushima Daiichi in March 2008 for Unit 5 and April 2009 for the other five reactors. NISA accepted TEPCO's interim findings but found the scope of their work thus far to be too narrow.

22 Of course, the provisions in the US are not perfect either, but in 2011 the Fort Calhoun nuclear plant in Nebraska survived relatively unscathed after being surrounded by water when the Missouri River burst its banks during freak weather.

or "voice from the heavens." In 2000, Maeda, a multi-billion-dollar construction company founded in 1919, won a contract auction to build the Kido Dam, 10 kilometres south-west of Fukushima Daini, on the heels of work on Kashiwazaki-Kariwa and the Hong Kong International Airport. As the contract progressed, Maeda instructed subcontractor Mizutani Construction Co in 2002 to overpay for a piece of land occupied by a menswear factory owned by Koriyama Santo Suit – a company founded by Eisaku and Yuji's father, where Yuji was president and Eisaku the principal shareholder. Koriyama Santo Suit agreed to sell it for ¥870 million but ultimately took payment of ¥970 million ($9.5 million in 2020). Yuji donated ¥30 million (around $280,000) of the profits to Eisaku's winning 2004 re-election campaign, with the rest going towards restructuring the company.[cg]

The Tokyo District Public Prosecutor's Office came upon the sale while investigating Mizutani Construction for tax evasion and, though the statute of limitations for bribery had expired, decided to pursue the case anyway. While in custody, Yuji was apparently asked why his brother was "against nuclear power plants," said that Eisaku was "not good for Japan" and told him that any governor opposing nuclear power would be removed.[ch] In court, prosecutors argued the elder Satō knew Mizutani were working for Maeda and, even though he had no involvement in the general running of his brother's company, that he must have known about the sale and it was therefore bribery because the price exceeded the market value of ¥800 million by almost 10%. This in itself was odd, since the value of land constantly fluctuates. Nevertheless, despite lacking proof, the court convicted Eisaku of accepting intangible bribes via the bizarre crime of *kankin no rieki*, or "making a profit," handing him a suspended prison sentence in 2008.

He appealed in October 2009. Mizutani Construction had resold the land for an even higher amount during the intervening time, thus forcing prosecutors to drop the main charge. The Tokyo High Court, however, upheld the decision of accepting an "intangible bribe" for a nonsensical zero yen and declared that simply selling the land constituted bribery.[ci] The judgement caused some puzzlement but seemed to be based on Eisaku Satō having signed a confession under interrogation shortly after his initial arrest, though he claimed this was purely to speed up the bail process and relieve pressure on his supporters. The act of confessing during questioning, then claiming innocence during trial, is not uncommon during high-profile cases in Japan. Satō asserted that the entire arrest and trial was character assassination orchestrated by TEPCO and the month-old Shinzō Abe government over his continued refusal to permit MOX fuel in Fukushima (Satō was, after all, the only person left standing in their way), but the Supreme Court upheld the sentence again in 2012.[cj] Whether there's any provable truth to his claim remains to be seen. I have my doubts, it seems just as likely that the prosecutors found the sale suspicious and decided he must have been involved. That said, the weak evidence is also unusual, as Japanese prosecutors enjoy a 99% conviction rate precisely because they don't move forwards unless victory is assured, so something strange went on. Satō covered the saga in a 2009 book, which became a best-seller following the 2011 disaster.

iv. Earthquake at Kashiwazaki-Kariwa

At 10:13 a.m. on 16th July 2007, the Okhotsk tectonic plate[23] deformed along a previously unknown fault, triggering a magnitude 6.6 earthquake 17 kilometres offshore from the village of Kariwa. The Niigata-ken Chūetsu-oki Earthquake, as it came to be known, lasted almost 20 seconds and collapsed hundreds of mostly old, wooden buildings, resulting in 11 deaths. Tens of thousands of homes went without gas, electricity or water for days, and small 30-centimetre tsunami sloshed against the shore. Though the temblor originated out at sea, it still occurred close to the world's largest nuclear power plant: TEPCO's Kashiwazaki-Kariwa (KK). Combined with the quake's shallow focal depth of only 8 to 10 kilometres, a rupture length of 30 kilometres – all underwater – and ground failures up to 50 kilometres away, the shaking was far more violent than might be expected for a magnitude of 6.6.[ck] It was the strongest quake to ever hit a Japanese nuclear plant.

Three of KK's seven boiling water reactors were offline for periodic inspection and maintenance at the time. Of the remaining four, Units 3, 4 and 7 were under normal load, while Unit 2 was

---

23 The Okhotsk tectonic plate is a minor plate sandwiched between the Eurasian, North American and West Aleutian Orogeny plates to the north, the Amur plate to the east, a sliver of the Philippine Sea Plate to the south, and the Pacific Plate to the west. The joint between the Okhotsk and Pacific plates is a subduction zone that was the source of the 2011 megathrust quake that caused the Fukushima Daiichi disaster.

commencing start-up. All four shut down automatically the instant KK's array of seismic monitors detected ground motion exceeding 100 Gal.[cl] Owing to unusual local geology, the direction and intensity of shaking across the 2.25-kilometre-long site varied enormously, with one academic paper describing it as "highly incoherent and impulsive."[cm] Japan's plants are built to survive the design-basis event, or, more specifically in this case, a theoretical largest possible local temblor called the S1 "maximum design earthquake," derived from historical activity and known faults and defined as 200–440 Gal. The S1 reference is called the S2, or "extreme design earthquake," a devastating quake conceived as the absolute limit of what regional tectonic structures could summon, with a minimum magnitude 6.5 right beneath the plant.[cn] An S2's range is 370–588 Gal. The shaking at KK far exceeded even the latter, with ground motion up to 2,058 Gal at its most frenzied spot around the base of Unit 3's turbine hall – over twice that exact location's rated limit of 834 Gal. Other parts of the plant encountered similarly severe shaking. Unit 4, with a design-basis of 194 Gal, was hit by 492 Gal. Ground acceleration at Unit 2 reached 606 Gal, but it was only guaranteed to withstand 167 Gal.[co] The highest value at a reactor building was 680 Gal at Unit 1. This pattern repeated across the site, where shaking knocked workers off their feet, causing nine injuries.[24]

With all that in mind, the plant performed remarkably well, suffering no damage to safety-critical equipment, no loss of external power and no station blackout – the nightmare scenario. Most important of all, almost no radiation escaped into the environment. Less vital components, however, experienced widespread failures, with up to 2,555 "non-conformances," according to TEPCO.[cp] Most problems were immaterial, such as buckled roads, cracked concrete and damaged turbine blades at one unit, but a couple are worth mentioning. Unit 6 had just been refuelled, and an enormous overhead gantry crane was about to replace the reactor's 700-ton outer lid when the earthquake struck, damaging the crane. Over at Unit 7, a control rod became stuck inside the reactor, while Unit 3's blowout panel fell off after its hinge bent. The blowout panel is a section of the reactor building's external wall which is designed to detach (blow out) by itself when internal pressure reaches a certain

---

24 The new September 2006 seismic guidelines replaced the S1-S2 system but S1-S2 had applied to any plants built prior to this (i.e. all of them). The replacement system defined a single design-basis earthquake.

point, to prevent a build-up of steam or flammable gases. This panel failure meant the building lost its negative pressure, thus allowing for the potential release of radiation.

The worst problems occurred when soil deformation short-circuited a large electricity transformer beside Unit 3, which burst into flames. The unit's dedicated fire team responded quickly, locating the blaze within two minutes. Suspecting an oil fire, they first searched for chemical extinguishers but found none, then tried to douse the flames with water from a nearby hose, but the water pipe had been damaged by the quake.[cq] Out of options, they ran to use KK's dedicated hotline to the local fire station but couldn't get into the emergency response room because shaking had buckled the door frame, forcing them to call the congested national emergency number – 119. Owing to the local devastation, firefighters took an hour and a half to arrive, and then another 30 minutes to extinguish the blaze.[25] Thankfully, a tall firewall contained the blaze, but a thick plume of black smoke terrified local residents who thought the reactors must have been damaged. They were lucky.

TEPCO president Tsunehisa Katsumata visited the plant after two days and called it "a mess." Mohamed ElBaradei, Director General of the IAEA, asked for an investigation and offered to send expert inspectors to help, but the central government declined. They soon changed their minds under pressure from Niigata Prefecture officials, who argued that IAEA inspectors could provide neutral information to the public, who did not trust TEPCO.[26,] [cr] The Minister for METI, meanwhile, ordered TEPCO not to restart Kashiwazaki-Kariwa's reactors until repairs were completed and gave each power company just seven days to prove they were tightening their safety procedures, including giving each plant its own fire station.[cs] The *Yomiuri Shimbun*, a right-wing outlet normally banging the drum for nuclear power, also expressed concern: "That the Kashiwazaki-Kariwa nuclear power plant was hit by a massive

---

25 What caused the delay is unclear. Plant manager Akio Takahashi said in a 20th July press conference that nobody had called the fire brigade prior to his arrival at the plant at 11:00 a.m. and that he had not explicitly instructed them to earlier because he assumed someone had already done so. When someone finally did try, he said, the earthquake had disabled the hotline to the local fire station. During a 30th August press conference, however, TEPCO mentioned none of this, but instead claimed the shift supervisor had contacted the fire brigade at 10:27 "but was asked to use [the] in-house self-defence fire brigade" (presumably meaning the fire team that did initially try to fight the fire). TEPCO claimed the staff had called the fire brigade again an hour later, which was when they responded to the call.

26 The IAEA had submitted the report from its first-ever comprehensive review of Japan's nuclear regulatory system just a few weeks before the earthquake. Though phrased in diplomatic language, it highlighted innumerable problems. Hence, the government did not feel like entertaining further IAEA inspectors.

earthquake cannot be dismissed as somebody else's business by any of the nation's nuclear power plants."[ct] The paper encouraged NISA and NSC inspectors to be frank about any problems to help restore public faith that safety was a priority, and called for the gap between public and official perceptions of accidents to be addressed.[cu]

TEPCO received considerable bad press in the following weeks for their untimely accident reports. For example, 1.2 cubic metres of water had sloshed out of Unit 6's spent fuel pool and then leaked down through cable ducts before finding a drain into the sea. TEPCO did not report the leak in their first press conference, then declared it soon after but claimed a smaller amount than what had actually leaked; they then corrected the mistake before declaring another minor release via a chimney stack. They also took 12 hours to report spilled low-level radioactive waste barrels (typically containing ashes from incinerated overalls, gloves and masks) which had toppled during the earthquake. The facility contained this extra leak.

The mayor of Kashiwazaki was furious with TEPCO, more so after learning that the IAEA had warned them about their inadequate fire-prevention measures two years earlier. The plant experienced a further nine fires within two years of the quake while it remained idle, with the ninth coming less than two weeks after the Kashiwazaki City Fire Department lifted a ban on flammable materials at KK based on a TEPCO fire-prevention plan.[cv, cw] Even Prime Minister Shinzō Abe weighed in. "I believe that nuclear power plants can only be operated with the trust of the people," he said. "For this, if something happens, they need to report on it thoroughly and quickly. We need to get them to strictly reflect on this incident."[cx] Katsuhiko Ishibashi chimed in, too, saying the earthquake "could have been predicted, and should have been predicted," and that the new seismic guidelines "are very insufficient and have loopholes."[cy] TEPCO president Katsumata apologised for "all the worry and trouble we have caused," but defended his company and the plant, saying, "I think fundamentally we have confirmed that our safety measures worked.... It is hard to make everything go perfectly."[cz]

Given the plant's immense size and the need to first find and then analyse the situation and identify how much radioactivity had been released, some delay in announcing spills and other environmental-impact issues was to be expected. Unlike with many other instances, there doesn't appear to have been a concerted effort to hide what happened here. TEPCO provided daily updates as workers combing the site discovered more problems, yet each time they found something new, there was a collective "ah ha!" from the media, which no longer trusted them. The amount of radioactivity was extremely low (less than one ten-millionth of the annual ambient background level from the spent fuel pool), but that didn't stop headlines such as "Japan Admits More Nuclear Leaks" from even reputable outlets like the *BBC* appearing worldwide.[da] TEPCO didn't "admit" to the accident: KK held 22,000 barrels, and it took a while to check them all. When workers found that lids had fallen off 40 barrels and the contents spilled out, they reported it.

Nevertheless, given the company's recent history, TEPCO's tardiness was understandably perceived to be another cover-up, albeit not as blatant. That's also not to say the utility wasn't back to its usual PR tricks – it was and did its best to downplay the string of failures that had allowed the fire to burn for two hours. TEPCO also used subcontractors to mop up the contaminated water by hand using paper towels, which, on the one hand, isn't as alarming as it sounds because the radiation levels were so low, but it's also an archaic image for such a cutting-edge industry.[db]

The real issue was not that TEPCO took time to report small and inconsequential waste spillages, but that the plant had been hit by an earthquake far stronger than it was designed to survive and they were unprepared for the aftermath. Instead of responding with a palpable hit of knee-trembling relief that KK had scraped through what could have been a catastrophe relatively unharmed, TEPCO, in public at least, patted themselves on the proverbial back and demonstrated this survival as proof that plants could withstand anything. Yes, the reactors had passed the test, in that they didn't shake themselves to pieces, but how close had they come to failing?

The Minister for METI poured scorn on regulators he claimed hadn't properly reviewed TEPCO's initial pre-construction geological survey back in the 1970s and complained that the utility had underestimated the size and threat of nearby faults. Sitting on the committee which had signed off on that survey, incidentally, was none other than Yoshihiro Kinugasa – the government seismologist who approved the Shika NPP site before it endured its own quake. TEPCO should have done better, too. "Not finding the fault was a miss on our part," said Toshiaki Sakai, head of TEPCO's nuclear plant engineering division, with an unironic deadpan, "but it was not a fatal miss by any means."[dc] In fact, TEPCO informed NISA in 2003 that the fault was far longer and more active than initially believed, but the company took little action. METI did what they always did when faced with a public backlash: increased their promotional spending and host-community subsidies, the former by ¥199 billion ($1.96 billion in 2020) or 12%, the latter by 5% and around the same value.[dd]

∗ ∗ ∗

Nuclear village complacency had gone on long enough for Professor Katsuhiko Ishibashi. He penned an op-ed in the *Asahi Shimbun* on 11th August 2007, warning that the worst was still ahead and that the time to act was now. "In the 40 years that Japan had been building nuclear plants," he wrote, "seismic activity was, fortunately or unfortunately, relatively quiet. Not a single nuclear facility was struck by a big quake. The government, along with the power industry and the academic community, all developed the habit of underestimating the potential risks posed by major quakes."[de] That period of inactivity, he said, was over, and since the 1995 Kobe earthquake, Japan had experienced several medium-strength quakes. Ishibashi highlighted three which struck close to nuclear plants from the previous two years alone, saying that "in each case, the maximum ground motion caused by the quake was stronger than the seismic design criteria." Because even perfect seismic research could miss faults large enough to cause a magnitude 7.3, he argued, the minimum tolerances should be raised to reflect such strong shaking: around 1,000 Gal. Instead, they accounted for considerably less. Ishibashi worried, as he and others had for years, that Chūbu Electric's Hamaoka NPP was at particular risk. The plant was old – commissioned in 1971 – and right on the Shizuoka Prefecture coastline, where seismologists had long predicted "the Big One" would hit. He urged its closure.

Turning to Japan's institutional problems, Ishibashi wrote, "The most serious fact is that not only are the new design guidelines defective, but the system to enforce them is in shambles. Much of the blame for the underestimation of the active fault line near the

Kashiwazaki-Kariwa plant rests with the shoddy examination of TEPCO's design for the plant that overlooked the problem." He urged the Diet to consider "a radical reform of the government approach to ensuring the safety of nuclear power plants" in order to prevent disaster. "Otherwise," he warned, "there can be no viable future for Japan's nuclear safety." Part of his wish came true in December 2008, when Chubu Electric announced their intention to retire Hamaoka's first and second of five reactors.[df, dg] Due to their age, installing upgrades in line with the new regulations proved too costly.[27] Instead, Chubu revealed plans to replace them with a single state-of-the-art reactor, though this plan fell through after the Fukushima disaster.

Though the damage to Kashiwazaki-Kariwa was not serious, the incident symbolised every dysfunction prior to the looming disaster at Fukushima. Yet, even as the most prominent foreign media outlets – the *New York Times*, the *Guardian*, *Der Spiegel* and others – sang a chorus of open criticism, Japan's largest mainstream media groups still shied away from harsh commentary. Under the circumstances, after a decade of accidents and scandals, you would expect more, so why the silence? Journalists from top news organisations such as the *NHK* television network and the *Asahi* and *Yomiuri* newspapers have, for over a century, enjoyed a cosy relationship with government officials and the nation's largest companies via *kisha kurabu*, or "press clubs." There are hundreds of separate press clubs for each government ministry, political party, major industry sectors, sports and entertainment organisations, and so on, each with their own journalist members who are ordinarily stationed within the heart of the area they cover, such as the Diet building. Few private companies have their own club, with some exceptions (TEPCO is one).

Contrary to similar concepts like the White House press pool in the US, the clubs are an exclusive closed clique with special access to the best sources and information, which, as a rule, they do not share with those outside the clubs – even within their own news agencies. You can't simply pass a security check for the Diet building and automatically be added to the club, even if you belong to a news organisation; the club must grant you permission to join. They also

---

27 Incidentally, eight months after this announcement, Shizuoka Prefecture and Hamaoka were hit by a magnitude 6.5 earthquake, which automatically shut down Units 4 and 5 at the plant. Chubu Electric issued the somewhat smug statement that "as the reactors were suspended as planned, we see no problems in their quake-resistance."

self-regulate, punishing anyone who disobeys the rules. One famous example occurred in 1990 when the *Asahi* was forced to submit a written apology to Diet club members and was almost ostracized when one of its journalists printed a high-level government source's name after the members had agreed not to attribute a sensitive quote. The journalist responsible was reassigned.[dh] Prime Minister Kan's press conferences were strictly *kisha kurabu*–only affairs during the Fukushima disaster, meaning no freelance or foreign journalists, which is the norm for high-level sources. Individual reporters within a club have tiered access to these sources, with the most intimate tier reserved for relatives of powerful people or graduates of the same university as the source: *gakubatsu*, discussed earlier. Advocates justify this system with the paper-thin argument that club members know not to jeopardise the access they enjoy and therefore gain more information than would otherwise be possible, but, in practice, club members won't challenge sources with difficult questions, regardless of the circumstances.[28] This scenario naturally leads to situations where non-members – smaller, independent news companies, weekly and monthly magazines, tabloids, or freelancers – who have no such privilege to lose, break most major scandals.

Still, the problems were widely recognised, despite the reticence: Japan, sitting atop the world's most earthquake-prone region, had gone all in on nuclear power, on which the country now depended. Media groups therefore ignored the dangers. This unwillingness to change did not go unnoticed. During the December 2008 G8 gathering in Tokyo, for instance, an IAEA representative highlighted the troubling fact that an earthquake had now exceeded a plant's design; the representative explained that "safety guides for seismic safety have only been revised three times in the last 35 years" but assured delegates that the agency was now re-examining them.[di]

* * *

TEPCO staff presented new tsunami research at a nuclear engineering conference within a week of the Kashiwazaki-Kariwa earthquake. As the company's own engineers admitted, "We still have the possibilities that the tsunami height exceeds the

---

28 Several of the largest news companies also do not name the individual journalists behind each article, unlike common Western practice, as the collective voice is valued over those of individuals.

determined design height due to the uncertainties regarding the tsunami phenomenon."[dj] Their research identified four earthquakes close to Fukushima Daiichi and Daini of magnitude 8.0 or higher in the last 400 years, and concluded that Daiichi's coastal defences stood a 10% chance of being breached within the next 50 years. Despite this latest warning coming from inside their own company, TEPCO's executives did not act. Amazingly, NISA, under pressure once more, announced they would *extend* the time between reactor inspections, from 13 months incrementally up to 24 over the next decade.

The world's largest nuclear plant was still offline at the end of December, costing TEPCO over $1.4 billion in lost revenue and prompting a national panic over energy shortages, since the plant provided 13% of the company's power capacity. A net loss of ¥95 billion ($935 million in 2020) in that fiscal year pushed the company into the red for the first time since 1979.[29, dk]

Sixty-seven-year-old Tsunehisa Katsumata, who had complained of feeling old and tired even before the earthquake, had quietly stepped back from the day-to-day operational stresses of leadership in June 2007 and seized the losses as an excuse to retire as TEPCO's president in January 2008. Executive Vice President Masataka Shimizu, who rose through the Materials Division, took his place. A numbers man and expert in procurement who'd joined the company where his father worked at age 23, Shimizu once explored a TEPCO plant on foot with a local town chairman after he noticed that the number and location of street lights did not match the blueprints. Interestingly, he became the first-ever graduate of a private university (Tokyo's Keio University, the same as Yasuzaemon Matsunaga) to assume control of the company. He is also the husband of Katsumata's daughter, her senior by 35 years.[30] Under Shimizu's leadership, TEPCO upgraded KK's earthquake resistance to 1,000 Gal in October 2008 and restarted the first reactor – Unit 7, the newest, strongest and least damaged – on 9th May 2009.[31] Units 6 and 1 followed within a year.

"We judge that [our disaster-prevention procedures] worked properly," commented NISA Deputy Director General for Nuclear

---

29 TEPCO announced an expected customer rate increase of 12% the following summer (the largest since the oil crisis), to pass on the cost of imported oil and LNG, until METI told them to lower the rate.

30 At the time of the Fukushima disaster, Katsumata was 71, Shimizu was 66 and she was 31.

31 This was raised to 2,300 gal for Units 1 to 4 and 1,200 gal for Units 5 to 7 after the Fukushima disaster.

Safety Hitoshi Sato, a few weeks after the July 2007 earthquake. "But some may attribute it to luck, because the quake was indeed stronger than the maximum level estimated in the plant's design.… There is always an unexpectedly strong earthquake. We are dealing with nature. It would be a lie if we say our guideline is sufficient. The bottom line is we should learn from the experience of larger-than-expected quakes." His comment is disheartening in hindsight: it was almost too late. Still, perhaps because KK had weathered the storm without a major catastrophe, an August 2008 *Asahi* poll showed that more people now believed nuclear power was safe than unsafe for the first time in 12 years – a trend reflected across a variety of other polls. This appeared to clash with what were now enormous lead times of 20-plus years for new plants, in part because enthusiasm for nuclear power in the abstract did not transfer into support for proposals in local communities. This distinction typically resurfaces with each new accident. Some people remain unfazed, but even supporters of nearby plants have doubts. "I thought I had made peace with the power plant," remarked Yonin Kondo, a fishmonger and village official living close to Kashiwazaki-Kariwa in 2008, "but now I can't keep it out of my mind."[32, dl] This sentiment was reflected in a local poll on trustworthiness conducted in Tōkai-mura, in which fewer than 20% of respondents now considered the government to be reliable in an emergency.[dm]

v. A Matter of Time

Japan was better prepared than ever for a nuclear disaster by the tenth anniversary of the JCO criticality, owing largely to changes implemented in its aftermath. The government spent millions of dollars in the intervening decade establishing 22 "off-site centres" nationwide, where private and public officials would converge to coordinate between any local and national response. Each of the 19 nuclear-host prefectures designated their own hospitals for patients exposed to varying doses, while 13,000 medical workers attended training programs on radiation treatment. These preparations, while good on paper, had one flaw: they assumed an accident

---

32 I have deliberately avoided the temptation to use quotes like these throughout this book. It isn't hard to find comments from people attacking nuclear power for any reason, but they usually feel anecdotal. For every person who hates or feels scared by it, there will be someone who does not. I felt this quote, however, was interesting because it came from someone who previously accepted the plant.

would be a single, confined, internal event in an otherwise fully functioning society.

TEPCO made improvements too. In addition to reinforcing various components, Fukushima Daiichi gained a highly advanced, earthquake-proof seismic-isolation building (SIB) in July 2010.[33] The SIB came from experience gained during the earthquake at Kashiwazaki-Kariwa, where the emergency response office sustained damage and could not be used, forcing management to coordinate the early recovery efforts from outside. Daiichi's two-storey SIB was designed to, in TEPCO's own words, "house an emergency response centre and other important communications and power supply facilities that would provide the necessary foundation for disaster prevention activities."[dn] It sat on the inland side of the main administration building, 35 metres (115 feet) above sea level, on an untouched section of the original cliff between Units 4 and 5. Flexible rubber dampers cushioned any movement between the building's foundations and the earth below, allowing it to ride out even severe shaking.[do] With its narrow porthole windows and own generators and air filtration system to protect occupants from outside contamination, the SIB would become critical to Daiichi's survival less than a year after completion.

Rural communities now regularly fought legal battles against nuclear power and those elected officials who supported it. Tying up utilities in years-long litigation became activists' weapon of choice, though higher courts invariably quashed any initial victories. On the flip side, many people born and raised within some of those same communities felt bound to the plants. "Nuclear power is the economic lifeblood of Tsuruga, and hosting the plants has provided us with a far better lifestyle than we otherwise would have enjoyed," a mother of two young children told a *Japan Times* reporter after the 2007 earthquake. "Yes, I worry about a severe earthquake damaging the plants and I don't believe that nuclear power is as safe as we're often told by the government," she conceded. "But I'm more worried about what would happen to our community if the plants were all shut down and dismantled."[dp] After everything

---

33 Kashiwazaki-Kariwa took priority and received its own seismic-isolation building first, in January 2010. Fukushima Daiichi and Daini both gained theirs at the same time. These buildings are referred to by several different names depending on the source, the most common being "quake-proof building." TEPCO call them seismic-isolated buildings, but I'm using the government's term of "seismic-isolation building," since it sounds better in English. I'm abbreviating it to SIB, but I don't think this is a recognised shorthand. Seismic-isolation buildings at nuclear plants have been in use since France first built one at the Cruas NPP in 1985.

that had transpired during the preceding years, after the accidents and the deaths and the lies, the view that Japan should rid itself of atomic power remained rare. Instead, this young woman's feelings may have reflected the majority view. After all, nothing truly devastating had happened since Chernobyl almost 25 years before, and even that had occurred in a cost-cutting, dystopian, secretive society, not one of the world's bleeding-edge economic and technological powerhouses.

Regardless of any individual feelings, nuclear power was experiencing something of a renaissance (in public perception, at least, if not in terms of new builds) as world leaders pushed to drastically reduce greenhouse-gas emissions. Coal power produces, on average, 979 tons of CO_2 per gigawatt-hour of electricity generated. Oil produces 810 tons, natural gas 550, solar 94, biomass 50, nuclear 32, and hydro and wind both 29.[dq] Because electricity generation produces around one third of global emissions, the push towards renewable energy was becoming a real policy driver. The JAEC reaffirmed its belief in the aims first set out in 1956 when, in 2008, it encouraged the government to stay the course with existing nuclear plans and the breeder program, in part because of their benefits in the fight against global warming.[dr] METI agreed and said it hoped to approve 14 new reactors by 2030.[ds] Japan was ahead of the curve: as one of just a handful of nations with over 50 years of unwavering support for nuclear energy, its low-carbon nuclear technology and expertise was arguably the best in the world.[34]

* * *

Monju was on track for an October 2008 restart until early that year, when implementing the revised seismic guidelines revealed a 15-kilometre-long active fault nearby.[dt] Further inspections of the plant showed that around half of 403 brand-new sodium leak detection sensors were defective due to poor workmanship.[du] At the same time, the JAEC revealed that taxpayers had been paying

---

34 Japanese companies continued to introduce little innovations. In December 2009, KEPCO installed a water wheel and turbine on the water outlet at Sendai NPP, with plans to use the power it generated to charge electric vehicles. Plans to become the centre for future Asian nuclear power development were also well underway, including jointly organised programs through government and industry encouraging students from Indonesia, Thailand, Vietnam and others to study to become the next generation of nuclear experts at Tōkai University.

to keep Tōkai-mura's Recycle Equipment Test Facility (RETF) idled for eight years while waiting for Monju to resume operation. Scientists hoped to use the RETF to improve the reprocessing method used on Monju's spent fuel. Media outlets expressed their frustration over the delay in bringing Monju back online with no end in sight, with the *Yomiuri* complaining about "defects in the inspection plans, poor discipline among workers and a low level of awareness of safety concerns" among senior officials.[dv]

Ultimately, almost 15 years passed before low-power tests recommenced in May 2010, in part because of legal wranglings over inadequate precautions and a 2003 High Court nullification of the facility's original planning permission. Another crippling accident occurred just four months later. A flaw in a custom gripping attachment for the overhead gantry crane, used to lift Monju's unique three-ton re-fuelling In-Vessel Transfer Machine (IVTM), caused a vital screw to loosen over time. The screw popped out on 26th August, just as the crane was withdrawing the IVTM – a long cylinder, 46 centimetres wide and 12 metres tall –from the reactor, where it had spent the week exchanging 33 fuel rods. The gripper opened, dropping the refuelling device into the core, where it deformed and became lodged inside where two sections of the IVTM were joined by metal pins. All efforts at extraction failed and the fuel could not be removed.[35] The device remained stuck as tsunami water pounded Fukushima Daiichi five months later.

Stalling the breeder program yet again was a huge blow, but a backup plan remained: MOX. The government had always intended to use recycled MOX fuel in fast breeders (its experimental use in light water reactors years earlier was more of a stopgap to reduce plutonium stockpiles while engineers worked out Monju's kinks), but using MOX in the country's existing reactor fleet would serve much the same purpose. Back in 1971, bureaucrats had set aside a 5,000-hectare area of flatland in the remote far north of Aomori Prefecture, near the village of Rokkasho-mura and not far from Mutsu, for future industrial use as part of the "Shin zenkoku kaihatsu sōgō keikaku" (usually shortened to Shinzensō), or Comprehensive National Development Plan. The site was originally

---

35 Monju has an unusual design. In an ordinary reactor, even if one fuel channel becomes blocked for whatever reason, the remaining channels can be unloaded. Monju uses the refuelling device to insert fuel perpendicular to the main fuel group. A revolving rack then positions the new fuel with the rest in the middle of the reactor vessel.

intended for petrochemical, refining and other purposes but was eventually designated for Japan's pre-eminent, catch-all nuclear fuel enrichment, reprocessing, MOX production, spent fuel storage and disposal facility.

JFNL first built a low-level waste storage facility and uranium enrichment plant, both of which came online in 1992. The following year, construction began on the Rokkasho Reprocessing Plant for a 1997 opening, but after extensive trials it still had not commenced full operation two decades later because of constant legal and technical challenges. The main problem arose from a decision to use an incomplete glass vitrification process under development at the prototype Tōkai Reprocessing Facility, rather than the processes long-used at Areva's La Hague and BFNL's Sellafield facilities, to much consternation from some quarters. "The nuclear fuel cycle, a pillar of Japan's energy policy, has again run into trouble," a furious *Yomiuri* journalist wrote in October 2010, under an article titled "Nuclear Fuel Cycle Delay Is Intolerable." The journalist noted that Rokkasho had been "delayed for two years until 2012. This is the 18th time this project has been postponed."[dw]

In contrast to the government, the utilities had always opposed breeders and MOX. "In public we will tell you that this is taking the long-term perspective," one unnamed senior TEPCO official remarked back in the early 1990s. "But what everyone knows, and won't say in public, is that this will cost consumers a fortune."[dx] A 2011 JAEC estimate suggested that domestically produced commercial MOX fuel would be twelve times as expensive as regular uranium fuel – a cost paid for by consumers.[dy] For years, Japan had been stockpiling and sending spent fuel to Europe for reprocessing as a temporary measure, though much of it returned to Japan with nowhere to go. Most spent fuel storage pools at individual nuclear plants were (and are) almost full, though a process of re-racking allowed some to exceed their original capacity. Building additional on-site storage was politically impossible, so the chosen option became reducing the volume of spent fuel via reprocessing and re-using, then storing it long-term at Rokkasho, lest plants be forced to cease operating.

In June 2009, the government delayed its goal of 16 to 18 reactors using MOX by five years to 2016. Nevertheless, by the eve of the Fukushima disaster, over 50 years after first announcing their intention to create a closed loop fuel cycle, a few plants loaded large volumes of MOX fuel for the first time.[dz] Unit 3 at Kyushu Electric Power Co's Genkai NPP came first, in November 2009, followed by Shikoku Electric's Ikata NPP in March 2010. Six months later, Fukushima Daiichi's newly refurbished Unit 3 became the third. The Fukushima region had become Japan's largest "power-supplying prefecture," with an abundance of thermal, hydro and nuclear power plants distributing energy throughout the nation.[ea]

Yet, despite all the upgrades and the crucial importance of Fukushima's power plants, the prevailing safety myth meant Japan remained unprepared for a natural disaster. During a meeting with TEPCO back in August 2002, NISA had pointed to a then-recent government report stating that a major east-coast tsunami could hit at any time, with a 20% likelihood in the next 30 years, and asked the utility to conduct tsunami simulations. TEPCO personnel "resisted for 40 minutes," according to one NISA official, and ultimately declined to investigate.[eb] July 2009 telegrams from the US embassy in Vienna about then-upcoming IAEA executive transitions complained about the organisation's Japanese head of safety and security, Tomihiro Taniguchi. The former deputy director-general of MITI had been, as noted in the telegram, "a weak manager and advocate, particularly with respect to confronting Japan's own safety practices," indicating little had changed among the country's bureaucratic elite.[ec]

TEPCO did conduct belated tsunami simulations in 2010 after the magnitude 8.8 Chilean earthquake inflicted ¥6.3 billion of tsunami damage on Japanese fishing businesses.[ed, ee] Here, again, they produced the old 5.7-metre maximum wave height. "The whole exercise," according to a Royal Society analysis of the work, "[was] nothing short of baffling to any reader with a minimum amount of scientific competence in the field and, most importantly, of common sense."[ef] That same year, according to Yoichi Funabashi and Kay Kitazawa of the distinguished Rebuild Japan Initiative Foundation, the government of Niigata Prefecture (where Kashiwazaki-Kariwa is located), "made plans to conduct a joint earthquake and nuclear disaster drill. But NISA advised that a nuclear accident drill premised on an earthquake would cause 'unnecessary anxiety and misunderstanding' among residents. The prefecture instead conducted a joint drill premised on heavy snow."[eg]

On 7th March 2011, just four days before the imminent disaster, TEPCO shared a three-year-old internal report with NISA, containing the results of a simulation applying the 1896 magnitude-8.5 Sanriku Earthquake (mentioned in chapter 3) to the Fukushima coast's Japan Trench, not much farther south. The bone-chilling result produced wave heights exceeding 10 metres (33 feet) and flood heights of almost 16 metres (52 feet) at Fukushima Daiichi. TEPCO had long dismissed this result as "mere assumptions with no actual basis" and later blamed their inaction on NISA, saying the agency had "not instructed [us] to immediately implement counter-measures" during the meeting, just as they hadn't in September 2009 when TEPCO shared calculations placing the 869 Jōgan tsunami at 9.2 metres at Daiichi.[ch, ei] Withholding this latest information for so long suggests a clear fear of being forced to pay for more robust defences.

On 9th March, at 10:45 a.m., an earthquake struck 160 kilometres off the east coast. Quakes are logarithmic, meaning the effects scale up, with each whole number on the scale releasing 32 times more energy than the number below. At magnitude 7.3, it was two and a half times larger and four times stronger than the magnitude 6.9 that destroyed Kobe in 1995, but the Onagawa, Higashidōri, Fukushima Daiichi and Daini plants barely noticed its 60-centimetre tsunami waves.[ej] The quake was a foreshock, a harbinger of the disaster ahead. Precisely 1,142 years had passed since the earthquake that triggered that great 869 tsunami – the last major east coast temblor. Warned for years of another long-overdue "Big One," Japan's time was up.

CHAPTER 7

A Complex Disaster

It's important to remember while reading this chapter that we benefit from hindsight and are tracing the course of events one step at a time. Throughout their ordeal, the people at Fukushima Daiichi faced constant uncertainty and took many actions not discussed here. At no point did they do nothing. Any gaps in the narrative exist to streamline the story.

i. Friday, 11th March – Day One

The Pacific tectonic plate, stretching from Japan to San Francisco, accumulated incalculable potential energy over the centuries, straining beneath its Eurasian cousin at an indifferent, inexorable nine centimetres per year.[1] This subduction dragged a thick layer of surface clay between each tectonic behemoth, lubricating the joint until, on 11th March 2011 at 2:46 p.m., the tension became too great. A 500-kilometre section snapped back 50 metres (164 feet) – by far the largest-distance fault slip ever recorded – and drove a colossal volume of water upwards.[2] The magnitude 9.0 Great Tōhoku Earthquake released over 350 times more energy than the 7.3 two days earlier as it shook, first with lateral motion, then vertical, at a peak acceleration of over 1g for an exceptionally long six minutes. It was felt up to 2,000 kilometres away and the entire coastline moved a few metres east. Automatic warning systems broadcast alerts via television, radio and text messages, while bullet trains applied their brakes and surgeons stopped their delicate work. Ferocious spasms caused widespread damage across

---

1 Some geologists consider the Eurasian Plate to be a fragment of the North American Plate.

2 The lubricating clay is generally thought to have enabled such a powerful movement; normally there's far more friction. Source: "The 2011 Japan tsunami was caused by largest fault slip ever recorded." *National Geographic*, 7th December 2013: https://news.nationalgeographic.com/news/the-2011-japan-tsunami-was-caused-by-largest-fault-slip-ever-recorded/.

the nearby Sendai Plain and beyond, but most buildings – long reinforced – fared well and the earthquake produced just 5% of total fatalities: 145 from burns and around 650 crushed by falling objects. In addition, there were 6,142 injuries compared to 43,792 from the Kobe quake.[a] Landslides and other failures, however, devastated local infrastructure by twisting and tearing apart roads, railways and bridges, and disrupting gas, electricity and water lines across a vast area. As such, though emergency services leapt into action, the widespread damage hindered their response.

Masao Yoshida, 56, clung to his desk for dear life. Like all Japanese, he had experienced countless small to medium earthquakes in his life and yet, at Fukushima Daiichi, 180 kilometres from the epicentre, he could only watch helplessly as his spacious office shook itself apart. Born on 17th February 1955, Yoshida joined TEPCO straight out of university in April 1979 as an engineering graduate, and, apart from a short stint at the FEPC, dedicated his entire career to its nuclear plants. He headed TEPCO's Nuclear Division for three years prior to earning the general manager position at Daiichi in June 2010. Tall, with large, round glasses, and a heavy smoker who loved to bet on horses, Yoshida stood out, according to those who knew him – in the same vein as Yasuzaemon Matsunaga and Yanosuke Hirai – as a reliable, principled leader. "He was a man who evaluated others by their character," recalled a close friend, "not by whether they could do their job well, regardless of job title."[b] When at last the shaking subsided, Yoshida picked his way through offices strewn with toppled computers, chairs and cabinets, wind-swept blinds fluttering in broken windows, and marched out of the administration building to the emergency assembly point. There, in the chilly March air, he instructed his subordinates to account for all personnel, then strode into the new seismic-isolation building (SIB) next door.

Up in the SIB's second-floor emergency response centre (ERC), damage and injury reports flew in from department heads, but practicing their emergency drills the previous week meant everyone knew what to do and the situation appeared stable. Sensors had recorded a maximum ground acceleration of 550 Gal, surpassing the tolerances of each reactor by around 20%, but all critical hardware seemed to be undamaged. Units 4, 5 and 6 were, thankfully, already offline for periodic maintenance: Unit 4 sat empty, its fuel cooling in the spent fuel pool, while 5 and 6 were loaded with fresh fuel and water in preparation for restart. Control room chiefs for Units 1, 2 and 3 confirmed that these reactors – which had been at full power – had automatically scrammed and were being cooled. Their main steam isolation valves (MSIVs) slammed shut during the quake, cutting off steam from the spinning turbines, which were now decelerating. The closed MSIVs meant that each reactor's main feedwater coolant pumps, which are powered by their own steam-driven turbines, also began to coast down.

The quake quickly isolated Daiichi from the outside world. All seven external power lines failed (six national grid transmission lines and one standby line direct from Tōhoku Electric Power Co), either from damaged lines, falling towers or shaking and subsiding land at the off-site Shin Fukushima Substation and on-site switchyards. This disconnect tripped the plant's fleet of 13 bus-sized emergency diesel generators (two at each unit, plus a third at Unit 6), each capable of cooling a single reactor.[c] They roared into life. With external AC power unavailable, the facility now relied on these generators to operate almost all water pumps. The DC batteries provided electricity for instruments and valve control, but if the diesels kept running they wouldn't have any further problems, given their secondary role as battery chargers. Despite the alarm bells and bruises, relief settled around Yoshida's ERC: everything had performed as intended and the plant had survived a mammoth earthquake.

* * *

To the south, shaking at Fukushima Daini bounced a 54-year-old subcontractor around in the cramped cabin of a tower crane he was using for earthquake-reinforcement work on the exhaust stack. He became trapped and died of his injuries before anyone could rescue him. The earthquake knocked out three of Daini's four off-site power lines, but all four of its reactors automatically scrammed without issue.[3, d] The lone Unit 2 at the Tōkai NPP, which had lost all external power amid 225 Gal shaking, had also scrammed.

To the north, at half Fukushima Daiichi's distance from the fault slip, Tōhoku Electric's Onagawa plant faced "the strongest shaking that any nuclear power plant has ever experienced from an earth-

---

3   The 500 kilovolt (kV) Tomioka #1 line remained intact, while #2 was damaged. The 66 kV Iwaido #1 was already offline for maintenance. The Iwaido #2 line was intact but its transformer was damaged, so it was taken offline to prevent further damage. The line resumed operation within 24 hours.

quake," according to the IAEA.[e] (Despite this famous quote, the peak shaking at Onagawa of 567 Gal did not reach the maximum experienced at Kashiwazaki-Kariwa in 2007, but I had to include it.) The entire peninsula had subsided one metre into the ocean, but Onagawa itself rode out the shaking with admirable tenacity.[4] Four of five off-site power lines fell, so the plant just barely retained external AC power, but it was enough. The three BWRs scrammed, but a breaker short at Unit 1 severed off-site power to that unit and set fire to the turbine hall basement.[5] Sensors detected the power loss within seconds and activated emergency generators to pump coolant water, which they did without problem for the remainder of the disaster. Earthquake (and soon tsunami) damage to the nearby roads meant fire engines could not reach the site; the breaker fire raged until 11 p.m. before the plant's own firefighters could extinguish it.[f] Otherwise, according to IAEA inspectors, "The structural elements of [Onagawa] were remarkably undamaged given the magnitude of ground motion experienced and the duration and size of this great earthquake."[g] Even farther north, Rokkasho also lost all external power, as did Higashidōri, though its sole reactor was already offline for maintenance.

* * *

Fission at Daiichi ceased and all six reactors were now offline, with no indication of any coolant leak. From here, preventing a dangerous temperature and pressure rise in Units 1, 2 and 3 required each reactor to be cooled with 20–30 tons of water per hour for 10 days, followed by 5–10 tons per hour for a month to combat the decay heat.[h] This occurs when fission products (i.e. waste) continue to produce heat through radioactive decay even after fission itself has ceased; new fuel contains no fission products, but spent fuel is full of them.

The Mark I containments of Units 1 to 5 all had safety relief valves (SRVs) to vent steam out of the core and down into the donut-shaped pressure suppression pools below, plus an array of cooling systems, two of which were about to play a major role. At

---

4   Onagawa was not completely undamaged but, like Kashiwazaki-Kariwa before it in 2007, no safety-critical equipment was damaged.

5   Unit 1 was in the process of restarting after being offline; the other two were at full power.

Unit 1 – the plant's oldest reactor and its only Series 3 design – the isolation condenser (IC) directs radioactive steam from the core, along a pipe and through a tank of cold water. Here, the steam inside the pipe condenses back into water before falling back into the bottom of the reactor, while the uncontaminated tank water gradually boils and its own steam vents harmlessly into the atmosphere. This system operates using convection and gravity alone. Unit 1 had two independent IC loops with enough water in each tank for up to eight hours of use before they needed to be refilled. The IC was, however, considered a weak link in the emergency cooling system for various reasons, hence the Series 4 BWR used a reactor core isolation cooling (RCIC) system instead.

Available in Units 2 to 5, the RCIC works similarly to its baby brother the IC, but instead of steam passing through a pipe and being cooled, the RCIC directs the steam through a small turbine then into the pressure suppression pool, where it condenses. To offset the lost reactor water, that small turbine drives a pump that pulls more water from a separate tank and injects it at high pressure (though not at high volume: only 1.5 cubic metres per minute) into the core. A separate spray system cools the pressure suppression pool, and both that system and the RCIC can also recycle water from the suppression chamber if needed. The RCIC is normally used for less than 15 minutes at a time but can operate for "at least four hours," according to the IAEA, and far longer if necessary.[i] It is designed to supplement other cooling methods, however, and is not for pressure control of a reactor by itself, nor for use in a loss-of-coolant accident. Neither the IC nor RCIC require AC main power, only DC battery power to open and close their four valves (two inside the containment vessel and two outside it) to ensure they could operate in an emergency.

A highly trained team of 11 operators shared a control room for reactors 1 and 2 inside the joint Units 1/2 service building, jutting out from the centre of their shared turbine hall on the ocean side. Likewise, two more teams of 11 controlled Units 3/4 and Units 5/6 from their own service buildings. These operators started the Unit 2 RCIC moments after the shaking subsided, but the system stopped itself one minute later because sensors detected sufficient water already in the core. Two minutes after that, Unit 1's IC started automatically, but the operators, following procedure, soon disabled it to avoid exceeding a 55°C per-hour temperature drop, which could have damaged the steel pressure vessel. The system started and stopped multiple times over the next 45 minutes.[6] Next, they restarted Unit 2's RCIC, which now stayed on, while operators in the shared control room for Units 3/4 manually started the third reactor's RCIC.

The first Japan Meteorological Agency (JMA) tsunami warning came within minutes of the quake: a wall of water, hundreds of kilometres wide, was hurtling towards the east coast. They expected it to reach land in 25 minutes and to be three to four metres high for the Fukushima coast, thus posing no danger to the plant.[7] The earthquake, however, had maxed out JMA's dense network of teleseismometers, preventing accurate estimates of tsunami heights using the standard moment magnitude scale. Quick backup formulae for prompt evacuation warnings also proved misleading because they only work well for small to medium temblors, with more intense shaking throwing off the results. By this time, workers at Daiichi had fanned out to inspect for damage, but Yoshida ordered the evacuation of low-lying areas around the harbour as a precaution.

Masamitsu Iga, a supervisor from Unit 3, along with two of his young protégés, Takuya Ara and Takuma Nemoto, were walking towards the auxiliary service building behind Unit 4 to check on a generator when they heard the warning. Unhurried and safe up at the 10-metre level, they split up as planned and each opened a separate windowed outer door to the building's set of two-stage secure entrance lobbies. Set up like airlocks, the outer doors closed and locked, but the inner security doors wouldn't verify the identification cards carried by either Iga or Ara.[8] "I was trapped inside the security gate compartment," the 23-year-old Ara recalled.[j] The pair had earlier exited another building through doors left open after the quake (the usual strict security being disregarded under the circumstances) and were thus considered by the logging computer to still be inside. It refused them entry as a precaution. After calling out to each other

---

6   This was TEPCO's explanation and is written in the regulations, but the NAIIC later interviewed control room personnel, who claimed they'd stopped the system to check for IC water leaks because the reactor pressure was dropping.

7   The JMA faced heavy criticism in the aftermath of the 2011 tsunami because of this low first prediction. Only 58% of people moved to high ground after the warning (August 2011 government survey), in part because of that initial warning.

8   These doors are not true airlocks in the traditional sense, as they are neither air nor water-tight. They are low-leakage doors, which allow for controlled air intake to maintain a slight negative pressure in the buildings.

through the walls and learning of their shared predicament, they tried the security office intercom in their respective rooms. There was no response.

At 3:14 p.m., the JMA revised their tsunami alert to six metres. The first wave reached Daiichi at 3:27 p.m., but, at four to five metres (13–16 feet) tall, broke harmlessly against the breakwater's thousands of concrete tetrapods. Minutes later, as ever more precise information beamed in from a sophisticated network of ocean-surface buoys, JMA meteorologists frantically warned of water up to ten metres (33 feet) high – double the breakwater height and the same elevation as the Units 1–4 reactor buildings. They were still short by up to five meters.

When it came, the second wave destroyed almost everything along hundreds of kilometres of coastline and up to ten kilometres (six miles) inland. Cars, buildings, entire towns and villages were wiped out with dispassionate ease. At Kesennuma, a city in Miyagi Prefecture, driving water carried the 330-ton fishing boat *Kyotoku-maru* three-quarters of a kilometre (half a mile) inland, decimating everything in its path. Propane gas canisters, burst oil and fuel tanks, and other explosive or flammable items started fires, which spread as the tsunami carried clusters of burning debris across wide areas.[k] Large portions of Kesennuma, Otsuchi and other coastal towns and industrial facilities burned into the night. Twenty-two thousand men, women and children unable to outpace the water or seek refuge on firm high ground were either crushed or drowned.[9]

The second wave reached Onagawa just as the first wave hit Fukushima Daiichi. Onagawa's massive sea wall embankment, built at the insistence of Yanosuke Hirai 30 years earlier, held almost all the water at bay and almost certainly averted a second nuclear disaster. Nevertheless, some water collapsed a heavy oil storage tank and ran down through cable ducts into the service building basement for Unit 2, penetrating a water-tight door and flooding electrical components that helped supply the diesel generators with cooling water. This flooding, in turn, tripped out two of the reactor's three diesels, but operators were still able to receive main AC power for functional components of the cooling system, so cooling was uninterrupted.[l] With fortuitous timing, Unit 2 had commenced operation after an extended shutdown less than an hour before the

---

9 Approximately 2,500 of this number are still listed as missing.

earthquake and thus required almost no cooling anyway – it achieved a cold shutdown prior to the tsunami's arrival.[m]

The town of Onagawa itself was not so lucky. Water tore trees from the ground and flattened almost everything not made of thick concrete. Some of the weaker structures were lifted whole from their foundations by the water's hydrodynamic and buoyant force, to be swept away with a mighty surge that reached an astonishing 18 metres (59 feet) above sea level. Most of the town's 10,000 residents made good use of their 30-minute advance warning and fled to the surrounding hills, but almost 1,000 Onagawans perished. Post-tsunami photographs resemble those taken at Hiroshima.

That second unstoppable wave rolled over Fukushima Daiichi's 40-year-old sea defences as if they weren't even there. Thundering across the harbour at 3:36 p.m., it annihilated all six sets of seawater inlet pumps – vital for cooling the reactors – and picked up everything not embedded in concrete.[10, n] The harbour's enormous twelve-by-nine-metre heavy fuel oil tank was lifted whole and pushed 150 metres (500 feet) inland, where it came to rest beside Unit 1's turbine hall, blocking the road and, with it, the easiest access to the site's coastal side. The 10-metre level on which most of the 3.5 million square-metre plant sat, long considered to be an insurmountable safe haven against coastal ingress, offered no such refuge as water gushed into open doorways, service hatches, ventilation louvres, cable trenches and all other penetration points, flooding the ground and sub-levels. Weak points, such as the tall roller-shutter door of Unit 4's Large Equipment Service entrance, buckled, then burst apart as the wave hurled a truck straight into it. Down in the basement of Unit 4's turbine building, as water surged into stairwells and corridors, Kazuhiko Kokubo, age 24, and Yoshiki Terashima, age 21 – two of six young equipment operators sent to inspect the building – were both killed by head trauma after being repeatedly flung against machinery by the deluge.[o] Firefighters used a digger to clear out the truck-clogged service entrance and drove a fire engine inside to pump out the water, but it was too low to reach.[p] Divers also tried to find them, but the events to come forced an end to the search after three days. Their bodies were not recovered until 30th March.[q] Despite the calamity, many of those indoors – in the bunker-like service buildings or high up on the 35-metre level, where

---

10 The timing of the second wave differs depending on the source, as the exact time is unknown. This time comes from the IAEA.

the administration and seismic-isolation buildings sat – had not the slightest inkling of the events outside.

Takuma Nemoto entered and then exited the auxiliary service building, having earlier swiped out of another building and thus encountered no problems with the logging computer. As he casually chatted with his two trapped colleagues, debating whether to try breaking the reinforced glass windows, Iga spied the water rising behind him. Iga yelled for the younger man to run inside as it raced past the reactor buildings towards them. Nemoto scrambled through both sets of doors with seconds to spare.[r] To their horror, though both outer doors held firm, water pressure overwhelmed the air seals and began to pour all around a narrow gap on both door frames as the pair tried desperately to escape. The large pane of glass in Iga's outer door collapsed inwards seconds later, tossing him around the small space like a ragdoll. Each man fought for several minutes as the water rose, almost touching the ceiling before it finally lingered. Iga scrambled out, shredding his hands on the broken glass, then grabbed a floating chunk of wood to hammer on the smaller glass holding Ara. "I was minutes away from being drowned when my colleague smashed opened the window and saved my life."[s]

Twelve of the plant's thirteen diesel generators spluttered to a stop as water variously swamped the generators themselves, their dedicated seawater coolant pumps, their nearby power distribution and switchgear equipment, or the whole lot.[11] Unit 6's third diesel, air-cooled and on higher ground, was the only survivor. DC battery power failed at Units 1, 2 and 4 over 15 colour-draining minutes as water seeped into electrical components. Suddenly, every light inside those reactor and turbine buildings' thick walls flickered out. In the windowless shared control rooms of Units 1/2 and 3/4, the last remaining slivers of light from the wall of lit dials, switches and indicators slowly, randomly faded until they, too, went out, plunging the operators into near-total darkness, spared only by the faint glow of emergency lights.[12] Station blackout.

* * *

---

11 Ten of the twelve generators drowned, while water rendered the other transmission equipment of two air-cooled units useless.

12 Unit 2's emergency lights were also out; only the emergency lighting from the Unit 1 side of the room continued to work.

**Tsunami water hits
Fukushima Daiichi**

Source: Tokyo
Electric Power
Company Holdings

Down the coast, the second wave rode up Fukushima Daini's sloped harbour road beside Unit 1 and swept inland passed the reactor and turbine buildings all the way to the plant's own seismic-isolation building. Here the wave hit a tall, steep embankment on two sides and backwashed north into the grounds surrounding Units 2 through 4. Daini's reactor buildings sat 12 metres above sea level rather than Daiichi's 10, helping somewhat to prevent water from overrunning everything, but Unit 1 flooded up to 15.9 metres above sea level: far higher than elsewhere at the site.[i] Water inundated the basement, ground and first floors of the enormous spent fuel building, the SIB behind it (which lost power when its power distribution panel drowned) and all but one of the eight seawater pump buildings, caking equipment in seaweed and sludge and depositing an unsuspecting shark outside Unit 1. The Units 2 through 4 reactor buildings avoided serious incursion, though some parts of Unit 3's reactor building basement also flooded, as did the turbine building basements of Units 1–3, with water penetrating through underground cable ducts.

Farther south, Unit 2 at Japan's original Tōkai nuclear plant also experienced flooding from a 5.4-metre tsunami, despite recent construction to increase the height of its seawall from 4.2 metres to 6.1 to protect the two emergency seawater pump pits.[13] The wall stopped the wave but work to watertight its piping penetrations towards the north end was incomplete. Water gradually poured through, flooding the north pit up to 4.9 metres, which in turn swamped a seawater inlet pump used to cool one of the plant's three backup generators, which then failed at 7:25 p.m.[ii] Each of the off-site power lines did, too, but with two remaining diesel generators, plus another 500-kilovolt generator on the SIB roof, the ageing reactor was in little danger of overheating.[v]

* * *

Fukushima Daiichi lost all remaining AC power in Units 1–5 when the diesels cut out, including crucial reactor cooling at Units 1–3, the high-pressure coolant injection (HPCI), two independent core spray systems and the ability to remotely open or close any valves at

---

13 Numerous sources say that work on this wall, and the closure of a large watertight gate in the middle of it, was finished mere days earlier, but I have been unable to verify these claims.

Units 1 and 2. Unit 3's DC battery systems escaped the tsunami, so its HPCI and some instruments continued to function. Cooling for each reactor's spent fuel pool also stopped. This fuel was nowhere near as hot as that still inside the reactors, but when water stopped circulating in the pool, it began to warm up. Yoshida and his staff in the emergency response centre were stunned when they heard that emergency power was offline.[14, w] Control room managers had called the ERC via the plant's own fixed internal landline phones because the Personal Handy-phone System (PHS), basically a private cell-phone system, was inoperable. Cell- and public landline telephones near the plant continued to sporadically work until noon the next day, when batteries depleted at the telephone company's Ōkuma base station.[x] Systems to display plant parameters inside the ERC were also offline, forcing people to scribble information on whiteboards for everyone to see.

They had trained for a loss of AC power of up to eight hours and were therefore prepared for this, but no contingency plan existed for the total destruction of AC and DC power sources; no reassuring words lay in the handbooks to guide them, and nobody was coming to help any time soon. The impossible disaster scenario had become a reality. As Akio Komori (Yoshida's immediate predecessor as manager of Daiichi and then-managing director of TEPCO's Nuclear and Plant Siting Division) put it: "We were entering into territory that rather exceeded what we had considered."[y]

Staff in the darkened control room of Units 1/2, under the steady command of 52-year-old shift supervisor Ikuo Izawa, broke out a set of flashlights and shone them across the bank of lifeless instrument panels.[15] Displays for water level, temperature and pressure offered no hint as to the state of the cores. Were the passive cooling systems – Unit 1's isolation condenser (IC) or the reactor core isolation cooling (RCIC) system in Unit 2 – even functioning? They had no way of knowing. On the plus side, Unit 4 was already offline and Unit 3 was okay for now because some of its battery systems had survived and its operators had kept the RCIC online. Unit 5's batteries also survived, but its main heat removal systems were inoperable because they relied on steam power; initial attempts

---

14 Yoshida later expressed his great surprise about the flooding. Though he had anticipated the harbour potentially being flooded, nobody at the site had considered the possibility that water could reach the 10-metre level.

15 The control rooms are on the second floor of each service building. I have to wonder how things would have gone if they'd been on the ground floor.

to depressurise the core for low-pressure systems failed, even as the pressure crept upwards. Unit 6 had a working generator with full power and was fine.

Consequently, various procedures were implemented over the next several hours. At 3:42 p.m., 56 minutes after the earthquake, staff in the ERC notified TEPCO of a "special event" under the Nuclear Emergency Act, which they raised to a full-blown "nuclear emergency" an hour later. TEPCO and Tōhoku Electric both dispatched high- and low-voltage generator trucks to Daiichi from multiple locations within hours, but with so many roads destroyed, when – or if – they would arrive was unknown. Though the government was initially reluctant to do the bidding of a private company, the Japan Self-Defense Forces (JSDF) also sent more than 40 generators in convoys on various routes to ensure that at least some would get through. Attempts to airlift a generator truck using JSDF and US military cargo helicopters failed because the truck was too heavy.

In Tokyo, TEPCO, METI and the Prime Minister's Office all activated their own ERCs to focus specifically on Fukushima Daiichi, in addition to those centres already monitoring the natural disaster. TEPCO, however, was somewhat paralysed by the absence of its two most senior executives. Chairman Tsunehisa Katsumata was in China on business with two fellow TEPCO executives and a gaggle of unidentified members of the press (likely of some seniority, based on comments by Katsumata) who were enjoying a most-expenses-paid trip.[z] "We probably paid more than our share [of the bill]," he later conceded, while refusing to disclose who was with him.[aa] The Chinese offered him a plane, but with Tokyo's airports closed, he was left stranded. TEPCO president Masataka Shimizu was in Japan, at least, but he was also away on a business trip with his wife to Nara in south-central Honshū in his capacity as chairman of the all-powerful FEPC.[16] Aware of his urgent responsibility to take command, he tried to return to Tokyo but found the Tōkaidō Shinkansen bullet train closed and the Chūō and Tomei expressways impassable. Frustrated, he boarded a local train towards Nagoya Airfield, hoping to use an affiliate company's helicopter.

---

16 Explanations for why Shimizu was in Nara vary. The trip appears to have been for "business and pleasure," with his work with the FEPC taking him there but his wife accompanying him for a short holiday while they were away. Katsumata claimed in his government testimony not to have learned of Shimizu's initial absence until after the disaster had ended.

NISA activated the nuclear emergency off-site centre (OFC) just down the road from Daiichi at Ōkuma. Around 40 representatives from the emergency services and regional and national governments were meant to assemble there in times of crisis, but due to an overlap in responsibilities, most of those same people were tied up dealing with the earthquake and tsunami response. Of those remaining, NISA had enormous difficulty contacting the relevant government employees due to inoperable or congested telephone lines, while many of the rest could not reach the OFC because of the local devastation. The fire, police and ambulance services already had their hands full. TEPCO executive vice president Sakae Muto was among those who began to make his way there, along with METI senior vice minister Motohisa Ikeda and six of his staff, plus personnel from JAEA, NIRS, the Japan Chemical Analysis Center and several others. Ikeda's group did not get far – traffic congestion blocked their departure from Tokyo, forcing them to turn back and take a JSDF helicopter from the Ministry of Defense at 9 p.m. that evening. Several other ministries and agencies were supposed to send staff to the OFC, but most did not. The Ministry of Health, Labour and Welfare, for example, failed to send a chief medical officer for ten days.

The first arrivals found the OFC itself to be without power, radiation air-filtration systems for protecting its occupants, or working lines of communication except for a single satellite phone. NISA was told of the latter short-sightedness in February 2009 but had done nothing. The site's emergency diesel generator worked for a time but had suffered earthquake damage and could not be refuelled. With no electricity, the OFC couldn't be used for the local nuclear emergency response, so most personnel soon moved to the Environmental Radioactivity Monitoring Center next door.[ab] By law, proposals made by senior staff at the OFC were intended to be passed up the chain of command for approval from the PM's Office, but this proved impossible. Fukushima's prefectural government had also planned to establish a headquarters for disaster control at their main office building in Fukushima City, but shaking rendered the site unusable because seismic reinforcements had never been carried out, forcing them to shift operations to a backup building which only had two radio lines, rather than the 47 they'd planned for, causing serious communication problems.

* * *

At Daiichi, amid constant aftershocks and the threat of more tsunami, a snaking line of vehicles carrying thousands of people – mostly subcontractors – slowly left the plant, while off-duty shift supervisors and other senior staff began to trickle back towards it to offer aid, leaving around 400 people on-site.[ac] With no pumps sending cooling water into the reactors, the existing inventory evaporated into steam, driving the pressure upwards. Any enclosed system will rupture if the pressure gets too high, so when it reached a pre-set point, the reactor's safety relief valves (SRVs) opened to discharge steam into the suppression chamber below, where it bubbled through water and condensed, scrubbing some of the most harmful radioactive and water-soluble particles such as iodine. This process is known as a wet vent. A dry vent goes through the drywell (the containment vessel, where there's no water) and releases more harmful gases into the environment. As such, wet venting is preferable. The SRVs close following either process until the pressure again reaches that pre-set level. Over time, with each successive internal venting, the amount of water covering the fuel in Daiichi's reactors decreased. When Unit 1's water gauge briefly came back to life an hour after the tsunami, the operators saw for the first time how much water remained: they estimated just one hour until the fuel became uncovered and started to melt.[ad]

With things deteriorating by the minute, Yoshida pondered the likelihood of a Chernobyl-like disaster, but he had experienced emergencies before and knew power plants like the back of his hand. His mind racing, he figured they had around eight hours before the IC and RCIC systems stopped supplying water – assuming they were working – but he knew that wouldn't be enough.[ae] "I was in despair," he remembered later.[af] Cooling the reactors was their top priority, but they had no way to pump water into any of the three previously online reactors, and no instruments to monitor temperature, water level or pressure even if they could. Reviving the instruments may not be too difficult – there were bound to be batteries somewhere – but how could they get water into the reactors? With all emergency backups potentially offline, he decided the fire suppression system was their only option.

Each reactor and turbine building came with fire protection lines connected to fire hydrants, but they were intended to fight fires and cool the enormous concrete containment vessel, not inject water directly inside the pressure vessel. As backups to the electrical motor-driven fire pumps, which were out of action, each Unit also held a single stationary diesel-driven fire pump (DDFP) in their turbine hall basements (though Unit 2's was flooded), all with their own network of hydrants and hoses. Many outdoor pipes, water inlets and hydrants sustained damage from the earthquake and tsunami debris, but internal systems were rated for stronger seismic activity and worth a try. In addition to laying kilometres of hoses, they needed to lower the core pressure since the pumps could only push water at up to 800 kilopascals (kPa, equal to 116 pounds per square inch [psi] in imperial measurements), but Unit 1, for example, was pressurised at 7,000 kPa. What did they have that was live, mobile and capable of pumping water at a higher pressure, even if just a little higher? Fire engines – each turbine building had an injection port installed on its outer wall a year earlier to connect a single fire engine.[17] "I had no way of knowing what happened inside the buildings after that earthquake struck," Yoshida said later of this decision. "In the end, it was a gamble of whether the fire protection lines were sound or not. But, since that was the only option, my feeling was that I wanted to use that to pump in water."[ag] He gave the order just after 5 p.m.

Three operators left the control room and headed down to check on Unit 1's DDFP, passing a dead fish in a puddle of muddy water on the way. They found the pump inoperable but fixed the fault and left it on standby mode, as it would take time to align the necessary valves for core injection from the fire pump. Outside, more workers found one fire engine damaged and trapped among tangled wreckage, while another up beside Units 5 and 6 was okay, but they couldn't get past the debris. The final engine sat unharmed up on the 35-metre level, from where a driver began to inch it down the hill toward the harbour as people dragged obstacles from its path. The enormous oil tank earlier deposited by the tsunami blocked the route past the north side of Unit 1 and nobody could get past a de-energised security gate on its south side, so they broke through another security

---

17 Much has been made of Yoshida's near-faultless decisions under tremendous pressure during the entire Fukushima disaster, but his early foresight to bring in the fire engines deserves enormous admiration in my opinion. Still, this decision to prepare the fire engines is a little cloudy. Some reports imply they weren't considered until after the DDFP failed, while others – including the IAEA vol. 1, Annex 1, timeline – say they were all ordered together. This seems the most likely scenario.

The oil tank dragged up to Unit 1.
Source: Tokyo Electric Power
Company Holdings

gate between Units 2 and 3 and navigated the fire engine through. Lookouts stationed atop the service buildings scanned the ocean with each successive aftershock, searching for the tell-tale sign of receding water which would indicate an impending tsunami. Thousands of aftershocks pummelled the coast during the next couple of months: four above magnitude 7.0 and five hundred over 5.0.

Just after 6 p.m., several battery-powered instrument panels on Unit 1's side of the control room again inexplicably came to life for a few moments, probably because the batteries had dried. They revealed that the core's water level had dropped and the IC valves were all closed, even though one had been open as the power failed earlier.[ah, 18] Operators tried to open the valves remotely. The IC exhaust ports were on the reactor building's landward side, so Izawa – not trusting their instruments – contacted the SIB and requested that someone there go outside to check if the system was working. These observers reported to staff inside the ERC that misty steam was rising in the twilight, which they thought could indicate that the

---

18 According to IAEA report, Vol. 1, Annex 1: "Operators found that the valve indicator lamps were lit sometime before 18:18. Not only MO-3A [motor-operated, number 3], which was controlled by operators for IC activation/deactivation and was left closed before the second wave of the tsunami, but also the IC supply piping containment isolation valve (MO-2A), which was normally open was closed. Thus, operators inferred that an IC isolation signal was generated during the loss of control (DC) power, possibly by the IC pipe rupture detection circuit. It was corroborated that they were closed at the time of DC failure; thus, there was no shutdown heat removal since then."

IC had activated, being a passive system that needed no electrical power once the valves were open.

Satisfied that Unit 1 was now in less peril than Unit 2, staff in the ERC in turn reported this information to Yoshida, his senior management team and TEPCO headquarters in Tokyo. Everyone switched focus toward getting power to Unit 2. Back in the control room, Izawa noted the odd description of steam in the IC exhaust – it should have been blasting out, not gently drifting – and the corresponding lack of audible cues, especially when that steam dissipated and stopped altogether in under ten minutes.[19] Now concerned that the system might have been compromised, he ordered his operators to close the valves to prevent a possible contaminated water leak at 6:25 p.m. In doing so, they lost the ability to cool Unit 1.[20] As Yoshida later recalled in his government testimony:

> One of the things I now regret is that we got no information from the reactor group leader, nor did we know if the manager on duty was in communication with the reactor group leader. Information of that sort was primarily shared via the reactor group leader, and there was no protocol in place whereby the manager on duty would call me to provide that information. I should have personally confirmed at several stages whether the IC was truly okay. Partly, I was operating under a false apprehension, assuming that the IC was working because the [reactor] water level was, to a certain extent, regular.[ai]

With no confirmation that Unit 2's RCIC was working either, its operators assumed that it was not and calculated that the fuel would be exposed by 10 p.m.

* * *

Twelve kilometres away, Fukushima Daini faced problems of its own and, had it not been for the events at Daiichi, Daini would be the infamous plant written about in history books. Plant Manager

---

19 TEPCO's BWR training simulator did not have an IC function, it's lucky Izawa knew how it worked.

20 Workers on the fourth floor of the Unit 1 reactor building reported water all over the floor immediately after the earthquake and prior to the tsunami. The earthquake likely damaged the IC tank or its piping, leading to a loss of water.

Naohira Masuda, a 29-year TEPCO veteran, took a different approach to Yoshida. Comparisons painting his response in a far better light than Yoshida's are unfair in my view, given that Masuda's site suffered nowhere near as much damage, but he nevertheless did a textbook job. Before sending anyone outside, he first began to calmly chart the magnitude of each successive aftershock on a whiteboard in the ERC, thus demonstrating for his staff that the natural danger was decreasing. "I was not sure if my team would go to the field if I asked," he later admitted, "and if it was even safe to dispatch people there."[aj]

As he took stock, control room operators checked their systems and found the plant had narrowly avoided a nightmare scenario because the tsunami was, at 9.1 metres, four metres lower at Daini than at its sister plant, despite their close proximity. Geologists believe this happened because the shape of the tectonic plates caused two distinct tsunami peaks to converge at Daiichi – sheer bad luck. All power and therefore all main cooling appeared to be offline at Units 1, 2 and 4 because the tsunami had knocked out seven of Daini's eight seawater pumps, but only Unit 1 experienced significant flooding. There was another issue: even though a single external power line had survived the earthquake, the electrical switching equipment had not, meaning power couldn't be sent where it was needed. On the positive side, passive cooling systems like the RCIC were intact, along with many diesel generators, pumps and electrical components at Units 2 through 4. One of Unit 3's inlet pumps also somehow survived, as did one of eight residual heat removal (RHR) systems (the main non-emergency means of reactor cooling), meaning Unit 3 was safe. With the RCIC up and running, Masuda declared a special event at 6:33 p.m., having received word from his operators that the RHR and diesels at Units 1, 2 and 4 were offline.[ak] After darkness fell, and once the water outside receded, he sent people out for the first time.

* * *

Prime Minister Naoto Kan declared an official nuclear emergency for Daiichi at 7:03 p.m., hours late. He was a staunch supporter of nuclear power, and the export of nuclear technology and expertise – with a particular aim towards India and Iran – was one of his key economic growth strategies. Despite being a precondition of activating the government's emergency response system, he vacillated over the minutiae because nobody gave him timely explanations of, or justification for, the legal process.[al] Members of the NSC were supposed to form an advisory group, but with the city at a standstill and most communications unavailable, this did not happen. Kan, his advisors and various other bureaucrats had already migrated out of the chaotic Crisis Management Centre (CMC), a basement bunker in the Kantei building (officially the Sōri Daijin Kantei: the Japanese White House), where the government's emergency disaster response leadership teams had gathered. The cramped and frantic CMC – built in anticipation of military use – had no cell-phone reception and held just two hard-wired landlines which were in constant use for tsunami and earthquake recovery efforts. Some people also found they were not pre-registered on the bunker's biometric system and could not gain entry. Upstairs, in the PM's quiet and expansive fifth-floor office, they could closely monitor the unfolding nuclear emergency without distraction.

The 64-year-old Kan, who became PM nine months earlier, was a rare politician from outside the typical circles of political elites. He majored in physics at the Tokyo Institute of Technology, rather than receiving the typical Tokyo University economics or law degrees held by most of his predecessors, and rose to prominence in 1996 on the strength of his performance as Minister of Health and Welfare, when the ruling Liberal Democratic Party (LDP) formed a coalition with the short-lived New Frontier Party, the precursor of Kan's Democratic Party of Japan (DPJ).[21] As minister, Kan investigated a scandal in which around 2,000 Japanese haemophiliacs had contracted HIV through tainted blood packs, over 500 of whom died as a result. He cut through the normally impenetrable political bureaucracy with an unusually frank and fearless approach, while even his own subordinate officials tried to conceal information from him. His investigation revealed that ministry officials – knowing that people could die – actively avoided preventing the untreated plasma from being withdrawn from use to avoid a panic and to protect companies importing the blood. This cover-up ended with the first-ever criminal conviction for administrative negligence of a Japanese government

---

21 Tokyo Tech, not part of Tokyo University, is just four years younger and is considered one of the country's most prestigious institutions.

official.[am] Interestingly, the scandal revolved around a blood company founded and run by former war criminals from the Imperial Army's infamous and brutal Unit 731, which became the country's principal blood products and pharmaceutical company.

* * *

At Daiichi, a crew of senior operators from Units 1 and 2 spent half an hour poring over piping schematics, searching for valves to align so the diesel-driven fire pump (DDFP) could inject water to the core, then suited up and entered the reactor building at 6:35 p.m.[22] They finished two hours later but the pump then failed to start. Meanwhile, another team carried batteries and a small portable generator into the control room to revive some instruments just before 9 p.m. The DDFP came online after repeated attempts but, heartbreakingly, could not push against the high reactor pressure. Water level indicators suggested that water in Unit 1's core was still 200 millimetres above the top of the fuel – not great, not terrible – but Izawa's worried operators tried again to remotely open the IC valves at 9:30 p.m.[23]

The readings were wrong. This model of BWR measures water using a system called a reference leg. Its functionality is convoluted, but basically it compares differences in water pressure in different parts of the core, the problem being that when water boils down to a particular height outside the anticipated range, the pressure differences can indicate that water levels are higher than they really are. Knowing this, Yoshida and others did question the numbers it produced, but they had nothing else to go on.[24]

In reality, the tops of Unit 1's fuel rods were already exposed by 6 p.m. (at the latest; some analyses put the time at 4:30 p.m.) and the remaining water boiled down until the fuel became comple-

---

22 The government report states: "To establish a line capable of injecting water from the FP [fire pump] system line to the nuclear reactor, the shift team had to open motor-driven valves on connecting lines between FP, the MUWC [make-up water condensate] and the RHR [residual heat removal] (or CS for Unit 1) systems located at Units 1 and 2."

23 Two widely reported figures of 200 mm and 450 mm appear in different sources. The full government-appointed NAIIC report says 450 mm, for example, but the Rebuild Japan Initiative Foundation report says 200 mm, as do the JNTI and IAEA reports. As such, I've gone with 200 mm, although both are considered unreliable figures.

24 Yoshida is quoted by an eyewitness in the prologue to the Rebuild Japan Initiative Foundation's '*Investigating the Myth and Reality*' book as saying "Are those numbers really correct?" at least twice.

ly exposed 90 minutes later.[an] Air is a poor coolant compared to water, so the temperature increased exponentially as the water level dropped. It passed 800°C (1,472°F) and the internal stresses in each fuel rod caused them to swell and fail, releasing their deadly fission products. As temperatures quickly surpassed 1,200°C (2,192°F) – which only took a further ten minutes, according to TEPCO – the fuel's zirconium cladding began to react with steam to create an expanding bubble of highly explosive hydrogen gas.[ao] Extreme heat damaged the gaskets, flanges and seals around the vessel lid and penetration points, while the lid bolts softened and stretched, allowing the hydrogen to seep out and, being lighter than air, to rise up through the building.[ap] Just one hour later, the temperature reached 2,865°C (5,189°F) and both the zirconium and uranium fuel itself started to melt.[aq]

Workers outside once again observed steam rising from the IC vent, so this time Izawa dispatched a small team of his own to check on the internal machinery. Their pocket dosimeters bleeped off the scale the moment they stepped inside the reactor building, measuring 0.8 millisieverts within ten seconds, and they were forced to retreat.[ar] Entry into Unit 1 then ceased without express permission, while radiation levels in the control room began an ominous creep upwards. Meanwhile, operators asked the ERC to arrange for batteries to help open the valves, but this request was ignored for several hours, possibly because personnel there thought the IC was already working. To everyone's relief, the first generator trucks arrived at around 10 p.m., but their drivers soon found that Daiichi's electrical connections had been ruined by saltwater, the sockets were incompatible, and the required 480-volt supply was obsolete and could not be supplied by the trucks; the generators were useless.

* * *

After consulting with senior figures from NISA and the NSC, Kan ordered a precautionary evacuation of residents living within three kilometres of the plant at 9:23 p.m., but this order came hours after the first emergency declaration. The timing caused enormous confusion on the ground, since the local Fukushima prefectural government, concerned by the delay, had ordered their own two-kilometre evacuation just 30 minutes earlier without informing the central government.

The prime minister took an ever-greater interest in events, prompting his secretaries to begin scribbling progress updates on whiteboards about the generator trucks and other matters.[as] As the night wore on, however, his frustration at the untimely and incoherent information emerging from TEPCO, the NSC, JAEC and NISA grew, owing to the convoluted lines of communication. Yoshida and his staff at Fukushima would phone TEPCO headquarters to provide an update or ask questions; then TEPCO passed the message to NISA, which would, in turn, inform cabinet staff, who then told senior bureaucrats in Kan's circle. Conversely, senior government officials never really knew if their instructions were being relayed back to the plant. Eight NISA inspectors, permanently stationed at Daiichi and intended to act as the government's eyes and ears, fled the site, forcing the government to rely on TEPCO alone for updates. All communications sent to the Kantei's basement CMC required aides to run up five flights of stairs to Kan's office because the constant aftershocks meant the building's elevators were unavailable. This chain of "Chinese whispers" (or "telephone" in the US) caused continual confusion, compounded by the utility's two absent senior executives and the lack of technical knowledge among some of those sent to advise or relay messages to the cabinet building.

NISA director-general Nobuaki Terasaka, for instance, failed to impress, as Kan recalled in his memoir of the disaster. "It does not take long to discern whether a person who is explaining something really knows what they are talking about," he wrote. "I could not comprehend what he was trying to tell me, so I asked him, 'Are you a specialist in nuclear power?' To this he replied, innocently, 'I'm a graduate of the University of Tokyo's Economics Department.'"[at] Kan and his team knew that the generator trucks had arrived and therefore the crisis should have been over, but Terasaka and TEPCO's government liaison – 64-year-old executive vice president Ichiro Takekuro, a Tokyo University engineering graduate and veteran of the utility's nuclear division – could not adequately relay why the trucks weren't being used. "Frankly," Chief Cabinet Secretary Yukio Edano later remarked, "it was at this time that doubts began to grow concerning TEPCO."[au] Given the obvious confusion and apparent ineptitude of those charged with handling the unfolding events, Kan began to feel that he, as Japan's leader, should assume control of the situation.

TEPCO president Shimizu arrived at Nagoya Airfield after dark, having travelled through 150 kilometres of chaotic countryside, whereupon he learned that the promised helicopter was not permitted to fly because Japan's civil aviation law restricts private helicopters from flying after 7 p.m. Hearing of the deteriorating situation at the plant, he contacted government officials for help and received approval from the cabinet's emergency response team to use an empty JSDF C-130 Hercules cargo plane, which took off at 11:30 p.m.[25, av, aw] He didn't get far – Defense Minister Toshimi Kitazawa got wind of the plan and vetoed it, but his instructions to retain the plane for disaster relief didn't get through until ten minutes *after* lift-off.[ax] The plane returned to the airfield and Shimizu was forced to wait until dawn, when a helicopter flew him back to Tokyo. Katsumata, too, flew back to Tokyo on the morning of the 12th, but terrible road conditions meant he didn't return to company headquarters until 4 p.m., over 24 hours after the earthquake. All this meant he was absent while his company tried to avert a global catastrophe.

ii. Saturday, 12th March – Day Two

By midnight, efforts to scrounge every working portable generator and battery on site – even extracting some from parked buses – began to pay off as tired technicians, clutching circuit diagrams in the dark after poring through 10,000 pages of documentation because their computers wouldn't work, wired the power sources into control room instruments.[ay] To the horror of Izawa and all present, the data revealed that Unit 1's drywell – the enormous outer concrete containment vessel – had already exceeded its maximum design pressure of 528 kPa by over 10%. Faced with a potential explosion that could destroy the reactor and force Daiichi's evacuation, Yoshida – grappling with constant information and communication issues of his own – finally realised Unit

---

25 Even though Shimizu didn't ultimately make the trip in that C-130, it speaks volumes of the power wielded by TEPCO that its executive could ask to use such a massive plane and initially be granted his request. Shimizu received a great deal of criticism for trying to take the JSDF plane instead of just driving the rest of the way to Tokyo. "If the Tepco president had been stranded somewhere unable to move, I could understand asking for a Self-Defense Force plane, seeing as he had to handle the nuclear accident," said chief cabinet secretary Yukio Edano. "But between Nagoya and Tokyo, where roads were open for travel, I've got to wonder why he made such a request."

Top: A masked operator reads instruments in the dark.

Bottom: Batteries used to power the control room.

Source: Tokyo Electric Power Company Holdings

1's IC was not working and requested permission from TEPCO HQ to vent pressure straight out into the atmosphere.[az] He was later criticised for not noticing earlier, but, given everything he had to contend with, I hesitate to do so. That said, of the dozens of people in the ERC, *someone* should have realised far sooner. Izawa realised that something was wrong with the condenser hours earlier and thought he had conveyed this to the ERC. What exactly was said isn't clear, but that single miscommunication was arguably the trigger point that really set the disaster in motion.

The plant's severe accident management handbooks[26] offered no plan for hard venting, since the deliberate release of reactor core gases was never expected to occur. Doing so without electrical power was a complete unknown, so Yoshida instructed his staff to begin the laborious preparations just after midnight. At first, few knew if the required vent valves *could* be opened by hand, nor where they or the inlet sockets were, as working around the reactor buildings was not routine for TEPCO staff. They left that sort of legwork to the thousands of subcontractors, most of whom had evacuated, forcing Yoshida to send people to the tattered, pitch-black administration building in search of piping blueprints and valve schematics.[27] Teams then scoured these diagrams for the valves' physical locations (and how to reach them), but at least some diagrams hadn't been updated for years and no longer matched the actual plant.[ba] While they formulated a plan, others went to find equipment to open the valves or tried to locate subcontractors with the right skills.

All the while, another team entered Unit 2 to check on its RCIC. They descended through the crippled building, breathing through respirators, but discovered the basement space flooded with water almost to the tops of their boots. Unable or unwilling to risk opening the watertight RCIC pump room door but hearing faint mechanical sounds echoing in the dark, they returned to the shared control room of Units 1/2 to report that the system seemed to be working.[28] An hour later, at 2 a.m., a different group returned for another look but found the area flooded even higher. They instead went up one floor to check the non-electrical instrument rack,

---

26 Yoshida admitted afterwards to not using the accident management book during the crisis. Much of what he did was improvised and not in the book anyway, so this likely made little difference.

27 Around 30% of prime subcontractors remained, along with 15% of the subordinate subcontractors.

28 Some sources claim the RCIC pump room was flooded but this is likely a mistake, as there don't appear to be any credible sources that specifically state the men opened the door and stepped inside.

where a high pump discharge pressure signalled that the RCIC was working: a rare stroke of luck. They could not control the RCIC, but because it was active and its valves had been open when the tsunami hit, they'd stayed open when the power went out. Unit 3's RCIC also remained operational because its own power distribution panel had, by chance, escaped flooding. With confirmation that Unit 2 was not in immediate danger, attention in the ERC returned to Unit 1.

In Tokyo, Prime Minister Kan and his advisors considered the wider repercussions of venting. In addition to TEPCO's Takekuro and NISA vice director general Eiji Hiraoka, who was responsible for the agency's safety policies and had replaced his direct superior, Nobuaki Terasaka,[29] Kan was counselled by Edano, METI minister Banri Kaieda and several others, including 62-year-old NSC chairman Haruki Madarame. Madarame was the former nuclear research engineer and professor emeritus of Tokyo University who had dismissed Katsuhiko Ishibashi years earlier and, in a February 2007 courtroom deposition for Chūbu Electric, had become exasperated during cross-examination about the potential for total power failure at a nuclear plant, retorting that "if we took all these possibilities into account, we could never build anything."[bb] Madarame and Takekuro had forewarned Kan of the inevitability of venting hours earlier. They assured him that, though this would eject some radioactive particles from the core, most would be scrubbed by the pressure suppression chamber's water and it would not result in a localised radiation disaster. Takekuro estimated they could begin in two hours.[bc] Kan gave his consent and Edano and Kaieda scheduled separate press conferences for 3 a.m. to announce the plan, after which venting was expected to begin immediately. TEPCO relayed the instructions back to Yoshida, who had already been working on the assumption of approval. An eyewitness to this stage of the unfolding disaster overheard a desperate Yoshida, upon being told by someone at TEPCO's head office to inject water when none was available, pleading "It doesn't matter *what*, bring me whatever liquid you can find!"[bd]

While they waited, workers at Daiichi returned to the DDFP at

---

29 "I realized that deep technical knowledge would be necessary after such a severe accident and felt that someone with such technical expertise should stay at the prime minister's office," Terasaka later testified. "I did not study nuclear engineering, nor did I build my career on nuclear safety." It is ridiculous that the head of any highly specialised organisation would know virtually nothing about what the organisation does.

Unit 1, only to find it had died during the preceding hours. Their last sliver of hope now rested with the lone fire engine and on workers searching in the darkness outside among debris piled against the turbine hall for the fire protection intake port. Multiple hazards hindered their work, including open manhole covers and flooding in some lower areas, forcing a step-by-careful-step approach. They also needed to open a new piping path to direct water from outside the building through the fire protection line and into the core. The motor-driven valves weren't easy to use manually, and workers struggled to turn the mechanism's heavy wheel handle, which was in an awkward location above the pipes. The valves themselves varied in size (and hence time to open), from as small as a football to valves with wheels well over half a metre in diameter; some with manual wheel-handle bypass valves and some without.[be] To avoid areas with dangerous radiation levels estimated to be up to 300 millisieverts per hour and rising in the reactor building, still more workers spent hours laying heavy electrical cable (sometimes only to discover that it wasn't long enough) in an effort to revive the valves remotely.[bf] The urgency heightened at 2:30 a.m., when pressure inside Unit 1's outer containment rose to 840 kPa, 60% beyond its design tolerance of 528, before settling back to 800 kPa.[30, bg]

Radiation continued to climb inside the Unit 1/2 control room, mirroring the situation outside. It was most intense beside the ceiling and the door, which lay ajar to allow cable access to the room from the small generator near the building entrance. Several of Izawa's staff sat huddled on the floor, lit only by a few fluorescent strip lights resting on tables, while the more senior members discussed their next move. As Edano stepped into an apprehensive cabinet building press room in Tokyo to announce the vent, an equally apprehensive Izawa addressed his assembled operators – workers whose safety he was responsible for – to assign them their roles to implement it.[bh] In an emotional scene where he struggled to speak, the shift supervisor declared that only senior staff would be going because they were more familiar with the plant layout and could therefore navigate better in the dark. More importantly, radiation posed less risk to their health because cancer from radiation can take decades

---

30   The Mark I containments were apparently thought to be able to handle 300% of their design pressure, but this is now regarded as an over-estimate (and you don't ever want to test this in a working reactor). Regarding the increase in pressure, the SRVs may have lost their effectiveness after dumping too much hot steam into the suppression chamber. At least one report also mentions damage to the SRV nozzle gasket seals.

to form, and because it may make people infertile. After asking for a show of hands but being met only by scared faces in the gloom, he broke the silence and declared that he would go himself. The tension evaporated. One of his senior staff reminded Izawa that they needed him in the control room to coordinate and that he would go in Izawa's place. More and more volunteered until everyone rose to their feet, leaving Izawa speechless. "I was overwhelmed," he admitted later.

While the main group discussed a plan for reactor venting, another team from Unit 1 trekked across to the reactor building at 3:45 a.m. for a radiation check, but when the doors opened, according to the Japan Nuclear Technology Institute's report of the incident, "white haze was seen and hence the doors were rapidly closed and the workers abandoned the idea of measuring the radiation."[bi] It wasn't all bad news – the freezing work outside paid off at 4:00 a.m. as, finally, after more than 12 hours without cooling, freshwater from the fire engine flowed into the melting core of Unit 1. The core pressure had lowered on its own, indicating a leak that was confirmed by a corresponding increase in radiation, but this leak at least made low-pressure injection possible. Sadly, though, up to 50% of this water was redirected inside the web of pipes to various other areas.[bj] Thirty minutes later, another powerful aftershock rocked the plant and all outdoor personnel evacuated to high ground in case of a tsunami. Yoshida then ordered all work in and around Units 1 and 2 be done with full-face respirators and body suits, but each time anyone went outside the respirators became contaminated and could not be used again.[bk] Supplies soon ran low, in part because the inadequate disaster plan only required, for example, one stretcher and 50 Tyvek hazard suits, and individual components were soon being cannibalised to make new protective equipment.[bl, bm] Yoshida repeatedly begged TEPCO for additional supplies throughout the crisis but was often met with unhelpful responses.

As time wore on, the deteriorating apocalyptic conditions began taking their toll on Daiichi's staff as men and women crammed themselves into any free space and lined every corridor in the only true safe haven on site: the SIB. Some sat, staring into nothingness, while others lay on the floor under thin blankets trying desperately to grab a few minutes' sleep. All were exhausted, hungry, scared for their health (and, increasingly, their lives) and that of their loved ones in the wider chaos across Japan. Once local cell-phones stopped working, each person had the chance to make calls by routing them via TEPCO headquarters, and a platoon of personnel from Kashiwazaki-Kariwa arrived during the night, easing the burden to some extent, but resources and resolve remained stretched.

The government widened its mandatory evacuation radius from three to ten kilometres (six miles) at 5:45 a.m.: difficult to effect because of extensive damage to local communications infrastructure, which forced local fire engines to patrol the streets broadcasting the order via loudspeakers. At the time of the announcement, only 20% of residents from nearby towns were even aware of the ongoing nuclear crisis.[bn] Half an hour later, in an unprecedented move, Kan himself boarded a Eurocopter AS 332 Super Puma helicopter heading to Fukushima Daiichi. This decision – the head of a country crippled by an ongoing natural disaster, flying to a specific location just to assess the situation for himself – was met by incredulity from all who learned of it. He grilled NSC chairman Madarame about nuclear safety systems en route, making a point of discussing the possibility of a reactor building hydrogen explosion, which had concerned him for hours. "Even if hydrogen is released within the reactor containment vessel," the chairman assured him, "the vessel is full of nitrogen. In the absence of oxygen, an explosion is out of the question."[bo]

They landed at 7:12 a.m. at the plant's sports field/helipad, then rode a bus straight to the SIB. By now, most of Unit 1's reactor core had melted and was sagging down to the base of its pressure vessel, where the molten mass began dissolving the thick steel. Though some did suspect a core melt at this stage, none comprehended the true extent of the damage. When Kan hastened through the SIB's double airlock doors, he was momentarily shocked by all the people, likening the scene to a "war zone" in his book. He barged past the retinue of decontamination experts waiting for him, barked his presence at a line of weary workers queueing to be screened for radiation after returning from some vital mission outside, before squeezing past them and up the stairs to confront Yoshida.

When the two met, however, Kan was impressed. The plant manager – unlike Kan's advisors, who could only provide vague generalisations – calmly explained everything. In the same way that Kan was an atypical politician, Yoshida was an atypical TEPCO senior manager: frank, pragmatic and prepared to make tough decisions, even offering to assemble a "suicide squad" to go into high-radiation areas to speed things up. Kan learned that the two had attended the same university, instantly forming that *gakubatsu* connection, and that the reason the vent had not commenced was because they were waiting for confirmation of the evacuation of Ōkuma.[bp] When Kan left the facility less than an hour later, he felt reassured that a competent and level-headed person was in charge, and that he and his subordinates were doing all they could. Though he maintained his visit was the right thing to do, the prime minister would later weather fallout of a political kind over his decision to personally occupy Yoshida's valuable time, not to mention over his foul mood and conduct towards the other employees.

Kan signed an order to evacuate a three-kilometre radius around Fukushima Daini as he left. Daiichi's sister plant declared a loss of the pressure suppression functions being used to cool three of its reactors at around 5:30 a.m. The RCIC required no power to keep each unit cool for now, but as operators directed hot water into the suppression chambers, their temperatures each crossed 100°C. They appeared likely to have another major accident on their hands, but technical staff were working feverishly to dismantle the damaged seawater pumps while they waited for new motors and cables to arrive by truck.[31]

At 8:00 a.m., workers at Daiichi finished their work to link Unit 5 to the remaining diesel generator at Unit 6, granting the fifth reactor access to some of its emergency cooling systems and ending the need to repeatedly open and close steam relief valves, which had been ongoing for hours as water boiled. An hour later, another contingent of workers finished laying hoses from the fire engine outside Unit 1 to the firefighting freshwater tank which the fire engine had earlier returned to each time its own tank drained. The work took over five hours but now allowed for a continual water injection for the first time, just as two more fire engines arrived on site.

When, at last, word came through that Ōkuma had been evacuated at 9:02 a.m., Yoshida immediately ordered his staff to open the two valves necessary to vent Unit 1's containment vessel.[32] Izawa had prepared six men – the oldest and strongest – from among the volunteers and split them into three teams of two, then walked each through the task again and again. Each man carried a personal

---

31 Trucks arrived at Unit 1 at 5:22 a.m., Unit 2 at 5:32 a.m. and Unit 4 at 6:07 a.m.

32 The Ōkuma city office called the plant on the telephone and told them that evacuation was complete, but this was a mistake. In fact, the evacuation was not complete. Nevertheless, the vent began.

dosimeter and wore a full-face mask with a 30-minute air tank on his back, a Tyvek hazard suit, and a silver fire-proof suit on top of that to minimise exposure to contaminants.[33] With no means of communication, the teams would go one at a time, with Team 3 as a backup. Team 1 headed down a 200-metre corridor to the pitch-black reactor building. With only the beams of their flashlights to guide them through the silent labyrinth, shadows dancing on every surface, the pair climbed up stairs, ladders and along grilled walkways to the second-floor motor-operated vent valve. Already soaked in sweat and ever conscious of the radiation all around them, it felt like an eternity to turn the heavy valve wheel until it was 25% open. They returned to a jubilant control room exactly 20 minutes after they'd left.

Team 2 broke into a jog as they entered the hazy reactor building, this time heading around to the north-east corner before descending a flight of stairs towards the pressure suppression chamber access hatch. When they opened it, the needle on their bulky ionisation chamber survey meter leapt up to 900 millisieverts per hour: an amount that would surpass the 100-millisievert annual limit for emergency workers in less than seven minutes.[34] They hastened over the threshold, climbed atop the donut-shaped suppression chamber and began to walk along the catwalk around to the opposite side, but halfway to the valve their dosimeter needle swung off the scale. Terrified, the pair turned and ran, and they soon became the first to evacuate for exceeding the dose limit.[35]

Just as everyone's back-breaking work seemed about to pay off, things were going wrong. Descending to Unit 1's suppression chamber proved too dangerous from here onwards, and multiple attempts to remotely open the chamber's main vent valve, in the vain hope of enough residual air-valve pressure, all failed. Without both valves from the dry and wetwells, the pressure was insufficient to affect a vent by breaking the rupture disk – a last resort, single-use diaphragm, pre-set to burst open above the maximum design pressure. Soon afterwards, while operators debated their next move, Unit 3's RCIC system failed at 11:36 a.m. A 2014 TEPCO

---

33 I have seen both 20- and 30-minute capacity for the air tanks, sometimes recorded by the same organ-isation, so it's unclear which is correct. Because of this limit, the workers would not begin using their air supply until they reached the reactor building.

34 This meter measures the electric charge caused by ionising radiation passing through a gas-filled chamber inside the device; it is commonly used to measure beta and gamma radiation.

35 One of them received 106 mSv, according to the IAEA; the other is listed simply as ">100 mSv."

analysis concluded that the failure "was likely due to an electrical trip caused by high turbine exhaust gas pressure."[36, bq] Some additional battery power was now available, thanks to a supply of around 50 hefty 2-volt batteries, each weighing over 12 kilograms, that had arrived via JSDF helicopter from TEPCO's Hirono thermal power station 21 kilometres to the south.[br] Unit 3's high-pressure coolant injection (HPCI) system used this electricity to activate an hour later as water levels dropped. A squad of engineers reported that they'd fixed Unit 1's DDFP at 12:51 p.m. after working on it for ten hours. They immediately put the pump to use, but it only lasted 20 minutes before a motor fault knocked it out again.[bs]

After discussing a way of opening Unit 1's remaining valve, the operators and technicians decided to repeat steps they'd taken earlier with the fire protection line: force open the large pneumatic air-operated main isolation valve from outside the building. Unfortunately, TEPCO's operators found their portable air compressors were not powerful enough, nor did they have adaptors to connect any to the piping system, forcing them to hunt for equipment locked away in subcontractor buildings. By early afternoon, as a group of technicians toiled away at the task, Izawa and his staff felt they had no choice but to venture back into the reactor buildings. Team 3 – the original backup – suited up and set off towards the unknown radiation levels, but no sooner had they left than Izawa received word that white mist was escaping the ventilation stack, a possible sign of a containment breach. He yelled across the dimly lit space at two shift supervisors, who grabbed face masks and sprinted out the door to stop the men who had just left, catching them just as they neared the reactor building doors.

At some point, someone decided to pile lead-acid batteries in an external doorway to the Units 3/4 control room that wouldn't close because of a warped frame. The lead helped somewhat to protect against gamma rays. Whether from the meltdown or the air compressor, the reactor pressure finally began to fall at 2:30 p.m. Yoshida, Izawa and Daiichi's hundreds of other men and women all breathed a collective sigh of relief, but it was no time to celebrate.

---

36 This high exhaust pressure trip was designed to stop the system in case the exhaust path into the suppression chamber was blocked. At Daiichi, the chamber pressure was high but the path was not blocked, meaning the safety trip actually *caused* further problems because an RCIC's oil supply is cooled by the same water that the system pumps, hence causing the bearings to seize as the turbine span down. As one senior reactor operator told me: "We have explicit guidance now that if RCIC is operating beyond its design envelope, to defeat all trips (so it runs until true failure), and to never shut it off until you have a backup pump ready to inject, because it won't restart."

They knew by now that the core must have at least partially melted and that cooling it before it breached the steel pressure vessel was crucial, given that rising radiation suggested a compromised outer containment. Just before 3:00 p.m., word reached Yoshida that the 80-ton freshwater tank being used by the fire engines to cool Unit 1 had run dry. Faced with no alternative, he ordered his men to drop their hoses into Unit 3's enormous 6.6 metres deep, 9-by-66-metre-long backwash valve pit, which was full of ocean water from the tsunami. (This pit is where water valves for cleaning areas of the steam condenser with seawater are located.) The unfiltered saltwater would corrode everything it touched, guaranteeing the reactor could never be used again, but Yoshida knew it was already beyond saving. The best they could do now was contain the bubbling radioactive magma within.

As workers laid water lines to the pit, using three fire engines in series to increase the discharge pressure, another recovery team finished their gruelling work laying a reel of heavy-duty electrical cable weighing over a ton to re-energise an undamaged transformer and power centre at Unit 2 via one of over twenty generator trucks now on site.[bt] They could now send the required 480-volt AC power over to Unit 1, which would finally allow them to activate a standby pump to cool the reactor properly. Both teams finished their tasks at around 3:30 p.m.[bu] Six minutes later, almost exactly 24 hours to the minute since the tsunami hit Fukushima Daiichi and just as the fire engines began to draw up saltwater from the pit, Unit 1 exploded.

The fifth-floor refuelling deck – the building's entire top third – wrenched apart, sending shards of irradiated concrete and metal knifing through the air in all directions. The massive heavy-duty gantry crane bent like a twig and collapsed onto the refuelling floor control room, crushing everything that wasn't expelled in the blast. Below, the IC tank room ceiling caved in, wrecking everything nearby. Chunks of debris rained down outside on the people who, moments earlier, had reactivated the fire engine pump, injuring five[37] and shredding the hoses they had just laid into the pit. Among the injured was the plant's own 50-year-old fire chief, whose arm snapped when a piece of steel hurtled through the passenger window of his fire engine.[bv] The engine survived the beating, but the

---

37 Five people according to several sources, including the IAEA. Some others say it was only four.

electricity distribution panel connected to one generator truck and multiple sets of carefully laid power cables also sustained damage, including the one set finished minutes earlier to reactivate Unit 1's standby water pump. Everyone scrambled back to the SIB with its filtered air, halting all recovery work and reactor cooling for around two hours. The yellow domed lid of Unit 1's containment vessel now jutted out among what remained of the refuelling floor, clearly visible from outside.[38]

Izawa's heart leapt out of his chest as the explosion rocked his control room. Had the reactor exploded? The radiation would be fatal if it had. He quickly ordered all but the most critical personnel to don their hazard gear and retreat to the hilltop SIB with the others. Yoshida and his staff inside the windowless ERC felt a single sharp jolt but were momentarily unsure if it was yet another aftershock; it didn't *feel* like one. A static telephoto video-camera, perched atop a hill ten kilometres away and belonging to local news agency *Fukushima Central Television*, captured the explosion. They broadcast the footage locally within minutes, where it played on the ERC's giant screens to a stunned audience who watched as their reactor building exploded over and over again while a nervous television anchor tried to describe what she saw.

Yoshida felt sick. "When the first explosion occurred," he later recalled, "I really felt we might die."[bw] Certainly, at that moment, he couldn't imagine that anyone near the blast survived, but they were lucky: the core itself had not exploded. Up to 900 kilograms of hydrogen from inside the containment vessel – created by steam reacting with the fuel's zirconium cladding – had leaked into the building's upper floors, likely out of the domed vessel lid, via a compromised gasket or flange joint, where the explosive gas floated until some stray spark ignited it.[39] Unit 1's blowout panel could have prevented the explosion, but NISA ordered their reinforcement at all plants after the incident at Kashiwazaki-Kariwa. They were

---

38 Though several possibilities were ruled out in the government report, the definitive ignition source for the explosion is not known. Efforts to restore electric power to the unit may have inadvertently caused it.

39 The 21,000-cubic-metre fifth floor required 311.6 kilograms of hydrogen for detonation. In 2012, TEPCO and JNES each released separate analyses, the former predicting slightly over and the latter slightly under 900 kg of hydrogen produced by the time of detonation. Source: Investigation Committee on the Accident at the Fukushima Nuclear Power Stations of Tokyo Electric Power Company, "Chapter 2: The damage and accident responses at the Fukushima Dai-ichi NPS and the Fukushima Dai-ni NPS," pp. 58–59 (pp. 52–53 of PDF). *Final Report.* Government of Japan, 2012. https://www.cas. go.jp/jp/seisaku/icanps/eng/03IIfinal.pdf

redesigned with carbon steel fittings that warp under pressure. In a possible extreme overreaction, however, several anonymous TEPCO employees made unconfirmed claims that the panels at Daiichi Units 1 and 3 were welded shut – or, as the government accident report put it, "installed to avoid easy removal" – and thus did not function during the disaster.[bx] The plant was also fitted with devices called "autocatalytic recombiners" which use clever chemistry to convert hydrogen into steam. Passive recombiners require no power and became common after the Three Mile Island accident, but unfortunately Daiichi's versions required power and thus did not work. Workers had considered cutting the blowout panels open hours earlier but knew that doing so would create a spark, so they abandoned the idea. The 4.2-by-6-metre blowout panel on Unit 2's eastern wall, on the other hand, flew off from the force of the blast, despite theoretically requiring 352 kilograms per square metre of pressure from the inside to open.[by]

A photograph taken from the administration building after Unit 1 exploded. Units 2, 3 and 4 remain intact.
Source: Tokyo Electric Power Company Holdings

* * *

It took over an hour before Shōriki's old *Nippon Television* became the first national station to air the incident. This broadcast, at 4:50 p.m., was essentially the first time that Kan and his executive team learned of an explosion, having only been informed that smoke was seen rising from the facility.[bz] Kan demanded that Kakekuro explain the situation, but the TEPCO executive could not get satisfactory information from his company's headquarters. Madarame, who had that morning assured the prime minister that an explosion was impossible, now cradled his head in his hands. When Edano held a press conference an hour later, he was sharply criticised by the public (though not, at first, by the mainstream press) for calling it "some sort of an explosion event."[ca] Edano's remark has been translated in various ways, but he basically characterised the blast as something akin to an explosion but not necessarily an actual explosion, despite it being clear what had happened. He later claimed that, in the absence of any concrete information, to call it an "explosion" when it may not have been would've caused a panic. People understandably panicked regardless, but the government found itself unable to confirm the cause until almost 9:00 p.m. This bungling followed an infamous press conference just prior to the explosion wherein NISA deputy director general Koichiro Nakamura admitted that "a meltdown may be in progress" at Unit 1, attracting global attention. By the time of Edano's own post-explosion conference, however, Nakamura had mysteriously disappeared and did not return, prompting more questions about transparency. His replacement addressed the matter a day later, disingenuously saying that "the appropriate description is 'damage to the outer covering of the fuel rods.'"[40]

* * *

Everyone at Daiichi sheltered indoors as probes measured the situation outside. Radiation had increased around the plant but not to levels immediately hazardous, so, after 90 minutes, jittery workers

---

40 Nakamura was replaced at the behest of Edano and others at NISA. He "voluntarily" resigned in August that year, but the Rebuild Japan Initiative Foundation considered that he had likely been pressured to leave. Source: Independent Investigation on the Fukushima Nuclear Accident, *The Fukushima Daiichi Nuclear Power Station Disaster: Investigating the Myth and Reality*, Routledge, 2014.

returned to remove the contaminated explosion debris and replace several hundred metres of hosing between the linked fire engines and the pit.[cb] A contingent checked the mobile generator truck at Unit 2 and replaced the damaged cable, but when they switched it on, the truck's electrical overcurrent protection system tripped, preventing it from working.[cc] Work to replace the hoses finished at 7:04 p.m. after several hours with no water injection, but a mere 15 minutes after cooling resumed, TEPCO liaison Ichiro Takekuro called Yoshida to check on the seawater injection preparations. Yoshida informed him that injection had already begun, but his superior ordered him to stop. When asked why, Takekuro shot back, "Shut up! The Prime Minister's office keeps on pestering me," then hung up.[cd] The possibility of a fuel recriticality – a new burst of fission enabled by saltwater – concerned Kan, and Takekuro had just told him that injection would not begin for over an hour. Thus, Kan thought they had time to consider the ramifications before commencing. Madarame had already assured him that recriticality was almost impossible, but Kan no longer trusted his expertise. He asked for a fresh analysis while they waited for preparations to be completed. Yoshida knew none of this, but political micromanagement – fuelled by Kan's fears and frequent temper tantrums – had returned to hamper things at the Kantei.

Irritated, Yoshida contacted TEPCO headquarters and its executive vice president, Sakae Muto, at the makeshift OFC, who then seconded the command to cease saltwater injection. "However, I didn't have the slightest intention of halting the process," Yoshida commented later. "I told myself that since there was no guarantee about when the seawater injection would be resumed, I would do it at my own discretion."[ce] After conferring with a trusted deputy in the ERC, Yoshida pretended to acquiesce, making a show of issuing an order to stop on the ever-present video link with TEPCO HQ, but in reality he ignored the request. His defiance potentially avoided further catastrophe and received global praise in the aftermath, especially given the ordinary Japanese strict deference to authority. "I felt that, in the end, it had to be my decision," he said. "There was just no time for debate."[41, cf] Media organisations later incorrectly reported that Kan himself had ordered the cessation of seawater injection. In fact, when discussion of the matter resumed at his

---

41 Even ignoring everything else, Yoshida was correct in this regard: plant managers have ultimate discretion in the regulations for situations like this.

office a short time later, he immediately gave his consent without ever knowing what had transpired.

Owing to the deteriorating conditions, the government increased the evacuation radius a third time to 20 kilometres (12.4 miles), but this order caused further problems. The number of citizens abandoning their homes ballooned from just under 6,000 within three kilometres, to over 50,000 at ten kilometres, and now over 170,000 within a 20-kilometre radius. People who had earlier evacuated from the ten-kilometre zone stopped short of this new boundary and were forced to move again, while others – thanks to damaged communications infrastructure and widespread power loss – still weren't even aware of a problem at the power station and were completely unprepared to leave.[cg] As such, it was not unusual for people to arrive at evacuation centres carrying nothing at all on the belief they would soon return home.

The barely functioning OFC only contained maps and a plan for up to a ten-kilometre evacuation, as such an extreme distance had never been considered. According to a government report, personnel there found themselves "unable to designate the parameters of the mandated evacuation zone, and even when receiving questions from the relevant local municipal bodies [they were] unable to provide definitive answers."[ch] Rather than acting as a coordination point between national and local governments and the plant itself, the OFC often floundered in confusion despite the best efforts of those who battled to reach it. The wider government, too, notwithstanding annual (though non-practical) emergency preparedness drills, had not actually prepared for an extreme or complicated accident and was overwhelmed on almost every front. "In the wake of this accident," the NAIIC wrote in their report, "many participants indicated that they felt the drills were useless."[ci]

* * *

Unlike reactors 1 and 2, which had suffered a total loss, operators were able to monitor and cool Unit 3 during those first two days because some of its DC systems stayed dry. Things went well until its RCIC system failed at midday on the 12th, just as all efforts were focused on venting Unit 1, but as water levels dropped the high-pressure coolant injection (HPCI) system automatically came online to continue cooling. Despite its high input rate of 1,800 cubic metres per hour, this system is driven by a steam turbine like with the RCIC and requires no main AC power, though it is not designed to run for much more than a few hours as an emergency backup. To preserve battery power, the operators disconnected all non-essential systems and manually controlled the HPCI to prevent it from stopping and starting as water topped up the reactor.[42] Their efforts stretched Unit 3's battery supply until 8:36 p.m., at which point all instruments started to fail. Preparations were already underway for venting.

What remained of Unit 1's reactor core had now burned a hole through the steel pressure vessel – an almost impenetrable barrier – and begun to chew its way through the concrete containment vessel below.

iii. Sunday, 13th March – Day Three

A merciful period of quiet followed the explosion until 2:30 the next morning, when Unit 3's operators managed to take another pressure reading and realised they had a new problem. The HPCI's high-pressure injection had done such a good job of using up the available steam and flooding the core with cold water that the reactor's temperature (and, by extension, its pressure) had dropped below the working pressure range of the system's decay heat steam-driven turbine.[cj] Years later, a detailed analysis suggested the HPCI had actually ceased injection at around 8 p.m. the previous evening – just before the previous reading – after which the water level began a gradual decline, but no one knew this at the time.[ck] The operators grew concerned that the low pressure would cause the HPCI's high-pressure turbine to vibrate and break, creating an escape path for radioactive steam, but the dead batteries prevented the turbine from auto-stopping. With the fire engines already in use, they decided their best option was the low-pressure DDFP which, unlike at Unit 1, continued to function.

This was perfect, because core pressure was now low enough for it to work, but keeping things that way required an operator to remotely open and close the reactor's safety relief valves (SRVs) as

---

42 To be more specific, the operators were opening the valves on a test line and redirecting excess water into the suppression chamber each time the reactor water level became too high.

pressure rose and fell. They could do this from the control room, where the SRV switch light somehow still emitted a faint glow, despite the depleted batteries. Armed with a plan, a shift team suited up and entered the reactor building to realign the DDFP from wetwell-cooling to core-cooling just as the operators switched off the HPCI at 2:42 a.m. Unfortunately, poor communication meant they disabled the HPCI without confirming that the DDFP was up and running. Then they hit a stomach-churning problem: the SRVs wouldn't open. BWR valve solenoids require almost double their normal voltage to work when the drywell temperature exceeds around 150°C (300°F), and the HPCI's battery-driven auxiliary oil pump was draining what little power they had. That final trickle of electricity had been enough to light the control interface bulb but not enough to open the valve.

The rate of water evaporation indicated that the HPCI water tank had now run dry, but preventing the hot steam from exiting the core – where before it had vented out to drive the HPCI turbine – caused the existing coolant to evaporate into more steam and the reactor pressure to begin an uncontrollable climb. Though switching cooling methods had seemed like the right move, they made a critical mistake in not ensuring the SRVs opened first. Pressure quickly surpassed the DDFP's operating range, rising from 580 kPa to over 7,000 in under two hours, which forced the dismayed operators to first try restarting the HPCI and RCIC multiple times, then to send a team to check valves and other equipment inside the reactor building.

The top of the core's 94 tons of fuel became uncovered as water boiled away, at which time the temperature was only about 200°C (392°F). Within three hours, it had leapt to 2,800°C (5072°F) and was melting, triggering a release of hydrogen similar to what had happened at Unit 1 two days earlier.[cl] The gases leaked out of the reactor, mostly via the first-floor machinery access hatch, located on the widest part of the drywell. Rather than gather in the fifth-floor roof space, as at Unit 1, this hydrogen began to fill the entire building.

Neither Yoshida nor any of his team leaders in the ERC were informed of the HPCI shutdown or subsequent problems until around 4 a.m., but barely an hour later the deteriorating situation at Unit 3 forced him to declare another emergency.[43] Of the three

---

43 The problems were reported to lower-level staff in the ERC but did not go higher than that. Also, when Yoshida and his managers did learn of the shutdown, the noise in the ERC initially caused them to mishear the call as "automatically shut down" rather than "manually shut down" (the phrases sound

fuelled cores, Unit 3 was his greatest concern because of its 32 MOX fuel rods, the plutonium content of which would be devastating if released.[44] Yoshida instructed his staff to repeat the fire-engine cooling method already employed at Unit 1 and to open the containment vent valves up to the rupture disk while radiation levels remained low enough to enter the building, just in case. An engine from Kashiwazaki-Kariwa on standby at Fukushima Daini set off towards Daiichi while another at Units 5 and 6 moved to the Unit 3 backwash valve pit for a water source. Two men in full hazard gear and respirators left Unit 3's control room and descended into the pressure suppression chamber, then along a curved catwalk like the one the team from Unit 1 had failed to pass two days earlier. The space was sweltering after constantly recycling cooling water for two days. "In total darkness, I could hear the unearthly sound of [the] SRV dumping steam into the torus," recalled 24-year-old auxiliary equipment operator Satoru Hayashizaki.[cm] He was deeply concerned for his close friend Kazuhiko Kobuko, one of the two men still missing in the flooded Unit 4 basement. The pair had been among those checking the building for damage when the tsunami came, and Kobuko had just so happened to go down to the basement while Hayashizaki went upstairs.[cn] He spent almost an hour shouting for his friend but was kept off the search party because of their relationship, to spare him the distress.

Two days later, he reached the suppression chamber valves, but as the frightened Hayashizaki climbed onto the orange railing to brace himself on the chamber's outer surface, he slipped and almost fell. Shining a flashlight at his feet, he realised the rubber sole of his boot had liquefied the moment it touched the surface. Unable to approach the valve's display panel due to the heat, they returned to the unlit control room, where radiation levels were climbing. Hayashizaki, shaken and convinced they were doomed, wrote a brief letter to his parents. "Father, Mother, please forgive me for dying before I could fulfill my duty as a son. After the earthquake, I would have wished to hear your voices even once. I did not give up on life until the very end."[co]

Meanwhile, another team brought a generator truck as close as

---

very similar in Japanese), which was not a direct cause for concern as the system was designed to start and stop itself. They only discovered the mistake sometime later.

44 Realistically, it's difficult to gauge how much worse a MOX release versus a standard release would be, since standard fuel will have already generated plutonium by itself in minute quantities. As one person told me, it's like having 98% poison versus 99% poison.

Top: The ERC. Yoshida is seated on the back-right, wearing black.
Bottom: Fire engines and their hoses.
Source: Tokyo Electric Power Company Holdings

debris allowed to the reactor building's switchgear room to supply power to a set of backup water pumps, but they wasted hours trying to route cables past unending obstacles, including "an iron door to the passageway [which] was warped and could not be opened. The recovery team had to ask a partner company to cut the door with a gas torch to secure the route for laying cable."[cp] There were other problems. "Normally, laying cables requires one to two months," said one restoration worker, "however, it was completed in only a couple of hours. Also, we had to find the penetration seals in the darkness and splice the ends. With the puddles of water around, we thought we were going to get electrocuted."[cq] Another group helped connect the two fire engines together, but as they were about to begin water injection at 7:00 a.m., TEPCO headquarters ordered Yoshida to switch over to a freshwater tank and avoid using saltwater for as long as possible.[45] He reluctantly gave the order, thinking it came down from Kan himself, but by the time the team had rearranged everything, the reactor pressure had climbed too high for the fire engine pumps.

Operators spent the next few hours trying to reduce pressure by opening the SRVs, both remotely from the control room using ten scavenged 12-volt car batteries to power the valve solenoid (after first being mistakenly given more 2-volt batteries) and manually from inside the reactor buildings.[cr] Another team set off towards the nearest major town (Iwaki, 30 kilometres away) in search of stores that sold batteries. Navigating the destroyed country roads took six hours, but upon arrival they couldn't find anywhere with the correct type. Then, between 8:00 and 8:30 a.m., radiation monitors at the plant's main gate registered a 3,340% increase in radioactivity, from 35 microsieverts per hour (μSv/h) to 1,204 μSv/h, likely from the combined damage to Units 1 and 3.[cs] (The figure had been 0.069 μSv/h the previous morning.) Just after 9 a.m., drywell pressure in Unit 3 fell, with a corresponding sharp rise in wetwell pressure, though nobody could tell if the SRVs had opened or if the containment had been compromised.[46] Pressure inside the wetwell

---

45 This order to use freshwater appears to have been a misunderstanding. Staff from the Prime Minister's Office had asked a TEPCO executive if freshwater was available. He passed this question along to Yoshida, who took it as an order to use any available freshwater. The possibility of using seawater on all three reactors was discussed at length over the videoconferencing system during these first days, so understanding what was really an instruction from what was theoretical was not easy. Company executives have a vested interest in denying they delayed saltwater injection after the fact, as such actions could result in their prosecution.

46 The timing is in question, partly because the government's interim report stated that the batteries were connected at 9:08 a.m. and the SRVs were likely opened around then, though the operators were

climbed to a peak of 630 kPa (91 psi) – almost 20% beyond its normal limit – during ten agonising minutes before the rupture disk burst and containment pressure fell.[ct] This allowed the fire engines to begin injecting freshwater mixed with fission-dampening boron from the plant's firefighting tank after four hours without reactor cooling, though investigators learned later that backflow pressure and branches in the piping meant much of this water had been redirected away from the core. At some point during the subsequent two hours, the air compressor holding open the suppression chamber's air-operated vent valve failed and the valve snapped shut, once more blocking the pressure release. Workers found the valve closed at 11:17 a.m.

Yoshida, planning ahead, ordered the Unit 2 vent path opened to the rupture disk and fire engines connected to the building's fire protection system while the situation there remained stable. After struggling to get through locked doors with missing keys, operators opened the motor-operated containment vessel and air-operated suppression chamber valves by 11 a.m. The motor valve would stay open without issue, but they'd have to monitor the air valve. Yoshida grew increasingly concerned throughout the morning and ordered a team to begin connecting Unit 2 to the pit full of saltwater at midday, just in case its RCIC failed.[cu] The freshwater tank being used to cool Unit 3 ran dry at around the same time, forcing the exhausted firefighters to revert to saltwater from the pit here too. People at TEPCO HQ spent at least the next seven hours trying to prevent Yoshida from using saltwater at Unit 2 for as long as possible. There were legitimate safety concerns, such as salt clogging the fuel rods' debris filters, which could potentially damage the fuel itself and cause a radiation leak, but that does not appear to have been the issue. "Laying out the selfish thinking on our side," began one exchange recorded on the video link, "starting with seawater from the beginning will lead directly to decaying of materials, which would be wasteful. Can we have the understanding that there is also the option of waiting for freshwater as long as possible?"

"It is not about understanding," shot back an exasperated Yoshida, "we have already decided on the lineup that [will] be used, and since we chose to make the ocean the source of the water, we do

---

unsure. The final report changed this sequence of events and said the exact cause of depressurisation was unknown and that the SRVs did not open until 9:50 a.m., 71 minutes after the batteries had been brought to the control room.

not have the option of going with freshwater from now [on]. That would only lead to a loss of time, again.... What you want to say is using freshwater will, in the end, make it possible to use it again in the future because it will not have been damaged by the salt."[cv] Operators sent to Unit 3 recorded levels exceeding 300 millisieverts at the airlock that afternoon: the containment was compromised. Yoshida worried that this would lead to another hydrogen explosion and ordered all personnel to shelter indoors at 2:45 p.m. (halting the flow of coolant into Units 1 and 3 once more), but after two hours it looked like nothing dramatic would happen after all and work resumed.

By this time, someone had noticed steam rising above the open carcass of Unit 1's reactor building, coming from the top floor's spent fuel pool, which was now open to the atmosphere. Hundreds of depleted fuel rods in racks of boron-impregnated aluminium sat at the bottom of the massive pool – an average-size BWR pool is over 1,000 cubic metres in volume and around 12 metres deep – beneath eight metres of water.[cw] These rods typically stay there for months or years before being moved to a larger pool at the plant's dedicated storage building. Unlike fresh fuel, which is relatively harmless and emits no heat, spent fuel can stay hot for decades while radiating lethal fission products. One major reason the Chernobyl disaster was so severe is because it happened at the end of a fuel cycle, when the rods were at their most contaminated. Demineralised water inside Daiichi's Unit 1 pool began to heat up after the initial blackout, heated by decaying fuel at around 180 kilowatts, and was creeping towards the boiling point. Nobody knew if the water was evaporating or if the pool itself was compromised and leaking – or both. The fuel would ignite and cause an environmental catastrophe if it became uncovered, but they had no way of getting water up there.

Meanwhile, workers managed to open the Unit 3 suppression chamber's air-operated vent valve at 7 p.m., after struggling for hours to fill the air compressor. The drywell pressure then dropped over the next couple of hours as gases filtered down to the wetwell. Yoshida sent a group of operators to replace those in the Unit 1/2 control room and told Izawa and his team to get some rest, allowing the shift supervisor – spent of all energy, like his colleagues – to leave the service building for the first time since before the earthquake.[cx] The devastation left him speechless for a second time.

* * *

Over at Daini, the undamaged Unit 3 reached a cold shutdown while work to restore the proper cooling of Units 1, 2 and 4 continued. Masuda wanted to keep his lone operating diesel generator for cooling Unit 3, so he opted to direct power from the Radioactive Waste Building (RWB) beside the SIB, requiring workers to lay new cables. The working main external power line allowed operators to use their control room instruments and determine that, of the remaining three reactors, Unit 2 needed to take priority.[cy] Around 200 TEPCO and subcontracting personnel laid nine kilometres of heavy electrical cable within a day – a feat which normally would take 20 people over a month. They first connected Unit 2, but after several hours of laying cable, operators realised Unit 1 posed more of a threat, forcing Masuda to make the difficult decision to drop everything and start over again.[cz] Their work connected all four reactor buildings to the RWB and two generator trucks and was finished by around 11:30 p.m. on the 13th.[47] Main cooling for each remaining reactor was restored over the subsequent 16 hours.

iv. Monday, 14th March – Day 4

As a new week began, the backwash valve pit being used to cool Units 1 and 3 had become worryingly empty by 1 a.m., forcing workers to pause water injection for two hours while they lowered their hoses further. Twenty minutes after cooling stopped, operators noticed Unit 3's drywell temperature and pressure creeping up. Yoshida and his team considered Unit 1 a lost cause and unlikely to explode again by this point, so Unit 3 received cooling priority with what little water remained in the pit. Injection resumed there at 3:20 a.m., with pressure approaching 350 kPa (51 psi). Two hours later, as the pressure continued its inexorable climb despite the fire teams' best efforts, it became clear they had to vent Unit 3 to prevent a major containment breach. Operators sent remote commands to the vent valves but it was too late – their instruments showed that the fuel became uncovered just after 6 a.m. as the pressure crossed

---

47 One huge cable stretched from the RWB to Unit 1, which was also connected to a generator truck. Another cable went from the RWB to Unit 2, then shorter cables connected 2 to 3, and 3 to 4, where the second truck sat.

470 kPa (68 psi). That pressure dropped a little as an easterly wind carried the vented gases out across the ocean; lucky, because a drywell radiation monitor registered 167 sieverts per hour – enough to kill someone in about two minutes – just prior to venting.[da]

Fearing another hydrogen explosion, Yoshida again ordered all outdoor personnel to evacuate the area at 6:30 a.m., but after half an hour with no apparent change, the containment vessel pressure levelled off at 450 kPa, then began a slow fall. He was reluctant to send people back outside but executives at TEPCO headquarters were pestering him to resume and – more importantly – he understood the urgency. Unknown to all, the pressure vessel ruptured at 7 a.m. as molten fuel burned through steel. Work resumed 30 minutes later.[48] "Unit 3 could explode anytime soon, but it was my turn to go to the main control room," remembered one operator. "I called my dad and asked him to take good care of my wife and kids should I die."[db]

NISA officials, out of some misguided attempt to avoid a mass panic, began discussing a media blackout of ongoing events and by 8:00 a.m. had ordered TEPCO to halt updates to the press. Even Fukushima's local prefectural government was somehow browbeaten into silence after initially rejecting the idea. The unfolding disaster now gripped the entire world, so cutting off information was a rather audacious move.

Two fire engines, utilising a line thrown into the harbour, started pumping water back into the empty pit at 9:00 a.m. – one of four original proposals for refilling the pit, after abandoning the main idea of using water from the 2,000-ton filtered water tank when an apparent leak was found to have drained it – and were soon aided by the arrival of JSDF water supply tankers. Once the pit had been refilled, additional fire engines planned to resume cooling of Unit 1. At 10:30 a.m., a team of workers tried to enter the Unit 4 reactor building to check on conditions there, expecting little to report (remember, Unit 4 was offline and empty), but they encountered high radiation levels at the door and were forced to turn back.

---

48 According to Yoshida, "I realized there was a risk of a hydrogen explosion, so I called for us to decamp. However, when discussing this with headquarters, I was told, 'Just how long are you evacuating?' I told them that there was a risk of an explosion, and there was no way we could put personnel on the ground. They [headquarters] said to me, 'Can you get back to handling the site soon?' The pressure on the containment vessel had dropped a bit. But, when we sent personnel back, it exploded." Source: "Yoshida interviews: Strong words on Fukushima N-crisis from TEPCO's manager on the ground." *Japan News*, 10th September 2014. Archive link: https://web.archive.org/web/20140915210028/http://the-japan-news.com/news/article/0001553303.

With no warning, Unit 3 burst open in a deafening explosion far more powerful than the first, partially destroying the third, fourth and fifth floors and sending thick smoke and debris over 300 metres (around 1,000 feet) into the air. Rubble pummelled the crew laying fire hoses as they dove under their fire engines. Yoshida again thought they were finished. "Headquarters, headquarters! We have a big problem, we have a big problem," he yelled over the video link to Tokyo. "It seems there's been an explosion at Unit 3!"[dc] Racked with guilt for having sent people back outside when things were on a knife edge, his voice audibly cracked from stress as he declared the official time: 11:01 a.m.

When the dust settled, workers outside saw a thick line of steam from inside the containment vessel rising through the wreckage.[49] Staff up in the SIB initially reported dozens unaccounted for, which had a profound effect on Yoshida. "If that report were true," he said later, "and some 40-plus people were really dead, I thought I should commit *harakiri* [ritual suicide, also known in the West as *seppuku*]."[dd] Several workers who'd been outside near Unit 1 when it blew up and had been scared to return to the reactor buildings afterwards had the misfortune of being outside near Unit 3 for this second explosion. Forty-seven-year-old Mitsuhiro Matsumoto had just left Unit 2's turbine building and was approaching the car he'd been using to traverse the site when the explosion enveloped him in smoke and dust. Once it cleared, he found the driver's side of his car, mere metres away, flattened by fallen concrete. "I shuddered at the sight," he remarked later. "If I had been inside, I would have been dead."[dc] Seven subcontractors and TEPCO employees, plus four JDSF personnel – some of whom had arrived at the reactor building at that exact moment in a convoy of tankers – weren't quite so lucky and sustained injuries requiring hospitalisation, but again, by some miracle, nobody died.

Still, while most had escaped serious physical harm, the mental strain became difficult to bear, with many of those outside during the explosions, in particular, displaying signs of post-traumatic stress disorder, even as the ordeal continued. Some now jumped at any sudden sounds, like a closing door, while others were found curled up on the floor muttering to themselves. Yoshida recognised his staff's terror and contacted TEPCO again. "Perhaps I shouldn't be

---

49 TEPCO confirmed in 2013 that this steam did come from the containment vessel after ignoring the question for two years.

Left to Right: Units 3, 2, 1.
Note steam escaping from the detatched Unit 2 blowout panel.
Source: Tokyo Electric Power Company Holdings

saying this now," he said, "but still, uh … there have been these two explosions, and workers at the plant site have really been shocked, or whatever you might call that." Later, he put on a brave face. "I fear we are in acute danger," he said, trying to steady his voice. "But let's calm down a little. Let's all take a deep breath. Inhale, exhale."[df]

Much like the first, this second explosion shredded water hoses and wrecked the closest fire engine, while the seawater pit was now a jagged knot of radioactive debris, making it useless for cooling – they needed another water supply, and fast. Work to resume cooling of Unit 1 had just about finished, but much of it now needed to be redone. The blast also damaged a component on the air-operated valve of Unit 2's suppression chamber vent line, which had been opened 24 hours earlier, causing it to slam shut.[50] All efforts to reopen it and its bypass valve failed, making water injection their only means of reducing core pressure. With everyone expecting Unit 2 to blow during the coming hours or days, Izawa picked up a phone in the SIB and dialled through to the unit's control room. Though his frightened subordinate assured Izawa that they didn't require a rescue, he couldn't desert his staff, so he grabbed a team of operators and headed out to the control room while everyone else took shelter indoors.

After two hours, a bowing Yoshida apologised for putting everyone back in harm's way, but he explained that they couldn't stop now. As he said in his government testimony:

> I was really moved when everybody came out willing to go back to the front lines. I called for restraint, on the contrary, so they would not go out at random. We arranged things like, this team and that team should be doing this, construction guys should use backhoes to clear rubble. We worked out plans before they left. And most of them at that time got almost excessive radiation doses when they were, like, replacing hoses.[dg]

At the same time (1:00 p.m.), Izawa's team noticed that Unit 2's reactor pressure had risen and its water levels had fallen since they last took readings, at 7:00 a.m. Nobody had checked the wetwell's water temperature or pressure at all until 4:30 that morning, even

---

50   The blast somehow dislodged the electromagnetic excitation circuit. Source: "PCV venting and alternative cooling water injection preparation for Fukushima Daiichi Nuclear Power Station Unit 2." TEPCO: http://www.tepco.co.jp/en/nu/fukushima-np/interim/images/111202_12-e.pdf.

though the RCIC's constant use and the broken residual heat removal system meant that both were in a steady climb.[dh] With a sinking feeling, the operators realised that the RCIC, their last proper cooling system – which had somehow kept going all this time – had died.[di] They were incredibly lucky it lasted as long as it did, a testament to its solid design and engineering. According to one Nuclear Energy Agency analysis:

> It is theorized that two phase water carryover [i.e. water and steam] to the RCIC turbine resulted in reduced RCIC function and reduced water injection such that a self-regulating mode of operation was attained. Interestingly, the loss of DC power during this self-regulating operation period would also disable other shutdown interlocks such as high water level and high turbine exhaust, which may also have been a key to the sustained operation of RCIC for nearly 3 days.[dj]

At 1:25 p.m., operators reckoned they had about three hours before Unit 2's fuel became uncovered. The government raised the radiation exposure limit for emergency workers half an hour later, from 100 millisieverts to 250, out of a growing concern that they might soon run out of people.

Workers outside had already tried pulling up water for the fire engines from the water intakes in front of Unit 4's turbine building and the south shallow draft quay but couldn't get to either location because the ground had collapsed on the approach. They also tried using an underwater pump feeding through a hatch in the water discharge channel, but that didn't work either. They were now desperately trying to hook up the two surviving fire engines from the harbour but accepted after two more hours that the familiar high core pressure, the vertical difference of 20 metres from the water source and the engines' weak pumps made this impossible.

The portable generator powering the Units 1/2 control room lights and the wetwell's electromagnetic air valve tripped at 4:21 p.m. Operators disconnected the lights, restarted the generator and carried on in darkness, but the valve wouldn't reopen. By now, the pressure suppression chamber at Unit 2 was almost full of water from the RCIC running for several days, making it virtually useless for pressure relief. Faced with the imminent prospect of a third meltdown, their last resort was opening the pressure vessel's SRVs to the suppression chamber, even though doing so risked damaging the reactor. They spent 90 minutes trying to open five different SRVs, none of which moved, until at last a reconfigured battery setup allowed three valves to open at 6:00 p.m. Pressure dropped from 7,000 kPa (1,015 psi) to 650 kPa (94 psi). Such rapid core depressurisation caused the remaining water to flash-boil, completely uncovering the fuel, but enabled the workers outside to begin coolant injection. Water flowed for all of 15 minutes before the fire engine ran out of fuel. Nobody had checked its fuel tank in the chaos following the explosion; everything stopped again for another 34 minutes while they searched for more.

Much of it was stuck, along with many other incoming emergency supplies, at improvised staging grounds such as J-Village, the sports ground beside TEPCO's Hirono thermal plant. While a steady stream of fuel, tools, spare parts, generators and all manner of other useful equipment flowed in from companies such as Toshiba, it was stopped at J-Village, both because of uncoordinated efforts to keep people away from radiation and because few drivers were willing to approach the plant. The government disaster plan only outlined steps for getting people *out* of an evacuation zone, with nothing to aid bringing supplies and equipment *into* one, so police routinely stopped vehicles travelling towards Daiichi.[dk] Of the 2,048 batteries acquired by TEPCO, for example, only 348 had arrived by 15th March.[dl] J-Village itself did not fully reopen as a sports centre until April 2019, when it became the torch relay starting point for the 2020 Tokyo Olympics, itself postponed by the global COVID-19 pandemic.

Daiichi's only mobile fuel supply was in a tanker truck, but its tyres burst while driving over debris from the Unit 3 explosion, so the team had to carry small tanks of fuel to the fire engines by hand.[dm] When, at last, cooling resumed at 8:00 p.m., the water branched in several directions through the system, just as it had at Units 1 and 3. Some of the water also initially evaporated in the pipes before reaching the reactor, but it soon pushed through as the surrounding machinery cooled.[dn] Decay heat – still increasing by 700°C (1,292°F) per hour, even after three days – had boiled dry what little water remained inside Unit 2 in the intervening time, and the uncovered fuel rods had now reached 1,500°C (2,732°F).[do] They started to fail.

The operators continued trying to open the suppression chamber vent valve, eventually concluding that they had opened a bypass valve at 9:00 p.m., but for now the overall containment pressure was too low to burst the rupture disk. Over the next hour they opened more SRVs to allow cooling via the fire engines, while drywell pressure increased. When engineers restored the control room's atmospheric monitoring system just before 10 p.m., the readings horrified everyone. Drywell radiation had skyrocketed from around one millisievert per hour seven hours earlier to 5,360, indicating extreme core, containment and fuel damage.[dp] If this kind of intensity escaped the containment barriers, Yoshida would have no choice but to evacuate, as 5,000 millisieverts (5 sieverts) will kill most people in a short time. Though pressure inside the wetwell hovered around 300 to 400 kPa (44 to 58 psi), a pressure decoupling saw the value rise inside the drywell in parallel with water from the fire engine turning into superheated steam. This pressure surpassed the design pressure of 540 kPa (78 psi) at 10:55 p.m., while pressure inside the core also continued to climb, forcing panicked operators to try another SRV to enable further water injection. The valve wouldn't budge – its batteries had run out.

Quickly losing what little control they had left of the situation, Yoshida, his management team and those at TEPCO HQ now faced an impossible choice: pray the reactor barriers held firm against pressures far beyond their normal range, or perform a controlled dry vent of lethal radioactivity straight out into the atmosphere without scrubbing it through the wetwell, making the entire plant far more dangerous than it already was. At 11:25 p.m., just as the pressure reached 740 kPa (107 psi), the operators realised their earlier attempt to open the suppression chamber vent must have failed. Reasoning that anything would beat a core explosion, Yoshida gave the order to bypass the rupture disk and go for a dry vent. They sent a signal to open the drywell's air-operated vent bypass valve at midnight: nothing happened.[51]

v. Tuesday, 15th March – Day 5

Secret preparations for evacuating non-essential personnel from Fukushima Daiichi to Daini had gone on all throughout the

---

51 I must confess I don't entirely understand why they performed this vent now and not on other occasions. The purpose of the rupture disk is to ensure that the reactor can be depressurised before it reaches breaking point, but the design pressure of both the pressure and containment vessels had already been surpassed on multiple occasions in multiple reactors without anyone using the bypass valve. The risk of massive radiation exposure may have tipped the balance.

previous evening, having first been suggested by TEPCO's Akio Komori soon after 6:00 p.m. Most of the 800 people on site were now sheltering in the SIB with little to do. With conditions deteriorating all the time and an explosion expected at Unit 2, removing anyone who didn't need to be there became a priority.

Prime Minister Kan's executive assistant woke him from a brief nap on his Kantei office sofa at 3:00 a.m. Kan had barely slept since the 11th, having spent the previous few days helping to coordinate the natural disaster response, but METI minister Banri Kaieda brought troubling news that TEPCO president Shimizu had personally requested permission for a complete withdrawal from the site. When Kaieda refused, Shimizu then called Chief Cabinet Secretary Yukio Edano, saying "Can't something be done? The site simply can't hold out any longer."[dq] Edano, too, denied Shimizu's request, but they, along with Madarame and various other senior figures, felt Kan needed to be informed. "If No. 1 [Daiichi] goes, No. 2 [Daini] would go," Edano said later. "Then we'd see a devil's chain reaction that would also knock down Tokai. I was thinking, if something like that happens, common sense would tell you that would be the end of Tokyo."[dr] Together, they feared Japan's entire east coast region could turn into a radioactive wasteland. Kan was incensed.

TEPCO executives would later claim they had never meant for a total abandonment of Daiichi, but rather an evacuation of non-critical personnel, but key cabinet officials who spoke with Shimizu and others disputed this claim. At the very least, a lack of clarity caused enormous panic at the government offices, so the politicians proceeded on the understanding that Japan's largest utility intended to bail at the worst possible moment. Kan, still facing constant information problems, decided to create a joint command centre of cabinet officials based at TEPCO headquarters to better coordinate between each other. He installed his own special advisor, Goshi Hosono, to take charge, then summoned Shimizu to his office in a rage. But the 66-year-old president – speaking "in a small voice, almost whispering," according to Madarame – denied any intention of leaving Daiichi undefended.[ds] Apparently unconvinced, Kan remonstrated Shimizu for his company's untimely and misleading communications and their general mismanagement of the crisis, then announced his intention to visit TEPCO headquarters within the hour.

Back at the plant, Yoshida became so resigned to death after their last-ditch efforts to lower the pressure in Unit 2 had failed that he rose from his desk and slumped down cross-legged on the floor. "I was picturing the faces of all the people who would die with me," he recalled later, referring to those he could ask to stay behind to keep pumping water into the other two stricken reactors if Unit 2 exploded. He continued:

> At the time, I thought it was completely hopeless. I couldn't sit in my chair. So I moved the chair, got under the desk and – it wasn't Zen meditation exactly, but I crossed my legs and put my back to the desk and sat there. I thought, this is the end, or at least, there's nothing to do now but leave it to God or Buddha.[dt]

He still hadn't moved after ten minutes, watched in silent apprehension by the men and women who depended on his steady leadership, when suddenly he rolled over on his side and lay still. Though the venting had failed overall, later analysis showed that the valve did open for a few minutes and had caused one of the largest spikes in radiation.[du]

When Kan arrived at TEPCO's central Tokyo headquarters, an imposing grey tower topped by a tall red and white striped communications tower that bears an awkward resemblance to Chernobyl's iconic vent stack, he stormed straight up to its second-floor ERC. He was stunned to find a video-conference system on the wall, providing a real-time connection to the ERCs at Daiichi, Daini and Kashiwazaki-Kariwa. "I was really surprised," he said later. "There was this huge screen connected to the No. 1 plant. I wondered why information was coming so slowly to the prime minister's office given the existence of this system."[dv] The centre had been crammed with scurrying technicians, physicists, engineers, medical advisors and all manner of other support staff around the clock since the earthquake. Kan himself had not once returned home during this time – and would not for over a week to come – but knowing that TEPCO were even thinking about (in his mind) fleeing snapped his final shred of patience with them. After first confirming that no reporters stood within earshot, he lost it. The now-legendary tirade began in measured tones, wherein he explained to Shimizu, Katsumata and all assembled (plus everyone watching at the three plants) that he was assuming control, but it rapidly grew in intensity as Kan underwent his own meltdown. Shouting, "What the hell is going on?" and warning them that they had "nowhere to run" and that TEPCO itself would be destroyed if they did, he ordered every company employee over 60 – everyone beyond the age of all but the most serious radiation harm, including its executives – to do their duty and go to Daiichi, in effect to sacrifice themselves to save Japan.[52]

The speech was somewhat misplaced, and all he succeeded in doing was to anger, upset and demoralise the very people putting their lives on the line, most of whom had not seen (or even heard from) their families since the earthquake.[53] Yoshida, watching Kan yell and gesticulate via the video feed, felt disgusted. They weren't running away, he thought; they'd fought like crazy against impossible odds and were still going. Throwing more people at the problem without the heavy equipment they needed would make little difference at this stage anyway. Kan later defended his rant by saying it wasn't meant for the people at Daiichi but contained more general remarks on the entire situation. It's easy to sympathise with both sides. Kan felt, with good reason, that TEPCO – a world-class company with almost limitless resources – were doing a pitiful job of both bringing the situation under control and providing timely updates to the heads of government. The boots on the ground felt, also with good reason, that they were sacrificing everything for their country and that Kan was barging in and criticising them for it. Perhaps what the prime minister should have done, beyond setting up the joint command centre (which was a good move), was to address TEPCO's top executives and convey his dissatisfaction to them alone.[54] Regardless, they did not have long to dwell on it.

At 6:12 a.m., about an hour after Kan began his corrosive pep-talk and minutes after a new shift team had arrived to take over the Units 3/4 control room, a loud popping sound rumbled up from deep in the bowels of Unit 2. Then, within moments, Unit 4's reactor

---

52 While there is video of Kan's speech, taken from behind him, there is no audio, so everything we know about his words comes from those who heard them. Naturally, Kan's own recollection is more measured than everyone else's, who all say he was absolutely livid.

53 Days later, the *Mainichi Shimbun* quoted one unnamed TEPCO employee: "Saying, 'I won't allow you to pull out,' is like saying, 'Get exposed to radiation and keep going until you die.'" Source: "TEPCO wanted to withdraw all nuclear plant workers 3 days after quake." *Mainichi Shimbun*, 18th March 2011: https://web.archive.org/web/20110321030932/http://mdn.mainichi.jp/mdnnews/news/20110318p2a00m0na009000c.html.

54 Kan did, in fact, speak to them alone after his initial company-wide polemic, just as the next events were unfolding.

building exploded from somewhere on the fourth floor, annihilating the third, fourth and fifth floors and propelling concrete and steel in every direction. There were no eyewitnesses this time, but Izawa's crew felt the accompanying kick and rushed to connect batteries to check their instruments, which, alarmingly, showed that Unit 2's suppression chamber pressure had dropped beneath the scale. They were unsure what had happened – as, it appears, are many others, since certain details in the reports from this moment all contradict each other. For example, some people said both the Unit 2 and 4 events were heard *and* felt from the SIB, while others, including Yoshida himself, testified that nobody in the SIB heard or felt either event. Some reports say personnel close by witnessed the Unit 4 explosion, but in reality nobody saw it happen or were near any flying debris.[55] The strange popping sound had emerged from Unit 2 at almost the exact moment Unit 4 exploded, catching everyone off guard. Was Unit 2's containment barrier compromised? Why had Unit 4 exploded when its reactor contained no fuel? What of its open pool and the 1,535 extremely hazardous spent fuel rods?

Yoshida made up his mind: the time had come. With a possible major breach and the potential that Unit 2 might also explode, plus radiation monitors now displaying 400 millisieverts per hour *outside* Unit 3, he thanked his ERC staff for their help and – against the wishes of at least one TEPCO official in Tokyo – ordered everyone not directly involved in maintaining the reactors to take buses parked outside to shelter.[dw, 56]

The evacuation plan, along with locating the necessary buses, had been agreed upon the previous evening in anticipation of this moment. It was simple: they were to drive outside the plant to the nearest spot with low radiation levels and wait there. If radiation was high everywhere, then they should head to Fukushima Daini instead. However, when Yoshida issued his actual evacuation instructions, he said "Evacuate to an area *within* the plant's premises [meaning Daiichi; italics added] where the radiation level is low. If

---

55 In this case I trust the final government report, given that they interviewed hundreds of people.

56 At least one official thought that, since some parts of Unit 2 were still showing pressure, no breach had occurred and that the situation did not warrant an evacuation. Yoshida said later, "There was pushback that the containment vessel would surely not explode because there was still pressure, but I countered that the pressure gauges could not be trusted. So, thinking in terms of safety, or the lack thereof – in other words, things were in an extreme state at the site – I stated my intent to prepare buses and evacuate." Source: "Yoshida interviews / Strong words on Fukushima N-crisis from TEPCO's manager on the ground." *Yomiuri Shimbun*, 10th September 2014. https://web.archive.org/web/20140915210028/http://the-japan-news.com/news/article/0001553303.

Left to Right: Units 4, 3, 2, 1.
Source: Tokyo Electric Power Company Holdings

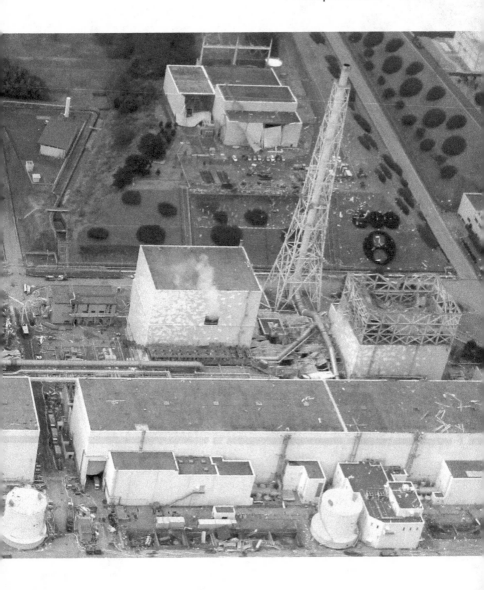

Looking north.
Note the concrete boom
truck leaning over
Unit 4 (bottom).
Source: Tokyo Electric Power
Company Holdings

Top: Unit 3. Bottom: Unit 4.
Source: Tokyo Electric Power
Company Holdings

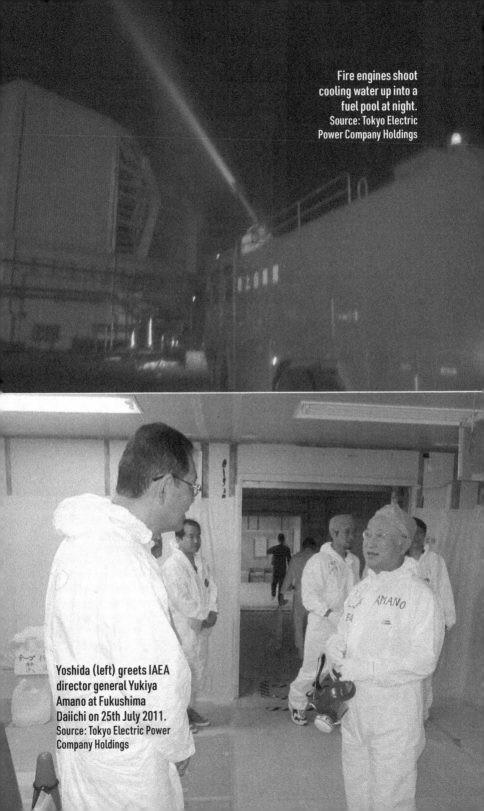

Fire engines shoot cooling water up into a fuel pool at night.
Source: Tokyo Electric Power Company Holdings

Yoshida (left) greets IAEA director general Yukiya Amano at Fukushima Daiichi on 25th July 2011.
Source: Tokyo Electric Power Company Holdings

headquarters confirms there is no problem, I want you to come back."[dx] This was a slight change from the plan, and the difference became lost as it got passed around.[57] Thus, hundreds of exhausted, hungry, unwashed and mentally drained men and women staggered out of the SIB and set off towards Daini after a brief stop by the main gates. Fukushima Daini achieved a cold shutdown of its final reactor just as Daiichi's staff boarded the buses, so their arrival did not hinder the efforts there. Though the altered evacuation plan caused some initial confusion when the buses tried but failed to contact the SIB soon afterwards, Yoshida ultimately decided they had done the right thing. He recalled:

> The No. 2 reactor was in the most perilous state. I mean, radioactivity, radiation levels. The quake-proof control center building stands very close to it. So I thought I was saying, "Get out of here and take temporary shelter to the south or to the north, wherever radiation levels remain stable." But if you come to think about it, everybody was wearing a full-face mask. So if you take shelter like that for hours, you'll be dead. I came to believe that going to [Daini] was by far the right thing to do if only you gave more thought to it.[58]

A fabled final few faced the danger alone. Each wrote their names on a whiteboard in case the worst came to pass. The press dubbed them the "Fukushima 50" after Edano said 50 people remained in a press conference, though in reality they numbered between 50 and 71, with the exact number unknown.[59] "They are the faceless 50," wrote the *New York Times* in a lengthy tribute that afternoon, "the unnamed operators who stayed behind."[dy] Mari Satō, 49, a manager from the Disaster Prevention and Safety division (responsible for firefighter training, among other things), said afterwards that she pushed up the stairs against a tide of people leaving, only to find Yoshida, Izawa and a few dozen senior staff in the ERC sitting in

---

57 Many people headed straight for Daini in cars but some of the buses did stop at Daiichi's main gate and took radiation readings. Even inside the bus, radiation was deemed too high, so they left.

58 Here is a translated quote from Yoshida's government testimony: http://www.asahi.com/special/ yoshida_report/en/1-1.html. Some of his colleagues later speculated that he'd changed the order deliberately to prevent accusations that they had abandoned their posts, but I suspect this was a simple misunderstanding.

59 There is no specific number because exactly who stayed is unknown. The number 69 is often reported in the media, but I'm unsure where this came from and it appears to be incorrect, as do all other exact figures. Various official reports state that the number is not known.

serene silence.[60] Some were lost in thought about the work ahead, while others composed emails to loved ones; all expected to die without ever seeing them again. After a time, Yoshida himself was apparently the first to break the extended quiet with a light-hearted, "How about something to eat?"[dz]

TEPCO took an hour to inform the Prime Minister's Office of the incident at Unit 2, despite Kan's rant and its appearance all over the news.[ea] The exact trigger for the mysterious sound has never been determined. Initial assumptions were that Unit 2's suppression chamber had ruptured somewhere, given the pressure drop to zero (i.e. atmospheric levels), but a robotic inspection in 2012 found no indications of damage.[eb] The government's cabinet report suggests the sound in fact came from Unit 4 moments prior to its explosion and that people mistakenly attributed it to Unit 2, which would make more sense.[ec] As for the suppression chamber pressure reading, it transpired that the readings were below the lower measurable limit (i.e. unknown; technically "downscale"), but that was communicated to shift leaders and the ERC as a zero reading. Pressure inside the outer drywell containment remained steady during this time, so the sensor may have simply failed.[ed] That said, the suppression chamber was damaged somewhere, because recovery workers found they couldn't fill it with water several months later.

Engineers first postulated that Unit 4's spent fuel pool had perhaps drained and boiled dry, creating a chemical reaction that had given off hydrogen from the overheating fuel, but the pool temperature and water level had been checked 14 hours earlier, ruling this theory out. It *was* a hydrogen detonation, but in this case hydrogen had backfilled into Unit 4 via the chimney stack it shared with Unit 3 during venting of the latter reactor.[61] Without power to the exhaust fans to guide the vented gases upwards, a cloud simply drifted over into Unit 4 until some stray spark caused it to explode. This theory was verified via a radiation survey of the gas filters installed at the reactor building boundaries, which showed that the Unit 4 filter had more radiation on the external side of its filter than on the internal side, which wouldn't normally be possible.[ee]

None of this was learned until much later, and the operators

---

60 Here is a photo of Satō, accompanied by a short quote: http://www.tepco.co.jp/en/challenge/csr/nuclear/stations-e.html.

61 This detonation never would have happened had the reactors not shared chimneys, but this little detail is usually ignored. The reason for the shared chimney stack, of course, was cost, plus people had never considered the possibility that something like this could happen.

didn't have time to ponder because someone spotted steam gushing from Unit 2's open blowout panel at 8:30 a.m., not long after everyone left. Half an hour later, a radiation monitor at the plant's main front gate registered airborne radioactivity at 11.93 millisieverts per hour (mSv/h), which, while low in comparison to standing inside one of the damaged reactor buildings, was by far the highest reading recorded thus far. For comparison, standard background radiation is around 0.00025 mSv/h or 2.19 mSv per year, so this was almost five million percent higher than normal. This moment, out of everything that had occurred so far, should have been the time to disseminate contamination maps produced by the government's System for Prediction of Environmental Emergency Dose Information (SPEEDI) network. After all, this exact type of situation was what the entire system had been designed for and practised with during annual emergency drills. In what would turn into one of the most controversial decisions of the entire disaster, however, the NSC and MEXT – the successor to the Science and Technology Agency – chose not to, just as with the Tōkai-mura accident.

A system malfunction prevented SPEEDI from acquiring emissions data about Daiichi from its linked network of sensors and so could not make accurate contamination forecasts. It did produce hourly estimates from partial or best-guess hypothetical data from 11th March onwards, but officials disregarded them even though they were the best (and only) projections they had.[62] Instead, the government assured people of safety outside the evacuation radiuses without any scientific basis, famously resulting in large numbers of evacuees – as many as 50% of the town of Namie's entire population, in one instance – moving *into* the path of serious contamination.[ef] The NSC did make some information public on 23rd March after reconstructing the diffusion path taken by airborne radiation, but by then it was too late. The graphs, which were reverse engineered afterwards once data from monitoring stations had been gathered, revealed for the first time the high radiation in areas people had fled to, but officials failed to clarify that the results were backdated and had not been known at the time. A 2012 government investigation could not determine if MEXT officials had even *con-*

---

62 This description is an oversimplification. SPEEDI is linked with another program called the Emergency Response Support System (ERSS) that pulls information from Japan's network of radiation monitors. SPEEDI pulls its information for contamination forecasts from the ERSS, but if the ERSS isn't working properly, then SPEEDI can't accurately forecast.

sidered whether to release timely SPEEDI predictions.[cg] Those who had remained in such areas for days were understandably upset. It's fortunate that nobody developed acute radiation syndrome, as the accident demonstrated that the medical establishment "did not anticipate the possibility that radioactive material would be released over a wide area and that many residents would be exposed," as the NAIIC wrote in their report. "Specifically, the accident clearly showed that most of the existing emergency medical facilities were incapable of fulfilling their intended purposes if many residents are exposed to radiation. The medical facilities were too close to the nuclear power plant, they had limited capacity, and the medical staff did not have sufficient medical training to treat radiation exposure."[eh]

The government expanded its warning radius once more at 11:00 a.m. because of the plant's alarmingly high radiation readings, this time issuing a shelter-in-place order to all residents living 20 to 30 kilometres from the plant. They were instructed to hide indoors to avoid airborne contamination, though many – forced to decide for themselves how much danger they really faced – chose to play it safe and evacuated instead.[63] Nobody was informed of the expected duration of this order, though internally bureaucrats expected it to last two days at most. Government and TEPCO personnel also abandoned the OFC because of rising radiation around this time, soon setting up a new centre at Fukushima City.

The brave souls back at Fukushima Daiichi worried that, irrespective of what caused the latest explosion, Unit 4's newly ventilated spent fuel pool, filled with an entire reactor inventory, might overheat. The deep pool water should have been enough to cope with decay heat, despite it still generating a whopping 2.26 MW of thermal energy (the greatest of Units 1 through 4), but if the pool lining had been damaged by any of the still-frequent aftershocks, then water might now be draining out. If that happened, the fuel would ignite and contaminate a huge area, its lethal particles carried by the wind. Daiichi as a whole, when including its dedicated spent fuel building, held several years' worth of used fuel because the Rokkasho plant was still not operating at capacity and there was nowhere else for it to go.[64] Without a working water level indicator,

---

63 The proportion of people in nearby towns and villages who chose to evacuate ranged from about 14% all the way up to 59% of residents from Minamisōma City, according to the National Diet of Japan Fukushima Nuclear Accident Independent Investigation Commission report.

64 Japan held 13,530 tons of spent nuclear fuel as of September 2010, according to the FEPC.

they were forced to send a team to investigate, but when workers tried to enter the reactor building at around 10:30 a.m., their dosimeters flew off the 1,000 millisievert-per-hour scale and they were forced to turn back.[65] Their fears materialised a couple of hours after the evacuation, when two senior staff spotted fire and smoke coming from the ruined building. "A fire has broken out at reactor 4," Yoshida called out via the video monitor. "[We] can't do anything. Please ask the Self-Defense Forces or US military for help." The fire engines could not stop cooling the reactors, but the blaze seemed to extinguish itself after two nail-biting hours (the fire was later attributed to burning lubricating oil) and the incident marked another shift in focus, this time towards keeping water in the inaccessible spent fuel pools.

While the remaining personnel brain-stormed ways to do that, they realised the radiation situation had not gone off the anticipated cliff. As such, Daiichi's next-most-crucial staff began a gradual return to the plant within hours (others provided technical assistance over the phone from Daini) and almost all would head back within days. Some went without hesitation, while others were torn, frightened by what awaited them. Several hundred people were on site by the 18th, but they remained the "Fukushima 50" to the public.

---

65 These levels contradict those found in the AESJ report, which says radiation around the building was fine, but the figures do appear in the IAEA and various other credible reports. The discrepancy is strange. I'm unsure if there was a misunderstanding somewhere.

CHAPTER 8

Aftermath

i. At the Plant

The situation at Daiichi would remain tense for weeks to come but, though they did not know it, the worst was now over. On the afternoon of the evacuation, a JSDF helicopter circled Units 3 and 4 recording a shaky video which appeared to confirm that water still covered the fuel.[1] It was not at all clear in the case of Unit 3, however, obscured as it was by debris and billowing steam, and the US government thought the video showed the opposite – that Unit 4's fuel was exposed to the atmosphere. The US Nuclear Regulatory Commission publicly urged all US citizens to stay 50 miles (80 km) away from the plant, prompting further mass panic among Japanese citizens who assumed the Americans knew something their own government wasn't revealing. Testimony by NRC chairman Gregory Jaczko later that day during a congressional hearing did not reassure anyone. "We believe that secondary containment has been destroyed and there is no water in the spent fuel pool," he said. "And we believe that radiation levels are extremely high, which could possibly impact the ability to take corrective measures."[a] Recovery teams at Daiichi made desperate attempts to spray water up from the ground using the fire engines, but little fell into the pools themselves.

Emperor Akihito, son of Hirohito, took this moment to deliver a rare video message to his people. "I am deeply hurt by the grievous situation in the affected areas," admitted the 77-year-old. He thanked emergency workers for their tireless efforts and drew particular attention to the ongoing crisis at Fukushima before saying, "I hope from the bottom of my heart that the people will,

1 Observers interpreted flecks of light as evidence of reflections glinting on a surface of water at Unit 4.

hand in hand, treat each other with compassion and overcome these difficult times."

JSDF Chinook helicopters returned the following morning, 17th March, flying four seawater drops over Unit 3 after radiation was deemed too intense the day before, but altitude and high winds meant most missed the mark. Special fire engines for combatting tower-block fires, Tokyo Police water cannon trucks and a colossal German 80-ton, 10-axle Putzmeister M58-5 concrete-pouring truck with a 58-metre boom arm (which happened to be docked at Japan en route to Vietnam) arrived over the subsequent days and weeks and were utilised for more direct spraying into the Unit 3 pool, to much greater effect.[2] Emergency workers arriving at the site for the first time during this period were shocked at what they found. "It was far worse than I expected," one firefighter recalled of his 2:00 a.m. arrival. "Everything was covered in rubble. There were concrete blocks everywhere, all the manhole covers had popped out, for some reason, and the road was impassable. We couldn't drive down to pay out the hose from the sea. So we had to run, carrying the hose, half a mile to the sea, in total darkness."[b]

Investigations later revealed that water still covered the Unit 4 spent fuel pool by sheer luck alone. When the reactor was shut down for maintenance, operators flooded the empty space above the containment lid (known as the reactor well) to enable submerged fuel extraction and transfer over to the pool via a double-gated channel. As water evaporated in the pool during the accident, it was topped up by water from the reactor well leaking around the tall steel gates.[c] Computer simulations by the US National Academies of Sciences, Engineering and Medicine concluded that if this hadn't happened, the pool's fuel would have been exposed by early April, at which time it remained inaccessible.

Yoshida managed the plant's emergency response for another week and was involved in everything from civil engineering work to acquiring more personnel and materials to radiation management. In fact, an analysis of the continual video feed connecting Daiichi to TEPCO HQ by the *NHK* network found that only 12.8% of his conversations were about the reactors. It's safe to say that, because he was so involved in the minute-by-minute accident response, he rarely slept. Tracking his brief periods of absence from the footage,

---

2   The US also flew in a Puztmeister M70-5, the largest concrete pump truck in the world with a 70-metre boom, carried inside a Russian Antonov An-124 on 8 April.

with the assumption he at least tried to sleep during these times –
because who knows how much he actually got, given the extreme
stress he was under – showed that he didn't close his eyes for more
than an hour or two each day until the 16th, and even after that he
almost never slept longer than three or four hours at a time.[d] By the
19th, eight days after it all began, he'd reached his limit and began
to complain of feeling unwell.

He wasn't the only one. Masataka Shimizu made headlines when
he disappeared for several weeks during the unfolding disaster, with
his last public appearance being a 13th March news conference.[e]
He remained present behind the scenes until his 15th March dress-
ing-down from Kan, but afterwards secluded himself in his personal
office for at least six days, not even attending crisis meetings with
other company executives and effectively leaving TEPCO without a
leader. The media noted his public absence days later, but TEPCO
would only say he felt overworked and had taken time off, to the
understandable dissatisfaction of everyone. Shimizu was admitted
to hospital on the 30th for hypertension and dizziness but did not
reappear in public until 13th April, after a month of silence.[f] In a
disastrous *owabi kaiken* apology press conference like the ones held
by his predecessors after the falsification scandals a decade earlier,
he expressed remorse for everything that had happened, but his
apology rang hollow and some attending journalists even shouted
criticisms – unheard of in Japan.

Engineers working around the clock at Daiichi eventually restored
external power to Units 1 and 2 on 20th March and to Units 3 and
4 on the 26th. Beginning on 21st March, many recovery workers
spent their down time between shifts aboard a brilliant white
four-masted tall ship used for training named *Kaiō-Maru* (*Sea King*):
110 metres (361 feet) in length, weighing 2,556 gross tons and
looking like something built 150 years earlier than its 1989 completion date would suggest.[3] The tall ship moored for eight days at the
damaged port of Onahama, 50 kilometres from the plant. Even
there, away from it all and with proper hot showers and food, most
sat in near-silent reflection. At least some anticipated their demise
within weeks, months or years because of their work, their tearful
mothers giving media interviews and saying things like, "My son and
his colleagues have discussed it at length, and they have committed

---

3   The ship was built, incidentally, by Sumitomo Heavy Industries: the giant *zaibatsu* that also founded
    JCO, the company responsible for the 1999 Tokai-mura accident.

themselves to die if necessary to save the nation."[g] In the outside world, the public hailed the "Fukushima 50" as heroes for putting their lives on the line.

Three subcontractors in particular drew attention in late March after being exposed to water "10,000 times higher than the amount normally seen in cooling water in reactors," according to the *Yomiuri*.[h] The trio had ignored their screeching dosimeter alarms' efforts to warn them about the dangerously radioactive ankle-deep water through which they were laying cables in Unit 3's turbine hall basement after transitioning from one area to another, believing the devices to be broken.[i] Two sustained injuries to their feet and each received doses ranging from 173 to 180 millisieverts. They ended up at NIRS, where Ouchi was treated in 1999, but were later discharged. This incident was the first public evidence that the reactor core had been compromised. Stories emerged from Daiichi, ranging in terms of altruism from workers' claims of only being there because they would lose their jobs otherwise, all the way up to a group of elderly company executives volunteering to go so that younger workers could be spared.[j] Ageing workers volunteering their services became a recurring theme, including one group of 270 specialists organised by former Sumitomo Metal Industries engineer Yasuteru Yamada. "I will be dead before cancer gets me," reasoned the matter-of-fact 72-year-old.[k]

During this period, when cooling the spent fuel pools seemed at times like a losing battle, Prime Minister Kan instructed JAEC chairman and professor emeritus of Tokyo University Dr Shunsuke Kondo to quietly consult with a team of specialists and produce a report on possible worst-case scenarios if the situation nosedived.[l] Dr Kondo's report, submitted on 25th March in the form of a 15-page PowerPoint document, terrified those few who saw it. If another explosion destroyed Unit 1's containment, Kondo thought, then Daiichi would have to be abandoned. Units 2 and 3 would melt down, ultimately forcing tens of millions of people up to 170 kilometres away to evacuate and for "voluntary migration" to occur as far as 250 kilometres away. "The content was so shocking that we decided to treat [the report] as if it didn't exist," recalled one senior government official.[m] The findings also haunted Kan, who later admitted that "it was a crucial moment when I wasn't sure whether Japan could function as a state." Kondo's report was buried, lest it cause a mass panic, with its existence only surfacing in January 2012.

Kondo himself later distanced himself from the report, believing that the science did not add up. There was no obvious source of another explosion besides more hydrogen, but the hydrogen-forming chemical reaction of melting zirconium had already exhausted itself. The report also ignored topography and assumed a strong, unwavering wind blowing inland towards Tokyo, as it was a rushed presentation intended more to show a theoretical disaster progression for politicians than to be a realistic scenario.

A forgotten story emerged about Unit 4 during this time. Babcock-Hitachi KK, Hitachi's boiler division, had fabricated the reactor's $250 million pressure vessel at its Kure foundry in Hiroshima Prefecture during 1974 and/or 1975 (sources vary), but braces used to maintain the vessel's shape inside the furnace during heat treatment to remove welding stress either slipped or may never have been installed at all. The massive vessel warped inwards. By law, as a component vital to reactor integrity, Hitachi should have started over, but discarding such a valuable item risked bankrupting them. Instead, Mitsuhiko Tanaka, an engineer who had helped design and supervise the vessel's construction, devised a computer model for salvaging the expensive component.[n] The solution had involved using hydraulic jacks to push the sagging area back into position, and neither the government nor TEPCO were ever informed. In 1988, frustrated by claims that something like Chernobyl could never happen in Japan because of its comprehensive safety regulations, Tanaka informed METI of what he'd done, but Hitachi argued their actions had not compromised the vessel's strength, and the matter was dropped.[4] When Tanaka drew fresh attention to the incident in late March 2011, Hitachi repeated their position. To be fair to Hitachi, they may well have been correct, as Daiichi's Unit 4 pressure vessel held firm when the reactor building blew apart. Had it been fuelled and online at the time, however, things may have gone very differently. Tanaka went on to join the NAIIC team investigating the disaster.

NISA raised the disaster's International Nuclear and Radiological Event Scale rating from 5 to 7 on 12th April, as much because of the widespread ramifications as the radiation, making Fukushima only the second accident in history to reach this maximum level. Some attacked the move for arriving too late and considered it evidence

---

4   Exactly what happened at this point is cloudy, as Tanaka told this part in several different ways, making his version unreliable. I've based this description largely on Hitachi's own comments.

of downplaying the severity. Others called it alarmist, given that only about 7% of Chernobyl's total radiation had escaped by this stage. The government quietly raised host-community subsidies on the same day, without a public announcement.

A gaping 27-gigawatt (GW) electricity shortfall emerged after the natural disaster knocked out eleven reactors and several thermal plants: 15.9 GW lost from six TEPCO plants, 5.56 GW from four belonging to Tōhoku Electric, 1.1 GW from Tōkai and 4.5 GW from smaller coal power companies. Rolling blackouts swept across east Japan as utilities sought to conserve and share as much energy as possible.[o] This caused inevitable tragedies, such as one man who died in a car crash because of inactive traffic lights. Intermittent commuter trains, meanwhile, meant people struggled to get to and from work, then faced problems caused by the unpredictable power supply at offices and factories. Matters were complicated by the peculiar old setup where the country's eastern side (including Tokyo and Fukushima) utilised a 50 hertz electrical frequency and the west used 60 hertz. Even running flat out, the frequency converter stations had a combined capacity of just 1.2 GW, though some hydro plants near the boundary helped by switching over to 50 hertz instead of their normal 60 hertz.[p]

Initial blackouts ended on 8th April but then returned during the summer. To compensate, utilities sped up offline maintenance of coal plants and reactivated mothballed sites, plus agreed with steel companies to utilise 900 MW of private generating capacity from steel mills. Many of the old mothballed power stations suffered frequent breakdowns, but while these measures did decrease the power gap, they could not meet the summer surge in demand. As such, beginning in July, the government, utilities and other large companies began a publicity campaign called *setsuden* ("saving electricity") and appealed for a blanket 15% reduction in electricity use across Tokyo and the wider Tōhoku region to prevent further blackouts. This was the first consumption restriction since the oil crisis.[q] People and businesses switched off their air-conditioners, room lights and other electronic items, and this strong public support and cooler-than-normal weather allowed the campaign to end early, at the beginning of September.

ii. Evacuees

The government revamped its procedures for evacuating the elderly and disabled after the Kobe earthquake, but though they were implemented in 2011, many plans only anticipated individuals living at home. Evacuating such a vast area – over 600 km^2 of earthquake-battered land once the radius reached 20 kilometres – was never foreseen and, unsurprisingly, most of those who perished were indeed the sick and elderly.

Hospitals and local emergency services, too, had new safety practices, including establishing their own evacuation procedures, shelters and transportation, but these only anticipated nuclear incidents of a scale similar to the JCO criticality accident. They had also trained for environmental disasters requiring the transportation of people under their care, and in most cases this now went smoothly, under the circumstances. Aerial medical units, in particular, airlifted well over 100 bedridden patients without any loss of life during transport.[r] But destroyed road and communications infrastructure, plus the far larger than anticipated scale, meant ground transport, via ambulance or bus, did not always go according to plan for the seven hospitals and seventeen nursing homes within the 20-kilometre zone. In the worst instance, almost 45 patients from Futaba Hospital died when no adequate food, water or health-care facilities could be located after the evacuation radius expanded to 10 kilometres.[s]

The hospital, which lies just under four kilometres south-west of Daiichi, was by far the largest nearby medical facility and among the slowest to react to the unfolding crisis. Its 76-year-old director, Ichiro Suzuki, ordered 209 of the 339 patients and – astonishingly – all nurses, doctors and support staff to evacuate in buses on the 12th, while he remained behind with the most critical patients who could not be moved.[5] "I sent off all the hospital staff," he said, "because I believed that SDF troops and police would arrive at any minute to rescue us."[t] Suzuki cared for the remaining 130 bed-ridden patients as best he could while they waited, occasionally leaving to attend to almost 100 residents of a nearby nursing home (which had at least

---

5   I have seen "all" and "almost all." "All" is more common by far and is the term used in the NAIIC report, which has the most weight of the various sources that discuss the matter. Whether anybody stayed behind to help Suzuki is unclear. Some reports say a skeleton crew of nurses stayed; others say he stayed alone.

retained its staff carers, some of whom came to assist him in the hospital). They numbered among approximately 840 patients who remained inside the 20-kilometre perimeter by the evening of the 13th.[u] Confusion arose among the authorities as to how many (if, indeed, any) patients had stayed behind at Futaba Hospital after the main group left. The anticipated rescue didn't arrive for two days, in part because miscommunications meant the JSDF unit sent to the region could not find the OFC in Ōkuma where they were to receive instructions.

During the final evacuation, the soldiers tragically missed four people whose bodies were discovered by police three weeks later, on 6th April. The rest headed via bus to the nearest designated health centre for radiation screening in Minamisōma, 35 kilometres to the north. Some members of this group became stranded at the health centre for over 24 hours without heating or medical supplies because all available facilities were full.[v] The rest were redirected to an evacuation shelter 70 kilometres south, bringing their travel total to around 230 kilometres (143 miles). The journey, along a roundabout route to avoid approaching Fukushima Daiichi, ultimately took almost six hours, during which time they were unaccompanied and without basic identification or medical records for their care. By the time they arrived at the designated shelter – the Iwaki Koyo High School gymnasium, which was unprepared to suddenly receive dozens of sick and vulnerable people – three more had died.

Matters did not improve after their arrival, and eleven more passed away within hours, forcing the school principal to appeal for help on a local radio station.[w] "The condition at the gymnasium was horrible," conceded one Fukushima government official. "No running water, no medicine and very, very little food. We simply did not have means to provide good care."[x] Futaba was one of four hospitals inside the zone to experience fatalities, partly because hospital staff were left to arrange replacement health-care without help from the Fukushima prefectural government, nor even from local municipal governments.[6]

---

6 These local governments "prioritized the transfer of governmental offices over the evacuation of the hospitals," according to the NAIIC (Chapter 4, p. 35). Only the JSDF came to their aid, but this took a few days in most instances. Amazingly, the NAIIC discovered that none of the staff in six of the seven hospitals realised that the pre-existing plan was for hospitals to arrange for their own evacuation. The one that did evacuate – Imamura Hospital – had not prepared for a disaster of the scale that had occurred, and their plan proved useless.

The systems breakdown was not confined to facilities within that evacuation radius, however. Forty kilometres south of Daiichi, at the major Iwaki Kyoritsu Hospital, medical staff under the age of 40 (around half of all personnel) were told to evacuate the city on 14th March after Unit 3 exploded, leaving the older doctors and nurses struggling to cope by themselves. They were unable to perform even emergency surgeries, patient wait times grew longer, there was little clean water, and, with Iwaki's pharmacies and stores closed since the earthquake, medicine, fuel and other supplies ran low.[y] Large numbers fled Iwaki as the situation at Daiichi deteriorated, prompting widespread rumours that the city was contaminated. This, in turn, scared away truck drivers bringing basic necessities, and that entire region of Fukushima in effect became isolated from the outside world. Things grew so desperate that one hospital director sent an email on 15th March begging for assistance from professionals outside the area because emergency services in the city were stretched to the breaking point. Iwaki was just outside the then-current 30-kilometre shelter zone, so designated disaster assistance teams, such as those airlifting patients out of Futaba, could not be sent to help.

In the end, a private group of medical professionals from Tokyo's Nippon Medical School (NMS) came to the rescue. The director of the NMS Advanced Emergency Center, himself a physician, drove to Iwaki the next day and helped to triage around 200 patients while his colleagues back home first tried to send resources via official channels, then, undeterred by their lack of success, arranged for beds in the wider Tokyo metropolitan area for the 15 most critical patients. Doctors drove the ambulances, navigating twisted and collapsed roads to eight hospitals over five days. Iwaki and Futaba Hospitals were two of Fukushima's six appointed radiation emergency hospitals, all of which failed for reasons similar to those described above.[z] It didn't help that three of the four hospitals within 20 kilometres of Daiichi had no evacuation plan.[aa]

In total, around 2,200 inpatients and elderly residents were evacuated from that 20-kilometre zone, but the road network and congestion were so bad that some remained on the move for over 48 hours. The overall operation improved starting on the 15th, once a system had been established for contamination screening, triaging and transporting huge numbers of patients to various hospitals, depending on their health. A 2012 report by Japan's Reconstruction

Agency verified 34 cases of deaths related to Fukushima Daiichi, while the NAIIC counted 60 by the end of March, over 40 of them from Futaba Hospital. The most tragic thing about it all is that, despite hundreds of these vulnerable people remaining within that 20-kilometre radius for several days after the evacuation order, doctors found no significant radioactive contamination on any of them. They died for nothing. Had the government been better prepared, or even taken the time to prearrange evacuee-receiving facilities before moving the most vulnerable, they probably would have survived.

Of course, things became as bad as they did because of the wider calamity. Most heating systems across the north-east were out of action, and stocks of kerosene to burn in heaters soon ran low at the roughly 2,400 shelters, where over 460,000 distraught evacuees slept with thin blankets upon freezing floors. "There are a lot of people with chronic conditions and today [18th March] it's cold so some people have fallen ill," said one Red Cross doctor in Iwate Prefecture. "We've had a bad stomach virus going around so a lot of people are getting diarrhoea and becoming dehydrated."[ab] Many of the elderly could not remember what type or dose of medications they were meant to take, and doctors couldn't acquire enough of anything.

The question of whether vulnerable people living near the plant should have been evacuated would come up time and again in the months and years ahead. Should the doctors and nurses have hunkered down – sealing windows and doors – and waited it out? Perhaps, but we benefit from hindsight. Most of those who died were from the second wave of evacuations, the 10-kilometre radius, when things were growing more desperate by the hour. Though radiation levels outside the plant never approached being immediately life-threatening, it seemed possible at several moments that the entire region might soon become uninhabitable, so rescuing people who could not rescue themselves was given a high priority.

* * *

Rather than the expected two days, those living within the 20- to 30-kilometre shelter-in-place radius were trapped indoors for ten, until Edano announced a government advisory on 25th March, saying: "Given how prolonged the situation has become, we think it would be desirable for people to voluntarily evacuate."[ac] Many then left, but those who didn't – unable or unwilling to leave out of fear, attachment or for medical reasons – were forced to hide in their homes for over a month, cut off from the outside world, until the shelter-in-place order was lifted on 22nd April. The government established "deliberate evacuation areas" on the same day for people residing in radiation hotspots.[7] "It was reported that evacuees have moved to hotels or inns, continue to receive relief supplies, return home once a week, and have brought supplies from their homes," complained one family. "Meanwhile, having sheltered in our home, we are unable to purchase any essential goods because the stores are closed. We have also been unable to drive because there's no fuel." Those in this position did not receive nearly the same support as people forced to evacuate their homes; some even claimed to have had none at all. The "voluntary evacuation" area was a major source of post-disaster controversy, with the damning NAIIC report going so far as to say: "We must conclude that the government... abandoned their duty to protect the lives and safety of the citizens."[ad]

A total of 99,205 citizens had left the region around Fukushima Daiichi by year's end, while a further 62,831 chose to bid the prefecture farewell and live somewhere else entirely. Most struggled to move on from the experience; the physical stress subsided, but damage to mental health did not heal easily. To assess their difficulties and determine the appropriate care, the Fukushima government conducted four detailed surveys spanning two years and tens of thousands of evacuees of all ages. Shinichi Yamashita, the head of the survey committee and professor of molecular medicine and radiation research at Nagasaki University, did not help matters during an initial misguided attempt to reassure people that they were safe. "The name 'Fukushima' will be widely known throughout the world.... This is great!" he said with a beaming smile to a hall full of worried citizens in Fukushima City on 21st March 2011. "Fukushima has beaten Hiroshima and Nagasaki. From now on, Fukushima will become the world number one name [associated with nuclear accidents and radiation]. A crisis is an opportunity. This is the biggest opportunity. Hey, Fukushima, you've

---

7   The enormous delay in setting up these new evacuation areas was another source of criticism. The government had possessed accurate radiation contamination maps for a month by this point, but bureaucratic red tape and indecision meant nothing was decided for weeks.

become famous without any effort!"[ae] He offered further helpful reassurance: "The effects of radiation do not come to people who are happy and laughing; they come to people who are weak-spirited." Later, Yamashita's comments on the incident were even more surprising: "I was really shocked," he said. "The people were so serious, nobody laughed at all.... [They were] really depressed."[af] As one of the world's leading experts on radiation medicine, his knowledge of the psychological effects of exposure was scientifically accurate, as shown by studies on rats, but he didn't know how to talk to people at all. "He did a great job of running the actual study and a bad job of managing expectations and communicating to the public," commented prominent German physicist Wolfgang Weiss.[ag] Yamashita stepped down from his government position after two difficult years.

The final survey results painted a sorry picture. Though limited in scope, it found that 18.2% of men and 24.3% of women suffered from probable post-traumatic stress disorder (PTSD), with 25% of all respondents indicating they felt a social disability because of symptoms arising from their experiences.[ah] Almost 40% now lived apart from those with whom they had shared a home before, while almost 80% had moved three or more times since the disaster. A depressing 30.9% had moved more than five times. The town of Namie, eight kilometres north with a population of around 21,000, was hit hardest, with over 30% of its population falling into the "six or more" moves bracket.[ai] The disaster affected children too: after twelve months, 24.4% of four-to-six-year-olds and 22% of six-to-twelve-year-olds had double the normal score indicating serious mental health struggles on the Strengths and Difficulties Questionnaire, a standard emotional and behavioural screening test for children and young people. In some positive news, this percentage fell to 16.6% and 15.8% respectively at 1.5 times the normal score after 24 months, showing signs of recovery.

Overall, people's ordeals caused a blanket increase in obesity, hypertension, sleeplessness, chronic anxiety, PTSD, and smoking and alcohol habits compared to those who did not evacuate, but even that group were not much better off.[aj] Some families were bullied for drawing attention to radiation dangers, while one child was refused entry into his local high school despite passing the entrance exam by a wide margin.[ak] Some of the people who received compensation from TEPCO and the government received similar verbal abuse as a result and now do not tell anyone where they were from. As of 2020, over 1,000 indirect fatalities have so far been attributed to the severe distress caused by radiation phobia, forced evacuation and the ongoing misery endured by those tens of thousands who still have no permanent home. "The consequences of the radiation accident is not purely about exposure to radiation," said Masaharu Tsubokura, a radiation specialist at Soma Central Hospital in Fukushima, to *Wired* in 2020. "It's also not purely psychological. It's changes in lifestyle, family issues, changes in society, hospitals closing, stigma, bullying, money. Hardly anyone here talks about radiation. Those people don't come back."[al] With the definition expanded, all have become *hibakusha*, like the victims of Hiroshima and Nagasaki.

iii. Clean-up Crews

The men and women working at Daiichi during and after the disaster, whether brave, desperate or duty-bound, faced their own agonising anxieties, often working long hours in hot, suffocating safety gear. They were also, inevitably, exposed to the highest radiation doses, though shortages meant not all recovery workers knew their own doses in late March 2011. "Since the number of monitors is limited, only one or two devices are handed to each group," revealed one labourer. "But sometimes you have to move away from that person and in that case you'll never know the level of your exposure."[am] Between March and April, six TEPCO employees exceeded the new, higher emergency safe dose of 250 millisieverts introduced on 14th March, while 146 TEPCO employees and 21 subcontractors exceeded the original 100-millisievert limit.[an] In the worst case, two operators from the Units 3/4 control room received 678 and 643 millisieverts, and one from Units 1/2 received 475.[8, ao] The average dose among TEPCO staff during this period was around 25 millisieverts, equivalent to two and a half full-body CT scans, but the legal limits meant the company quickly burned through most of its own trained and experienced workers, while the flow of support staff from major subcontractors proved insufficient to make up for the shortfall. In extreme cases,

---

8 The two men from the Units 3/4 control room received such a high dose (mostly internal) because there were not enough proper charcoal filter masks for everyone in the control room, so they wore simple dust masks.

workers claimed they took turns approaching a valve, turning it for just a few seconds each before bequeathing the task to the next brave soul.[ap]

TEPCO reported to the World Health Organization in December 2013 that 178 workers had by now exceeded doses of 100 millisieverts to their thyroid glands – tissue particularly susceptible to cancer – but were forced to re-check their numbers after the Ministry of Health, Labour and Welfare (MHLW) expressed doubts about the figure. The utility then conceded that the actual number was over ten times higher: 1,973, all of whom became eligible for certain free health services. Almost half of them described experiencing psychological distress in a 2011 study by Japan's military medical academy, the National Defence Medical College.[aq] With supply high but demand higher (88% of the 83,000 workers at Japan's nuclear plants were contractors in 2010), the same old players inevitably re-emerged to exploit the lucrative labour gap. Within months, workers from as far as Hokkaido in the remote north, down to the distant cluster of Okinawa islands 1,750 kilometres to the south, began to appear at Fukushima Daiichi.

The floodgates opened in August 2011 when Parliament approved a decontamination-funding bill that relaxed restrictions on contract bidding. Winning subcontractors were no longer required to receive government certification, nor even to undergo screening, meaning anyone – however unsuitable – could be sent there to work. The change made it more difficult to track how many workers came from legitimate, registered recruitment agencies with their own retained personnel (which are far more common now than during the 1970s and '80s) and which came from shadier sources. An area around the city of Sendai's train station, 90 kilometres north of the plant, emerged as a principal *yoseba* auction hub for labourers. Many there were employed by unregistered brokers who skimmed wages after lying about the pay, working conditions, job duration and any perils they would face. One man was told he would be monitoring the radiation exposure of people leaving Daiichi, only to find himself strapped up in full hazard gear and marched off to an area which would punch through his annual safe limit within an hour. When he protested, he was fired.[ar] He was, in a way, lucky, if the numerous reports of beatings and threats of murder for complaining about conditions are to be believed. In another case, the *Asahi* reported – and the responsible subcontractor admitted – that labourers "were ordered to cover their dosimeters with lead plates to keep radiation doses low enough to continue working under dangerous conditions."[as]

Incidents like these explain why the practice of independent brokers hiring labourers who are then managed by legitimate sub-contracting companies had been made illegal since the practice rose to prominence during the 1970s, but where there's a will – and money to be made – there's a way. Workers helping at Daiichi could receive much less than half the money allocated to them, taking them below the minimum wage, with some people driven into debt in the worst cases after the costs of their laundry, measly food and claus-trophobic accommodation were deducted, according to police data.[at] A 2012 TEPCO survey of 3,186 workers at the plant revealed that 47.9% of them were managed by a subcontracted company other than that which employed and paid them. A further 2.1% responded that they did not know who they worked for. Sometimes a legitimate subcontractor simply told the illegitimate labourers to lie to TEPCO about who they worked for, other times they falsified paperwork without the knowledge of individual labourers. Some brokers, according to a *Reuters* investigation, would even "buy" manpower by paying off workers' debts, then force their new "property" to work for them in and around Daiichi until the brokers were satisfied they had received their money's worth.

Unsurprisingly, this treatment, coupled with poor oversight and risk management, precipitated plummeting morale among workers, resulting in one clumsy, exhausted blunder after another. Losing consciousness in full-body protective suits from heat stroke was not uncommon during Japan's sweltering summer months, and several people died. One worker complained of feeling ill at the end of his shift in August 2015 and was sent to hospital but collapsed and died later that day.[au] A year later, another man collapsed in the entrance to the administration building and was rushed to hospital in an ambulance but was pronounced dead within hours.[av] The shortage of labourers became so acute by early 2014 that one subcontractor openly ran an online recruitment ad with the headline "Out of Work? Nowhere to Live? Nowhere to Go? Nothing to Eat? Come to Fukushima."[aw]

* * *

Police arrested yakuza boss Makoto Owada of the Sumiyoshi-kai, the country's second-largest syndicate, in late May 2012 for sending labourers to the plant via an approved subcontractor for several months in 2011, but yakuza members may have actually been at Daiichi the entire time. "The accident isn't our fault," said one mid-level member of the Sumiyoshi-kai. "It's TEPCO's fault. We've always been a necessary evil in the work process. In fact, if some of our men hadn't stayed to fight the meltdown, the situation would have been much worse. TEPCO employees and [NISA] inspectors mostly fled; we stood our ground."[ax] Tomohiko Suzuki, the author of *Yakuza and the Nuclear Industry: Diary of an Undercover Reporter Working at the Fukushima Plant*, claims three of the original Fukushima 50 were yakuza members. Organised crime involvement in Fukushima (and the wider nuclear industry) was not limited to labour either. It was considered standard practice, for instance, to pay local politicians and yakuza groups to win construction contracts, according to court testimony from executives of Mizutani Construction Co – the company indicted years earlier for allegedly bribing Eisaku Satō, lending credence to his own bribery accusations – after it again found itself in trouble over a decontamination contract.[ay]

Japan has a complicated relationship with the syndicates. They have plush offices in swanky districts of Tokyo, operate out in the open, have famous friends[9] and have been known to provide beneficial public services in times of crisis. They also opened their properties as public shelters and delivered food, water, blankets and other supplies to disaster-hit areas in March 2011, long before government help arrived.[az] As such, many people are somewhat agnostic towards them. Until October 2011, it wasn't illegal to pay yakuza for their services, nor, therefore, was employing them, but a change in the law put an end to decades-old revenue streams.[ba] Even then, TEPCO was reluctant to change their ways and did not update their contracts with new organised crime exclusionary clauses for quite some time. When pressed by journalists, TEPCO refused to disclose which companies they used for labour, background checks or general security, "because to do so would be in non-compliance with personal privacy information protection laws."[bb] When

---

9   For example, Shinsuke Shimada, one of Japan's most popular television hosts and comedians, was revealed to have yakuza ties in August 2011, effectively ending his career.

the MHLW finally waded in during early 2013, they found that almost 70% of subcontractors handling decontamination around Daiichi had broken the law. In one instance, a subcontractor took payment for the same several hundred workers twice, by having the company below them (i.e. the sub-subcontractor) claim to have supplied labourers who already worked directly for the first subcontractor.[bc] Multiple companies were forced to give back unpaid wages, yet none were prosecuted.

TEPCO, under political pressure, made some improvements once the extent of the problem became public, by scheduling lectures on employment law for contractors, establishing anonymous hotlines for unhappy workers and conducting random unannounced inspections of work areas. Matters did improve, but the behemoth utility sat atop an impenetrable pyramid of subcontractors seven or more layers deep. The first few layers of major contractors – conglomerates such as Kajima Corp. and Toshiba – do things by the book, but as you drop down among the hundreds of smaller companies, things become a muddled maze of manpower. It's unclear how any one company could solve the problem without changing the law and fundamental makeup of Japanese labour culture, even if it wanted to. Responsibility for what little enforcement did exist was kicked down the chain, but, at the end of the day, nobody checked because they simply needed the numbers and didn't care how they got them. Whenever government bureaucrats or company executives were questioned on this lackadaisical approach, most gave a variation of the same old excuse: if we vetted people that carefully, nothing would get done. Other countries with their own nuclear decontamination projects, many far older than Fukushima, such as in Britain and the United States, face similar obstacles, yet they seem to find a way. Admittedly, Japan is on a whole other scale, but changes could be made with sufficient willpower. They aren't because there isn't. As one homeless labourer summarised, "TEPCO is God, the main contractors are kings, and we are slaves."[10, bd]

10　Japan has much less of a homelessness problem now than it did in the 1990s, when numbers surged during the "lost decade." Changes to various welfare laws and the construction of homeless shelters reduced the number of people sleeping on the streets by 70% in the decade up to 2014.

iv. Political and Legal Consequences

The question of who to blame arose long before the immediate crisis ended. TEPCO, though striving to exonerate itself at every opportunity, understandably became the nexus of public outrage, with everyone from Shimizu at the top down to individual exhausted men and women at Daiichi held in varying degrees of contempt. Its executives received the most scorn, but an initial outpouring of respect and encouragement for the efforts of low-level employees wavered as the weeks dragged on. Incidents of harassment and abuse, for instance, prompted TEPCO to remove the company's name from all its dormitories across Tokyo "to protect our employees and their families."[be] The strain and social stigma of working for such a pariah became too much for some, and in 2011 the number of employees quitting – in a country where, if someone lands a job at a plush company like TEPCO, they typically stay for life – rose to 3.5 times its 2010 turnover rate, increasing to over five times in 2012.[11] TEPCO slashed all management salaries by 30% and all others by 20% in September 2012 to cut costs. When, in fiscal year 2014, the company offered its first-ever voluntary retirement program for employees aged over 50 to reduce staffing levels, 1,151 applied for the 1,000 slots.[bf] Senior executives realised they faced an exodus of personnel qualified to carry out the decontamination plan and reverted pay cuts for those involved to 7% a short time later, but, combined with the forced turnover resulting from radiation exposure at the site, finding enough skilled people to work there has caused occasional problems ever since.

The company refused to acknowledge that a meltdown had occurred until the release of a detailed reactor analysis on 15th May 2011, despite knowing on the very first day that Unit 1 had melted, with Vice President Sakae Muto admitting to the possibility at the time. A week later, just as IAEA inspectors arrived in Japan to visit Daiichi, TEPCO revealed that Units 2 and 3 had also melted. NISA director general Nobuaki Terasaka also admitted in August that he'd known by 12th March, even though his agency took over a month to concede as much.[bg, bh]

---

11 For comparison, the average employment turnover rate in the United States was 23% in 2018. In Japan, the rate was 7%. Among TEPCO's quitting employees, 134 quit in 2010, 465 quit in 2011 and 712 quit in 2012.

Five years later, TEPCO belatedly admitted it should have been clearer. Then-current president Naomi Hirose commented that "I would say it was a cover-up" after a report described Shimizu instructing company officials not to use the word "meltdown" on 14th March 2011.[bi] Shimizu later claimed this instruction originated from Kan and Edano to avoid confusing the public, which they denied.[bj] TEPCO had informed the government of an internal simulation predicting a 25–55% core melt while denying a meltdown in public, even though its Nuclear Disaster Countermeasures Manual (which they'd revised less than a year earlier yet somehow forgot existed until February 2016) defined the threshold as 5% or greater. "We sincerely apologize for failing to confirm the presence of the guideline in the manual for five years," came the almost laughable statement.[bk] "There were events where it may have been possible to issue notifications and reports more promptly."[bl] TEPCO's hiding the existence or delaying the release of information for an unreasonable length of time – even from the government – is unfortunately a pattern of behaviour that continues to this day.

Demonstrations against nuclear power, little more than a nuisance to bureaucrats in the past, erupted across the country after the disaster. Two national polls by *Kyodo News* and the *Asahi Shimbun* showed as many as 70% and 74% of people respectively favoured phasing out nuclear power, with that number higher in host prefectures.[bm, bn] And yet, few consequences troubled TEPCO's senior executives. Shimizu stepped down from his role as president in May 2011, landing a quiet job as an outside director at Fuji Oil Company in June 2012, where he has remained hidden ever since. TEPCO is Fuji Oil's principal shareholder, with a stake of 8.85% in 2019.[12] Tsunehisa Katsumata weathered the storm and was re-elected as chairman in June 2011 but only served for another year before he, too, stood aside. Most of TEPCO's other executives initially remained, and all, though quick to apologise, denied any personal wrongdoing. Early news reports, quoting unnamed public prosecutors, suggested a "not if but when" approach to holding companies and individuals accountable, but by the disaster's first anniversary, not one TEPCO or government employee had faced any legal repercussions.

---

12 Followed by the Kuwait Petroleum Corporation and the government of the Kingdom of Saudi Arabia, with 7.52% each. Source: "Fuji Oil Company annual report," 2019: http://www.foc.co.jp/en/ir/library/annualreport/main/011/teaserItems1/00/file/Annual2019.pdf.

TEPCO itself was another matter. The company practically defines the concept of "too big to fail," but its existence as a private entity came into question after its shares plummeted to their lowest level since being listed in August 1951, losing over 80% of their value by 5th April.[bo] Pundits, politicians and government officials alike openly discussed a possible forced nationalisation because of overwhelming public mistrust and the inevitable liability claims. The company announced a loss of ¥1.25 trillion ($15 billion) in May, one of the largest losses ever for a non-financial Japanese company, after a ¥130 billion ($1.6 billion) profit the previous year.[bp] The following year did not paint a much brighter picture as Japanese utilities struggled to recover, posting a collective $15 billion loss.

Then, on 9th May 2012, an event that would've been inconceivable 14 months earlier became a reality. TEPCO kept its name, but a government investment of over ¥1 trillion ($12.6 billion) across ten years – in addition to the ¥2.4 trillion ($30.1 billion) in taxpayer money it had already provided to pay for compensation claims – gained the government a 50.1% majority stake in the company. "Without the state funds," commented Edano, who had replaced Banri Kaieda as minister of METI, "[TEPCO] cannot provide a stable supply of electricity and pay for compensation and decommissioning costs."[bq] The government would choose board members but could not dictate management reforms without a two-thirds majority vote share. It intended to cede control back to the company once ministers felt satisfied that significant progress had been made towards repairing the physical and public relations damage. Though the deal required cost savings of ¥3.4 trillion yen ($42.2 billion) over the subsequent decade and for the company's entire board of 16 directors to resign, the hope was that TEPCO could resume control as early as 2017, provided it met certain targets.[br] This close oversight has not prevented poor choices; a 2015 audit found that TEPCO had wasted over a third of the allocated ¥190 billion ($1.6 billion) taxpayer decontamination fund, often on custom-made decontamination equipment and machinery which either didn't work, barely worked or was only used for short periods.[bs]

The company had paid ¥9.32 trillion ($86.7 billion) in compensation through the government's disaster dispute resolution program as of March 2020, much of which came from taxpayer funds.[bt] Despite this substantial number, the program offers pitifully low individual pay-outs, and its initial 58-page application form (and accompanying 158-page supplemental information book) required original paper copies of receipts, proof of income, etc. and seemed designed to be almost impossible to complete.[bu] The revised 34-page form, produced after public outcry, was not much better, and a year after the disaster, only one-quarter of the government funds allocated for compensation had been distributed. Japan is not a litigious country, but as of 2020 over 12,000 dissatisfied victims have filed over 400 civil and class action lawsuits on various aspects of the disaster.[bv] TEPCO's first loss came in August 2014 when the company paid ¥49 million ($472,000) to the widower of a woman who had committed suicide by setting herself on fire after being forced to abandon her home and livelihood.[bw]

A year later, Prime Minister Shinzō Abe prodded evacuees to return home and begin rebuilding their lives, granting a one-time payment of ¥100,000 (around $830 at the time) per household to those who did so, but he warned that all subsidies would be terminated by March 2017. When the time came, as promised, the 27,000 people who had fled the voluntary evacuation zone (at the government's urging) but never returned due to persisting radiation lost their housing subsidies.[bx] This left them in physical and financial limbo, unable or unwilling to return to homes where they didn't feel safe or settle down where they didn't want to be.

March 2017 also saw the first successful ruling against the government for its mishandling of events when a group of 62 compulsory and voluntary evacuees were awarded ¥38.6 million (around $340,000) compensation for their trauma.[by, 13] Though this was a paltry sum in comparison to the original ¥1.5 billion court filing, some plaintiffs were satisfied with the outcome. "The money is not a problem," said one man. "Even if it's ¥1,000 or ¥2,000, it's fine. We just want the government to admit their responsibility. Our ultimate goal is to make the government admit their responsibility and remind them not to repeat the same accident."[bz] Numerous similar compensation victories followed, but judges have occasionally sided with TEPCO on the grounds that established compensation channels already exist. During one such hearing in 2011, a TEPCO lawyer made the audacious argument that the company bore no re-

---

13 A group of 137 people sued the government, but the court only awarded damages to 62 of them, comprising 45 households.

sponsibility for decontamination because "radioactive materials that scattered and fell from the Fukushima No. 1 nuclear plant belong to individual landowners there, not TEPCO."[ca]

* * *

It took until 30th June 2017 before a frail-looking Tsunehisa Katsumata, now aged 77, along with former vice presidents Sakae Muto and Ichiro Takekuro, stood in court for what was to be the disaster's only criminal trial. Naoto Kan, Haruki Madarame, various other government officials and almost 40 TEPCO executives were dropped from the original filing, most conspicuously Masataka Shimizu. No concrete reasons appear to explain why Shimizu was not charged when, as the CEO, he was most responsible.[14] Muto and Takekuro were beneath him; Katsumata was above him in some respects, but theoretically his was more of a ceremonial role by 2011. Some have speculated that, though retired from day-to-day matters at that time, Katsumata remained truly in control, and he therefore stood trial on that basis.

Citing insufficient evidence, the Tokyo District Public Prosecutor's Office declined to proceed with the case in July 2014, despite multiple lengthy independent reports from the IAEA and others detailing TEPCO's poor safety culture as a root cause.[15] The prosecutor's office attempted to bury the story by leaking their decision on the same day in September 2013 when the International Olympic Committee announced that Japan would host the 2020 Olympic Games. Not for the first time since the disaster, which had seen unprecedented tens of thousands turning out to express their views, protests erupted on Tokyo's streets. An 11-person independent citizens judicial panel known as the Prosecutorial Review Commission (PRC, selected by lottery and similar in function to an American grand jury) ruled that the three senior TEPCO executives *should* stand trial. This decision compelled the prosecutors to re-investigate the evidence, only to again elect not to prosecute in January 2015

---

14 I have spoken with several Japan-focused legal scholars on this matter and none knew of a specific reason.

15 The office also declined to prosecute any government officials. Their involvement caused some angst among plaintiffs, because the complaints had been submitted to Fukushima's Prosecutor's Office, but jurisdiction was transferred to Tokyo. There were some benefits to this, but it appears the overriding reason was that the prosecutors simply did not want the case to be charged.

because they felt the case was too weak for a conviction. A different PRC overruled them again, triggering an extremely rare *kyōsei kiso* "forced prosecution" indictment in February 2016 on charges of professional negligence.

The arguments hinged on whether TEPCO's executives knew of the risk posed to Fukushima Daiichi by possible earthquakes and tsunami. TEPCO's head of tsunami countermeasures testified that he briefed Muto in February and June 2008 on internal studies indicating possible breaches of the sea barrier up to 7.7 metres, with a further meeting to discuss a revised 15.7-metre prediction in July with Katsumata (and Shimizu), followed by approval and then postponement of expensive barriers.[16, cb] "I simply thought," said Muto, who made the decision, that "it would be difficult to come up with a design for a strong sea wall straight away."[cc] Such a seawall would be expensive and inconvenient. With Kashiwazaki-Kariwa also offline, losing further power capacity was undesirable. All three defendants, through profuse apologies, countered that the data was hypothetical and in dispute, that they could not have foreseen the scale of the eventual tsunami and therefore did not approve any tangible course of action. There was also the sticky fact that, though the company provided a vital public service, its leadership was beholden to shareholders as much as anyone. Regulatory bodies, if they thought a tsunami posed a realistic threat, held as much (if not more) responsibility for inaction as TEPCO, yet only these three people stood trial.

Prosecutors requested five-year prison terms for the deaths of 44 evacuees (though not the two men killed by floodwater in Fukushima Daiichi's pump room) and the untold harm to thousands of others, but the panel of judges found all three defendants not guilty in September 2019. No doubt anticipating the inevitable global uproar, the judges spent over three hours reading the verdict and justification for their decision. They acknowledged that the men knew about data indicating a possible mega-tsunami, but accepted doubts about the projections' validity. Had the executives taken things seriously, the planning and construction of new defences would have taken Daiichi offline for an extended period, perhaps years, and, in any event, these upgrades may not have been finished on time. The

---

16 Muto said "Please leave it to the Japan Society of Civil Engineers for further consideration," according to the trial transcripts. He was told that this would not meet the deadlines for backchecking the plants but was unconcerned.

acquittal made headlines across the world. Not one person from any branch of government has been charged with anything as of 2021, and it is unlikely any ever will be.

The triple earthquake, tsunami and nuclear disaster proved to be the most expensive in human history, with the World Bank estimating in March 2011 that costs from the two natural disasters alone could reach ¥19 trillion ($235 billion).[cd] By 2020, however, the sum far exceeded that lofty estimate, with the government's Reconstruction Agency announcing it had already spent ¥31 trillion ($281 billion) on repairing damaged infrastructure and buildings and expected another ¥2 trillion ($18 billion) by 2025.[ce] The cost of decommissioning Fukushima Daiichi and decontaminating the surrounding area ballooned as well. It almost doubled METI's 2013 calculation of ¥11 trillion within three years, but the respected Tokyo-based think-tank, the Japan Center for Economic Research, estimated in March 2019 that the total sum would at least double even that number by 2050 and could potentially rise to an eye-watering ¥81 trillion ($728 billion: around the GDP of Switzerland) if certain conditions are met.[17] Their lowest estimate of ¥35 trillion requires TEPCO to abandon its plan of finding and removing fuel from the ruined reactors and instead encase the plant in a concrete sarcophagus like the one at Chernobyl, but the government has explicitly ruled this out. Despite (or rather because of) these astronomical figures, TEPCO has effectively avoided financial accountability for the true cost of the disaster.

* * *

Uplifting stories did surface among the tide of misery. Executives at Tōhoku Electric quickly decided to open their Onagawa plant to those left homeless by the devastation, even though it was not a designated evacuation shelter.[cf] "The general public isn't normally allowed inside," said a spokesman, "but in this case we felt it was the right thing to do."[cg] The gymnasium eventually held up to 360 people for three months, with power, meals (albeit, with shortages,

---

17 The reasoning behind these numbers makes for interesting reading. The full document can be found here:
"Accident cleanup costs rising to 35–80 trillion yen in 40 Years." Japan Center for Economic Research, 7th March 2019: https://www.jcer.or.jp/jcer_download_log.php?post_id=49661&file_post_id=49662.

limited to two per day), heating, phone lines, and an abundance of space and working toilets: a rare luxury at the time.[ch] The first residents even watched Daiichi's reactor buildings explode live on television.[ci] One woman had to be airlifted from the plant to Sendai, where she gave birth to a healthy baby.[cj]

Elsewhere, a diverse range of reports praised the tireless efforts of countless men and women from the government, JSDF and private industry during and after the crisis, both as organised groups and individuals, such as the case of a lone policeman who single-handedly evacuated the train he was travelling aboard to high ground after the earthquake. But these reports were equally withering in their criticism of the upper-echelon government response. The planned and trained-for emergency chain of command – Daiichi reporting to TEPCO HQ, which reported to NISA, which reported to and advised the Kantei – was bypassed almost from the start, with TEPCO soon talking to both NISA and the crisis management centre in the Kantei basement. From there, communications often became muddled on their way upstairs to Kan's office. Some information did not go up at all. Other times, TEPCO's Ichiro Takekuro bypassed the main arm of government entirely by relaying messages straight to Kan or his team. These senior officials were meant to concentrate on the broader earthquake-tsunami disaster and take a hands-off approach to non-legal events at Daiichi, thereby letting government contingencies take their course – *not* to micro-manage.

TEPCO, its executives, the wider nuclear village and opposition Liberal Democratic Party (LDP) members – including former and future prime minister Shinzō Abe – conducted a formidable blame-shifting campaign away from themselves and onto the besieged Kan. They argued his visit to Daiichi had distracted Yoshida and side-tracked recovery efforts, which, in turn, prevented the core venting that led to the Unit 1 meltdown and explosion. Further, they claimed that he ordered workers to halt seawater injection. Essentially, he was to blame for the accident becoming so bad. Upon analysis, however, most of these claims fall apart.[ck]

Kan's left-wing Democratic Party of Japan (DPJ) only gained power in late 2009, while Kan himself became prime minister on 8th June 2010. Prior to that, the LDP held almost unrelenting control for 57 years, and the entire structure of government – including its disaster preparations – was theirs and theirs alone. Kan's problematic behaviour was on him, but the broader government response was all LDP. Saying he ordered a halt to seawater injection is outright false, but this ranked among the most damaging accusations. The suggestion that his visit somehow prevented or even delayed reactor venting borders on the ridiculous; one TEPCO report even specifically says that "venting preparations in the field continued during Prime Minister Kan's visit. Therefore, the visit did not directly delay venting operations."[cl] Yoshida also said Kan's visit caused no delay.[cm] Unit 1 had already melted before he arrived and preparations had been underway for most of the night, plus the vents did not open for another six hours after he left. Preparing for his arrival, such as finding a suitable helicopter landing area, did take up valuable time, but upon arrival at the SIB, he only occupied the attention of one man. Kan felt the poisonous mistrust between the political leadership and TEPCO that was caused, in part, by extreme communication difficulties and the failing contingencies (such as the off-site centre) needed to be addressed before it deteriorated into outright animosity.

On balance, I think the reassurance he gained from his time with Yoshida proved beneficial. So, too, did his unprecedented decision to install his aide Goshi Hosono at TEPCO headquarters, as this resolved most of the communications issues. True, the small group of politicians at the Kantei did sow confusion by becoming overly involved in matters they should have left to TEPCO, compelling its executives to second-guess Kan's intentions, but TEPCO executives also aggravated Yoshida's difficulties with their own micro-management. Everyone should simply have provided the supplies Yoshida requested and let him make all the decisions.

As more reports appeared, each more damning than the last, with many claiming Kan made matters worse rather than better, public opinion won out and he resigned on 2nd September 2011. His performance may have been far from perfect, but I think he was unfairly maligned. Existing disaster preparations, contingencies and communication channels at all levels proved to be completely inadequate, and neither he nor his party had enough time after taking power to make significant improvements.[18] Incidentally, the 2020 movie *Fukushima 50*, based on the disaster and starring Ken Watanabe, portrayed Kan (played by Shirô Sano) as a villain: a hysterical meddler whose actions risked lives and made everything worse.

---

18 Of course, the question of whether they *would have* made such improvements will never be answered.

NISA, for its part, did not rise to the challenge of providing guidance for the country's leadership, with top government officials describing agency advisors as being "capable of only hemming and hawing."[cn] They "only stare downward, stiffen, and remain silent despite being criticized," acting "like school kids who haven't done their homework, being unable to look the Prime Minister and his team in the eyes"; the NAIIC report is filled with various similar condemnations.[co] Industry observers, journalists and government ministers quickly suggested splitting the organisation off from METI, but NISA were not alone in their inadequacy. The sectionalism and rivalry common among different branches of government caused many powerful groups to chart a passive and inflexible course through ongoing events. MEXT, for example, floundered and chose to await requests for help from organisations that were clearly overwhelmed, rather than be proactive. Science-based decision-making also flew out the window at an early stage. While the first, three-kilometre evacuation radius came about from discussions between NSC chairman Haruki Madarame and NISA vice director general Eiji Hiraoka, based on their knowledge and experience of existing and incoming nuclear safety guidelines, the expanded second and third (10- and 20-kilometre) radiuses that directly and indirectly caused so many deaths weren't decided using any proper scientific basis. The 10-kilometre figure arose purely because it was the maximum distance covered by existing disaster training, and the 20-kilometre value seems to have been plucked from thin air. The local Fukushima prefectural government also struggled, being unprepared for a triple-pronged disaster of violent earthquake, coast-eradicating tsunami and explosive nuclear emergency on their doorstep (though, in their defence, I'm unsure how anyone *could* fully prepare for that).

Haruki Madarame, chairman of the JAEC's Nuclear Safety Commission, who had once mocked the warnings of Katsuhiko Ishibashi, conceded in February 2012 that "though global safety standards kept on improving, we wasted our time coming up with excuses for why Japan didn't need to bother meeting them."[cp] He apologised and accepted his own partial responsibility as NSC leader for this complacency, while reiterating that he had held the post for less than a year before the disaster. Madarame instead believed most of the blame lay with TEPCO, arguing that "power companies have the fundamental responsibility of securing safety and they need to set their standards much higher than what the government suggests.... It is extremely outrageous if power firms are using the NSC's safety standards as an excuse not to raise them."[cq] This was a startling statement from one of the nation's top nuclear officials. As the JAEC's ambitious national goals and policies have gradually declined in number and scope since its inception in 1956, so the JAEC itself has fallen from favour – declining to just a chair and two commissioners – while the prime minister no longer has to respect its recommendations. The FEPC's influence has also been greatly reduced, and it no longer dictates government policy to the extent it used to.

* * *

It wasn't just TEPCO and government agencies which came under scrutiny; the nation's largest news media companies did too. A comprehensive 2019 paper by the Disaster and Media Research Group goes some way towards explaining the uncritical behaviour displayed by news organisations during the 2011 crisis. After two years of research and interviews with newsroom executives, the group found that media organisations believed "[their] role is to cooperate with the government to communicate during emergencies, even if this means sacrificing their watchdog role."[cr] As such, they prioritised evacuation warnings and tsunami and earthquake updates over independent verification of radiation plume directions or accurate health-risk assessments. They also shared the government's concern over avoiding unnecessary panic. This was a reasonable approach at first: with their lands devastated, people naturally look towards media outlets for advice. After the first explosion on the afternoon of 12th March, however, journalists should have demanded more information, or at least tried to deduce some basic facts.

A lack of expertise was a key problem. Of all the largest news organisations, the *Asahi* newspaper and *NHK* television network were the only ones with specialist nuclear science and energy correspondents. The rest, arguing afterwards (not unreasonably) that they couldn't retain staff specialists from every field, rounded up "talking heads" from predominantly pro-nuclear institutions and had them analyse events live on air, with the perhaps inevitable result that they defended TEPCO's information. This weak point can be illustrated with that first explosion. Everyone soon realised that the reactor itself had not exploded because radiation readings remained low. The industry's scientific community knew that the only probable source of an explosion of that magnitude was hydrogen, but there was none in the reactor building, so what can make hydrogen? The only answer is the reactor itself, which had not been cooled for 24 hours and therefore must have melted by this point, generating a massive amount of hydrogen as a by-product. Yet, lacking journalists with scientific expertise, media figures were powerless to question TEPCO or government spokespeople's reassurances or vague references to possible "core damage."

v. Health and Radiological Consequences

Masao Yoshida spent a brief period off from work to recover from exhaustion before returning to oversee plant operations. Unlike most (if not all) of TEPCO's executives, who acted like children caught with their hands in the candy jar when questioned about the accident, Yoshida had no qualms about accepting responsibility and gave repeated heartfelt apologies for his failings. He felt an enormous sense of guilt for not taking the threat of tsunami more seriously back when he'd headed TEPCO's Nuclear Division from 2007 to 2010. Like his peers, he had considered extreme tsunami to be impossible, for which he received deserved criticism. He was considering quitting TEPCO and forming his own company by late 2011. "I caused an accident," he confided to a close friend and colleague. "So I have to do something to revitalise this local area."[cs]

But on 1st December 2011, after apologising to his staff for leaving them once more, he entered hospital following a diagnosis of oesophageal cancer. He absorbed 80 millisieverts during the accident, but his cancer was likely due to his heavy smoking habit. Doctors declared him cancer-free in June 2012, following seven months of treatment, but on 26th July, during lunch with a colleague in which he expressed his desire to return to work, Yoshida suffered a brain haemorrhage and collapsed. He survived, but his cancer had returned.

He avoided public commentary during the long months of the various accident investigations but broke his silence upon their release and the subsequent criticisms of his staff, speaking passionately about their actions. "Some of my colleagues went to the scenes of the accident with a great amount of radiation and radioactivity a number of times," he said during a video seminar recorded from his home in August 2012. "They did all they could; all I did was watch. I didn't do anything. In fact, I really appreciate and thank every one of my colleagues who went to the scenes of the accident."[ct]

He fought on for almost a year, but cancer is cruel. Masao Yoshida, now widely considered a national hero, passed away in hospital on 9th July 2013, aged 58. "I bow in respect for his leadership and decision making," Naoto Kan tweeted upon hearing the news. A born leader and a man of pre-eminent honour, his loss is an enormous tragedy. I wish I had been fortunate enough to meet him.

* * *

Fukushima Daiichi released considerably less radioactivity into the environment overall than Chernobyl. The total amount of aerosol-based fission products is around ten times less, while the most harmful isotopes iodine-131 and caesium-137 were around six times less, or 40 times less in the case of plutonium.[cu] That said, six or ten times less than Chernobyl is still a devastating amount, and the disaster, as you can imagine, did not warm Japanese hearts towards their government, TEPCO or nuclear power. For a start, people quickly noticed a stuttering stream of vague and contradictory information about what was happening, the severity of it all, and whether residents of nearby towns and villages were safe to shelter at home or if they should evacuate. A few days after staff at the power plant had evacuated to Fukushima Daini, thyroid cancer–inducing iodine-131 was found to exceed safety levels by three to seven times in milk 30 kilometres from Daiichi, and in spinach growing on farms up to 75 kilometres away.[cv] People would have needed to devour several times their own body weight in spinach for it to have had any kind of measurable effect, but such news nevertheless incited panic across the region, despite Edano's public reassurances.

Trace amounts discovered in Tokyo's water supply and a government warning not to add water to baby formula milk failed to put anyone at ease. Bottled water was delivered to all 80,000 affected families, but stocks still disappeared from store shelves across eastern Japan within hours.[cw] The government instructed water-purification plants nationwide to cease using rainwater days later, after iodine exceeded safe limits for infants by up to 145 times at 18 purification plants in five prefectures.[cx]

Despite a smaller overall release than Chernobyl, Fukushima was the largest-ever "accidental source of radionuclides to the ocean in terms of measured radionuclide concentrations," according to Ken Buesseler of the Woods Hole Oceanographic Institution, surpassing Chernobyl by many times (which is unsurprising, given their relative proximities to the ocean).[cy] Indeed, more than 80% of radioactive particles released during the accident blew out over, or ejected into, the ocean.[cz] This was good for everyone and everything on land, but bad for the aquatic environment and the fishermen who depended on it. Having already seen their nets washed away, their ports swamped with debris and any docked vessels smashed to smithereens – the tsunami damaged 29,000 fishing boats and 319 ports, around 10% of each of the national total – every fisherman's worst nightmares came true as government officials banned the sale of east coast fish.[da, 19] Similar bans from international neighbours China, Thailand and South Korea followed in time, forcing at least some fishermen to find decontamination work in and around the plant to make ends meet. Tokyo's Tsukiji fish market, the world's largest, still stood almost deserted of customers weeks after the accident began as local, national and international demand for Japanese fish collapsed. Most people understandably regarded anything caught in the region as unsafe to eat, tanking sales of Fukushima fish from ¥11 billion ($133 million) in 2010 to ¥1.6 billion in 2011, and total losses to the fishing industry (including tsunami damage) reached ¥1.2 trillion ($15.6 billion in 2020) by early July 2011.[db] Some forlorn fishermen took to catching fish and delivering it to ports outside Fukushima Prefecture to avoid the affiliation.[dc]

As we know, radiation is everywhere, including in all food, and an average 70-kilogram (154-pound) man has a natural "activity" of around 8,000 becquerels. The becquerel's tiny scale often leads to huge and scary numbers, but I'll try to explain how it works. A single becquerel corresponds to a half-life decay rate of one nucleus per

---

19 The Fukushima Prefectural Federation of Fisheries Cooperative Associations (Fukushima FCA) had already voluntarily stopped fishing on 15th March, though they did not really comprehend the scale of the problem at this stage.

second and is a measurement of radioactivity rather than exposure dose, like sieverts. The becquerel scale does not distinguish between different types of radiation, such as alpha, beta and gamma, nor the way that different isotopes have different toxicities. For example, 100 becquerels per kilogram (Bq/kg) of the natural radionuclide potassium-40, found in potatoes and bananas, is considerably less harmful than 100 Bq/kg of plutonium, making becquerels a handy unit for determining how much of a radioactive material is present – just not its strength. Beef, fish and poultry typically contain around 100 Bq/kg, while eggs contain 44, potatoes 170 and most green vegetables around 150. Some foods range higher, such as white bread, which contains around 550 Bq/kg, but few items cross the 500 threshold.

Japan did not have legal safe radioactivity limits for food prior to the Fukushima disaster, other than to prohibit the sale of anything "poisonous," but in 2012 the MHLW established a limit of 100 Bq/kg for caesium in food, 50 Bq/kg for milk and infant food, and 10 for drinking water.[dd] This corresponded to a new, reduced maximum permissible dose of one millisievert per year.[de] For comparison, the European Union's long-established caesium limit is 1,250 Bq/kg, making the Japanese guidelines extremely conservative.[df]

Around 40% of bottom-dwelling fish such as flounder and cod were contaminated beyond the new safety guidelines and their level of exposure did not initially decline, in part because radioactive sediments settled on the seabed and then continually re-contaminated the fish.[dg] The discovery of caesium from a Fukushima reactor in Pacific bluefin tuna caught off the California coast made headlines all over the world in May 2012. Unfortunately, some ill-informed and irresponsible websites – a globular morass of conspiracy mongers, personal blogs, Facebook and even some mainstream news sites – took the opportunity to pass around old photographs of sick fish and claim Fukushima as the source. The California contamination was far beneath that seen during the era of nuclear weapons tests and posed no threat to health, with natural potassium-40 over 30 times higher than the radioactive caesium detected by researchers, but it marked the first time ever that migrating animals had carried radioactive particles across the ocean.[dh]

Three months later, two greenlings caught near the mouth of the narrow Ōta River, 20 kilometres north of Daiichi, were found to contain 25,800 Bq/kg of caesium-134 and -137, prompting some alarm, as you can imagine.[di] No other fish caught outside Daiichi's harbour since a few months after the disaster had exceeded 5,000 Bq/kg, and few contained more than 1,000. Scientists conducted a thorough analysis of beta-ray emissions and determined that the greenlings had been present within the harbour as the disaster unfolded, hence their high activity count.[dj] For comparison, other fish caught inside that perimeter ranged from 1,030 to 740,000 Bq/kg.

In 2019, when fishing of all species had resumed in specific areas on at least a trial basis and the Fukushima Prefectural Federation of Fisheries Cooperative Associations (Fukushima FCA) had implemented a strict policy against selling fish in excess of 50 Bq/kg, the cooperatives still caught just 14% of their 2010 volume.[dk] In September 2019, the Fukushima FCA took the important step of setting haul targets for the first time since 2011, as no farmed or fished produce that year had exceeded the safety limits. Still, the cooperatives remain concerned about the long-term public impact of dumping tritiated water into the ocean. (More on tritium below.)

Radiation did not just enter the ocean, of course. About 80% of contaminated land throughout Fukushima was related to agriculture, and farmers endured a destruction of their livelihood only slightly less severe than that of the fishermen. Crops were routinely sprayed with potassium fertiliser to prevent the absorption of radioactive caesium, resulting in zero contaminated rice for the first time in 2015. Vegetable crops also recovered within a few years, but there remains a ban on Fukushima's famous wild mushrooms as of December 2020.

* * *

Doctors found 186 incidents of thyroid cancer in children living in the affected areas, but this can, in part, be attributed to heightened vigilance and mass screening for the tell-tale cancer nodules that appear on thyroid glands, as similar numbers may well have gone undetected under normal circumstances. In October 2015, the MHLW acknowledged the first instance of cancer (leukaemia) caused by radiation exposure from the disaster, of a labourer in his late thirties who worked at the plant in November and December 2011.[dl] He had been exposed to just 19.8 millisieverts. The ministry announced the first death caused by radiation exposure from the disaster in September 2018, of an unnamed male clean-up worker in his fifties who developed cancer after being exposed to 195 millisieverts. Overall, however, the scientific consensus is clear: direct physical illnesses and deaths from radiation exposure among the general public are expected to be imperceptible, according to a United Nations report published in March 2021. Even the specific cases mentioned above were acknowledged in a legal sense and not a medical one, in an absence of evidence that their illness was caused by radiation alone. The Fukushima prefectural government reports 2,268 "disaster-related deaths" in the prefecture. While this figure includes people displaced by the earthquake and tsunami, two-thirds are thought to be tied to the nuclear disaster.[20, dm, dn]

Unfortunately, the government has, at times, gone out of its way to convince people there's no risk while preventing independent monitoring and analysis of public health in the region. In December 2013, it passed the draconian Act on the Protection of Specially Designated Secrets (Tokutei Himitsu Hogo Hō), which granted the government powers to imprison for up to ten years anyone revealing broadly defined 'secrets,' with no independent oversight. "There are few specifics in the law," complained one opposition Member of Parliament, "which means it can be used to hide whatever the government wishes to keep away from public scrutiny."[do] That same month, legislators approved the Act for the Promotion of Cancer Registration (Gan Tōroku Hō), designed to unify and streamline the hodgepodge of information recorded by individual hospitals into a reliable national database. While the act encouraged the promotion of information on cancer survival rates, it also outlawed the sharing of data on any radiation-related medical conditions, on threat of another five- to ten-year prison sentence or a ¥2 million ($18,500) fine. These and a number of other things caused Japan to plummet on the Reporters Without Borders 'World Press Freedom Index' from 11th in 2010 to 72nd by 2016, clawing back up to 66th place as of 2020. Partly because of the mistrust arising from this and other measures, a number of citizen-run radiation labs have cropped up in Fukushima, providing a service for people to test themselves for thyroid cancer or food for contamination.[dp]

---

20 This figure of 2,268 remains the most up-to-date figure as of January 2021.

vi. Decontamination

By late 2011, once the fear of imminent nuclear catastrophe subsided and weary operators had long since achieved a cold shutdown of the reactors, TEPCO and the government began the decades-long decommissioning process. The almost insurmountable task comprised several parts. First: plug the leaks. Ongoing reactor cooling meant water filtered down into the basement through cracks caused by the earthquake, meltdowns and explosions. This structural damage also enabled groundwater to gurgle up from below, where it all mixed with radioactive contaminants before 300–400 tons of freshly contaminated water leaked back out into the Pacific Ocean each day. This figure remained fixed even by late 2013, despite TEPCO denying any ocean leaks for several months at one point. With roughly the same amount of water in every building because of ducts and other penetrations connecting them below ground, nothing of substance could be done about the three melted reactors until the problem was resolved.[dq]

Stopping the leaks from inside would be a straightforward job under normal circumstances, and it's perhaps because of this that TEPCO and the government both initially underestimated the issue and refused outside appraisals, believing it would be possible to contain, pump up, store and decontaminate without anything reaching the ocean. Early attempts to plug the first dangerous leak from a 20-centimetre crack in a cabling shaft beside Unit 2 were unsuccessful. With the reactor water emitting one sievert per hour, TEPCO first tried to fill the crack with concrete. Then, when that didn't work, they threw in a high-tech concoction of water-absorbent polymer, sawdust and shredded newspapers. The water kept flowing; more polymer mix, more concrete.[dr] That specific leak stopped on 6th April 2011, but TEPCO couldn't stop pumping coolant into the reactors, so the overall volume of contaminated water grew until they were forced to dump 10,000 tons contaminated to around 100 times the legal limit into the ocean after running out of storage space.[ds] This was harmless once diluted in the ocean, but two years later TEPCO were overwhelmed.[21] "The water keeps increasing

---

21 The harmlessness of the water did not prevent outrage, even though, according to TEPCO, "these discharges would increase the effective dose to a member of the public by 0.6 mSv, if he/she were to eat seaweed and seafood caught one kilometre from the discharge point every day for a year. It should be noted however that the movements of all ships, including fishing boats, are restricted within a 30

every minute, no matter whether we eat, sleep or work," admitted one spokesman. "It feels like we are constantly being chased, but we are doing our best to stay a step in front."[dt] Once they recognised the severity of the issue, and with little time to plan, the utility quickly built underground clay- and plastic-lined storage tanks, but these too soon began to leak. Hundreds of above-ground water tanks suffered the same problem. After 300 tons of highly radioactive water leaked into the ground in August 2013 (separate from the normal daily leak of 300 tons), TEPCO was forced to admit it did not know how the water had breached its tank, nor where it had gone.[du] The following month, workers detected an extremely dangerous hotspot of 2.2 sieverts per hour near one tank, reinforcing the widespread perception that the company was in over its head.[dv]

TEPCO removed stagnant water in the basements of each building during the spring and summer by drilling holes in the outer walls of the Units 2 and 4 turbine buildings or via enormous pumps, where possible.[dw] These pumps forced liquids through filters to strip out radioactive isotopes before redirecting it to more storage tanks.[22] Scepticism over the water's safety led to a somewhat perverse episode where a clearly frightened government spokesman named Yasuhiro Sonada, his hands trembling, drank water from the area around Unit 6 in front of a gaggle of reporters to prove its cleanliness.[dx] Early filters only captured gamma-emitting caesium, but as the years passed they became more sophisticated, battling through occasional breakdowns, until only the beta-emitting hydrogen isotope tritium remained.

Tritium had replaced the far more hazardous radium as the glowing material of choice for watch dials once radium's dangerous nature became public knowledge. Though far less of a threat than radium or caesium, tritium's potential for ingestion and internal exposure to fish or humans if dumped into the ocean proved a PR nightmare. The problem was that nothing could cleanse the water of tritium: it literally bonds with the oxygen atom and becomes tritiated water – tritium oxide, or T_2O – instead of regular H_2O. By January 2020, the site's forest of tanks held an estimated 860 trillion becquerels of tritiated water. This astronomical number prompted alarm bells among media outlets and environmental groups, but part

---

km zone from the NPP."

22 An early proposal to store around 10,000 tons of water on a gigantic floating 390-by-97-metre shore tank next to the plant proved impractical.

of the reason it was so high, as we've discussed, is that the becquerel is a unit adjusted to the atomic scale, much like counting the grains of sand on a beach. Tritium has a radiologically short half-life of 12.3 years, making it highly radioactive, but its biological half-life (the time it takes for 50% of a consumed source to pass through the human body) is just ten days, and it emits a very weak beta particle with a decay energy of only 5.7 kiloelectronvolts (keV), orders of magnitude less than caesium-137 (beta decay of 512 keV, gamma decay of 661 keV) or iodine-131 (beta decay of 606 keV, gamma decay of 364 keV).[dy] When these factors are combined and the 860 trillion becquerels (though a lot) are averaged across the stored water and then diluted in the vast ocean over a planned duration of 30 years, the tritium is expected to be virtually undetectable. This is good because TEPCO already has governmental and IAEA approval to begin pumping it out into the ocean via an underwater tunnel at some point in spring 2023, but to some extent its threat to health is an unanswered question.

* * *

Kajima Corporation sunk a deep steel cut-off wall between Fukushima Daiichi's reactors and the ocean, stopping most radiation from reaching the coast, but TEPCO faced calls for a more all-encompassing barrier. Twelve wells dug on the inland side captured around one-quarter of the groundwater and redirected it around the plant, but that still wasn't enough. Kajima's engineers then concocted an ambitious plan to isolate Units 1 to 4 from the surrounding environment.[23] They proposed using 1,571 "freezing pipes," cooled to −30°C (−22°F) using a brine solution and buried 30 meters (100 feet) into the uppermost aquifer layer of water-bearing permeable rock, to freeze a solid ring of soil around the plant's perimeter.[dz] Earth-freezing technology is normally used for temporary structural purposes or water reduction in mining or construction projects, not for long-term, monitorable containment. This new project would be almost double the previous largest use of the technology in Japan, when engineers froze 37,700 cubic

---

23   Kajima has its own Kajima Technical Research Institute with an entire department dedicated to work on groundwork such as foundations and soil properties. Members of this department designed the system used at Fukushima Daiichi.

metres of earth during work on the Tokyo subway. As a result, Kajima encountered multiple setbacks, while critics pummelled TEPCO and the government for not utilising something cheaper, simpler and more reliable, like a wall of concrete. This was wishful thinking to a certain extent: building water-tight, deep cut-off walls is not easy or cheap and they aren't immune to leaks either.

Construction of the ice wall began in June 2014, with a planned activation date of March 2015, but in June 2016, after revealing that the partially complete, ¥32 billion ($275 million) wall had made just a 10% difference to the water capture, TEPCO admitted for the first time that stopping all water could not be done.[ea, 24] Still, they soldiered on and began a slow freeze of the wall's final section in August 2017, by which time the government had thrown ¥35 billion ($320 million) at the project, plus an extra ¥1 billion ($9.2 million) per day to keep it running. Though it occasionally leaks, the wall, wells and pumps combined have done a reasonable job of reducing the flow of water to around 100 tons per day.[eb] This, however, has forced TEPCO to erect evermore tanks across the surrounding landscape, surpassing a million tons of stored water in early 2018 and 1.2 million two years later. As the volume of stagnant water within the reactor buildings drops, "high concentrations of radiation are being found," according to a September 2019 report by the Nuclear Damage Compensation and Decommissioning Facilitation Corporation (NDF), a quasi-governmental organisation set up to oversee decommissioning operations.[ec]

* * *

This leads us to the second and most obvious element of the grand decontamination plan: radiation. Beginning in May 2011, workers spent five months erecting a hastily designed temporary cover atop Unit 1 just as their liquidator predecessors had with Chernobyl's sarcophagus 25 years earlier. Others sprayed a sealing resin on the ground for weeks on end to prevent the wind and rain from further spreading radioactive particles. Unit 1's enclosure was demolished

---

24 TEPCO was pumping up just 31 tons less water per day in June 2016 than the 321 they had reported in May. Source: Otake, Tomoko. "In first, TEPCO admits ice wall can't stop Fukushima No. 1 groundwater." *Japan Times*, 20th July 2016: https://www.japantimes.co.jp/news/2016/07/20/national/first-tepco-admits-ice-wall-cant-stop-fukushima-no-1-groundwater/.

and replaced with a more permanent solution between 2013 and 2016. Unit 4's cantilevered version was finished in July 2013, complete with cameras and mobile cranes to facilitate emptying its spent fuel pool, which concluded in December 2014. Unit 3 stands out with a distinctive arched enclosure, while work to remove its own spent fuel was completed in March 2021. These measures together have stopped most external radiation above ground, save for hotspots like puddles of leaking water, but explosion damage and radiation *inside* have presented an obstacle not encountered since Chernobyl, making even basic tasks difficult.[25]

As hard as day-to-day operations are, finding and removing the melted fuel is nigh impossible. Human beings were able, for instance, to supervise the removal of spent fuel from Unit 4 in person because of the low levels of radioactivity inside the building, but Units 1 to 3 required technologies that didn't exist. Japan's renowned prowess in the robotics field proved invaluable here, as companies and institutions across the land spent years designing, testing and iterating numerous remote-controlled, often alien-looking robots, each more sophisticated than the last. All were customised to individual buildings, routes and tasks and hardened to withstand the punishing radiation within. In 2016, the government opened the $100 million Naraha Center for Remote Control Technology Development to research, design, test and train with robots, located close to the Hirono coal plant and J-Village. "I've been a robotic engineer for 30 years, and we've never faced anything as hard as this," said Shinji Kawatsuma, the centre's director. "This is a divine mission for Japan's robot engineers."[ed]

The machines nevertheless repeatedly struggled to complete their missions, despite all the investment. One, a snake-like robot built by Hitachi to explore the dark depths of Unit 1, stopped working for unknown reasons three hours into its ten-hour mission in April 2015.[ee] A Toshiba machine for extracting fuel from Unit 3's spent fuel pool quickly succumbed to radiation after engineers lowered it into the water in March 2016. Another, sent in February 2017 to clear a path beneath Unit 2 for yet another robot to search for molten

---

25 New problems also crop up all the time. As recently as 31st March 2020, TEPCO revealed that two buildings near Units 1 and 3 that had been used "as substitutes for tanks to temporarily store a huge volume of heavily contaminated water" contained sandbags that were now emitting up to four sieverts per hour and could not be approached. Source: *New problem at Fukushima site; sandbags found to be radioactive. Asahi Shimbun*, Yu Kotsubo, March 31st, 2020. http://www.asahi.com/ajw/articles/13228942?fbclid=IwAR3iubtKgtuFtw8102gjNuqHdr6jrdpTempFQaaikhf4kakjBVvGTEgm0tg

fuel, died after it encountered a staggering but crudely estimated 650 sieverts per hour – a number that stunned the world – but the figure was corrected to a still-lethal 80 sieverts per hour in July.[26, ef] The next robot also failed (the seventh to do so) after becoming stuck in the debris-strewn building, but each failure meant lessons learned, and there have been numerous important successes.

Mini-Manbo (sometimes called "Little Sunfish") was a red and white compact underwater vehicle around the size of a small loaf of bread, with four propellers and a tilting camera. After months of practice in the Naraha Center's huge water tank, its pilot spent three edge-of-his-seat days in July 2017 navigating Mini-Manbo through cramped, tangled spaces inside Unit 3's flooded reactor compartment – constantly hindered by impenetrable clouds of sediment thrown up by the craft's tiny propellers – before finding a stalactite-like mass of fuel, control rods and other debris hanging beneath the core.[eg, eh] Another new robot extracted fuel samples from inside Unit 2 in February 2019, but nine months later officials admitted they weren't going to meet the target date of 2021 for beginning to extract the rest (a target nobody ever saw as realistic).[27] Mitsubishi have built a robotic arm on rails that hinges like an accordion to make the attempt.[ei]

* * *

Scientists considered radioactive caesium, with its half-life of 30 years, to present the main health threat on land as much as in the ocean. Materials contaminated with caesium were expected to poison the ground for 300 years unless removed. Beginning in 2013, a small army of machines and masked workers combed the landscape, decontaminating everything in sight. They mowed gardens, collected fallen leaves in wooded areas within 20 metres of houses, cleaned roads, pavements (sidewalks) and building roofs and walls, emptied ditches, and removed over 16 million tons of topsoil before sealing it in huge black sacks for long-term storage.

---

26 This figure is based on an analysis of camera noise and the speed at which the camera sensor failed. Source: Yamaguchi, Mari. "Cleaner robot pulled from Fukushima reactor due to radiation." Phys.org, 9th February 2017: https://phys.org/news/2017-02-cleaner-robot-fukushima-reactor-due.html.

27 TEPCO released a video of the robot removing fuel from the reactor: https://www4.tepco.co.jp/en/news/library/archive-e.html?video_uuid=vy9uep38&catid=69631&fbclid=I-wAR0CLeXpa_jHXxf9XmsNmopxVeR_uzQOUNJGlSYo9WUrHOgnR83O-Py65GM.

The Ministry of the Environment estimated in March 2018 that decontaminating the ground would ultimately require digging up and burying 22 million cubic metres of soil and rubble across more than 9,000 square kilometres (3,475 square miles).[cj]

Environment Ministry officials have long promised to bury the most contaminated soil at a permanent facility by 2045 – and, indeed, people from Ōkuma and Futaba agreed to a nearby interim storage facility on the condition that it would be removed in the long term – but, unsurprisingly, nobody wants such a waste dump near where they live.[ck] What will happen remains to be seen, but removing the topsoil has proven to be very effective at reducing ground and air contamination, as 96% of hazardous particles lie within the top five centimetres of soil.[cl] Less harmful waste is burned in specially constructed incinerators. Unfortunately, however, forests, which cover 75% of the affected area, have not been cleaned because of the difficulty of the work, accessibility problems in rough terrain and the unfeasible cost (estimated at ¥16 trillion / $140 billion).[cm] The majority of trees are evergreen, and their leaves trapped most of the radioactivity in their upper reaches, but as time passes and these leaves occasionally fall, they re-contaminate the soil. Heavy rainfall during typhoon season also causes landslides, soil erosion and floods, spreading harmful particles into the environment via river networks and making the long-term effectiveness of decontamination difficult to judge.

Whole libraries of easily accessible information have been written on the accident decontamination efforts, so I will end this brief summary here.

vii. New Oversight

Japan faced mounting pressure to restructure its collusive regulatory bodies. Public trust in NISA especially plunged to rock bottom in late July 2011, when Chūbu Electric revealed that the agency had asked them to instruct their own employees to pose as curious citizens during an open public hearing back in August 2007 and to stage neutral or positive questions, speaking favourably about MOX fuel at the Hamaoka plant.[cn] The goal was not just to promote plutonium fuel but to fill seats that dissenters could otherwise have taken. Chūbu Electric planned to load the controversial fuel into

Unit 4 at Hamaoka at the time and NISA wanted to shift public opinion, even though that was not their remit.[co] Chūbu, to its credit, invited its employees to attend but, after drafting questions, reconsidered the ethical wisdom of voicing them at the hearing. Days after this revelation, a former high-ranking NISA official revealed he had done exactly the same thing with Shikoku Electric in 2010.[cp] Both episodes came hot on the heels of news that Kyushu Electric did something similar just a month earlier, when, of its own volition, the company instructed employees to email their opinions to a live televised debate on gauging public support for restarting Japan's idle reactors. The final tally came to 286 messages in favour, 163 against, but at least 130 of those in favour were traced back to Kyushu employees.[cq]

Kyushu's actions, though clumsy, were unsurprising; private companies conduct similar surreptitious self-promotion all over the world. Had METI made such requests – tasked, as it is, with promoting nuclear power – it could have been construed as skirting just within the boundaries of acceptable behaviour. But for them to come from NISA, the supposedly independent body tasked with dictating nuclear safety, was as unacceptable as it was perplexing. The government agreed and abolished NISA in June 2012, along with the 57-year-old Nuclear Safety Commission. In their place, a new Nuclear Regulation Authority (NRA) was formed in September within the Ministry of the Environment to combine the functions of those and a few other organisations to monitor nuclear safety, security and radiation releases. The NRA gained its own staff of 500 (though most came from NISA) and an annual budget of ¥50 billion ($600 million). In addition, the government created a new five-member Nuclear Safety Investigation Committee to ensure the NRA does its job. "I always think of the people who live their everyday life with anxiety under the effects of radiation," said the NRA's chairman, nuclear engineer Shunichi Tanaka. "I have determined that all my experience and knowledge, such as the experience of the JCO criticality accident, will be used to ensure nuclear safety in the new regulatory authority." People at first worried that – as a former general manager of JAERI's Tōkai Research Establishment, vice chairman of the JAEC and president of the Atomic Energy Society of Japan – though highly qualified, he was exactly the sort of nuclear village insider they *didn't* want overseeing the new organisation.

The NRA nevertheless set about reviewing NISA's data on potentially active fault lines (i.e. any which displayed evidence of paleoseismological movement within the last 130,000 years) beneath power plants. The agency had "discovered" that Tsuruga was sitting atop an active fault back in March 2012, despite the nearby and well-known Tsurugawan-Isewan Tectonic Line and earthquake occurring just three years before the plant began construction in 1966.[er] In this case, however, it was the Urasoko (sometimes Urazoko) fault, just 250 metres from Tsuruga Unit 2, and a bedrock fault called D-1 which raised concern as new research suggested both combined could cause a magnitude 7.4 quake: larger than previously realised.[es]

Japan Atomic Power Co had considered Urasoka inactive during their original surveys, mainly because they did not account for oceanic faults.[et] They investigated in the early 1980s after applying for a license for Unit 2 but decided the fault posed no threat, and then strongly denied it was active when applying for licenses for a third and fourth unit in 2004, even though the evidence by this time suggested otherwise.[eu] NISA, remarkably, forced them to conduct new surveys in 2008, which confirmed the fault's activity, yet still JAPC refuted the evidence and even called the new 2012 conclusion about D-1 "totally unacceptable."[ev] The NRA reconfirmed the problem in December 2012, then again in March 2015, leaving Tsuruga Unit 2 in limbo. Mihama, being on the same peninsula, is also considered under threat. Days later, an NRA panel determined that the country's newest plant, Higashidori, was also likely on an active fault.[ew] Unit 1 at Hokuriku Electric Power Co's Shika plant joined the list in March 2016.[ex] Another fault lies beneath Monju, though it was declared likely inactive in 2015.[ey]

Despite their vigour, the same old culture of collusion clung on at the NRA at first. Director General Tetsuo Nayuki, the NRA secretariat's third-highest-ranking official (responsible for safety regulations on earthquake and tsunami matters), admitted to leaking an internal document about the Tsuruga fault line to JAPC officials in January 2013 after meeting them in private multiple times.[ez, fa] He was fired and, over time, matters improved. The NRA introduced strict new regulations in July of that year, with provisions for "the impact of earthquakes, tsunamis and other external events such as volcanic eruptions, tornadoes and forest fires ... [in addition to] countermeasures for severe accident response against core damage, containment vessel damage and a diffusion of radioactive materials, enhanced measures for water injection into spent fuel pools" and a myriad of other precautions.[fb] The new rules took a firm position on building new emergency structures at nuclear plants, with Chairman Tanaka acknowledging "we are not taking into consideration how much it will cost atomic plant operators." Notwithstanding the NRA's difficult position – the pro-nuclear parties urging haste so they could return to business as usual, while those opposed demand a permanent end to nuclear power – the new organisation has, for the most part, done an admirable job of charting a strict but sensible course somewhere down the middle. It has made promising gains in hounding TEPCO and the other utilities to take safety seriously and now has broad power – unanswerable even to the prime minister – to do things NISA never could. In March 2020, for instance, it ordered the shutdown of Kyushu Electric's Sendai plant because the utility failed to construct its own seismic-isolation building, in the first incidence of its kind.[fc] If anything, some neutral observers suggest the NRA's efforts to appear impartial make it *too* strict in practice.

In addition to decommissioning all six reactors at Fukushima Daiichi, TEPCO confirmed in July 2019 that it would also decommission Fukushima Daini.[fd] Losing both sites was and is a devastating loss and enormous responsibility for the company, as the decommissioning is projected to take 44 years.[28] With other utilities reluctant to form partnerships because of the company's near-crippling financial obligations, TEPCO's hopes for a nuclear (and profitable) future rest with its prestige Kashiwazaki-Kariwa (KK) plant.[fe] Shrugging off government ownership and control is contingent, in part, on resuming operations there, but KK has not run at full capacity since the 2007 earthquake. Resuming operations, in turn, requires the consent of both the new NRA and Niigata governor Hirohiko Izumida. The utility applied for an NRA safety assessment of Units 6 and 7 in September 2013 without Izumida's permission (within days of vowing not to do so), much to the surprise of at least one NRA commissioner, who had expected TEPCO to deal with the problems at Daiichi before turning to KK.[ff, fg, fh] The first assessment failed after the NRA found TEPCO had neglected to report insufficient tremor resistance for its seismic-isolation building, despite knowing about the shortfall for three years.

---

28 This estimate seems incredibly unrealistic to me, I'd be surprised if it takes less than 50. That said, Japan will be world experts on decommissioning within another decade – a crown probably held by the Russians and Ukrainians at the moment because of Chernobyl – so it may be faster than that.

Chairman Tanaka continued his spree and, in 2014, rejected JNFL's new application for a post-2011 operating license for their enormous Rokkasho facility, citing unsatisfactory emergency preparations. He also banned the use of SPEEDI, saying it could not fully eliminate exposure risk, bringing a disappointing end to a promising technology. KK's original hoped restart date of 2017 passed, as did Tanaka's five-year term. There was a notable softening of the NRA's position towards restarting KK during his final weeks – and, indeed, toward even TEPCO's basic eligibility to continue operating the plant, which was not a foregone conclusion – after four years of safety screening, including from Tanaka himself. His fellow commissioner, Toyoshi Fuketa, a doctor of engineering and former senior JAERI and JAEA official, replaced him as chairman in late September 2017. Under Fuketa, the NRA granted its approval for TEPCO to restart Units 6 and 7 (the first BWRs to be approved, though Onagawa Unit 2 is expected to be the first to re-enter service) leaving just the Niigata governor's blessing standing in the way.[fi]

In office since 2004 and enjoying his third term, Governor Hirohiko Izumida was a staunch opponent of TEPCO, despite being a member of the pro-nuclear LDP and a former METI official. He consistently refused reactor restarts but in 2016 announced that he would not run for re-election, setting off a fight between pro- and anti-nuclear factions to elect their own candidate. The anti-nuclear group won by 6% (TEPCO's shares fell a corresponding 8% at the news), with an *NHK* exit poll indicating that 73% opposed restarting the plant. Niigata's new governor, left-leaning doctor and lawyer Ryūichi Yoneyama, continued to oppose TEPCO and the LDP until being forced to step down in April 2018, following revelations that he gave presents and money to women he'd met on dating websites. Though Yoneyama was single and therefore not having an affair, he was considered to have been paying the women for sex – unbecoming of someone in his position – so he resigned.[fj] Investors rejoiced. This time, amidst an election revolving around the plant, the LDP-backed Hideyo Hanazumi won. Opposing the stance of his own party, however, he followed in his predecessors' footsteps and has remained cautious of restarting operations. To appease Hanazumi, Kashiwazaki mayor Masahiro Sakurai and local residents, TEPCO have agreed to consider decommissioning and/or mothballing (to be decided) Units 1 to 5 within five years of restarting Units 6 and 7: an incredible concession.[fk]

Kashiwazaki-Kariwa itself gained a 15-metre (49-foot) seawall, watertight doors, protected battery houses and a multitude of other upgrades, including heavy earthquake reinforcement throughout. Stupid incidents have continued to undermine the positive changes, however, like the time plant staff caused a brief panic at the office of Mayor Sakurai following a magnitude 6.7 earthquake in June 2019 by checking the wrong box on a faxed form, thereby saying they were having problems with a reactor.[29, fl] TEPCO remains optimistic. "We have to fulfill the mission of bearing the responsibility for Fukushima," said KK's plant manager Toshimitsu Tamai, conscious of the widespread opposition to restarting the plant, in 2019. "And part of that is to generate the revenue necessary to pay for the decommissioning of the Fukushima plant."[fm] Restarting even a single reactor is expected to boost TEPCO's profits by over $800 million a year.[fn]

viii. The Road Ahead

The disaster changed Naoto Kan's opinion of nuclear power, a change he made no attempt to hide. Members of the nuclear village became so concerned that even the mayor of Futaba – the then-abandoned town beside Daiichi – visited Kan on 4th April 2011, along with seven other host-community mayors, to implore him *against* ending its use, despite everything going on at the time.[fo] But Kan had made up his mind. "Japan has always had risk," he said in May. "It is sitting on this island with lots of earthquakes, and now there are plants sitting on them. We're at a point where all politicians must check the old policy and decide whether we should go on – or change."[fp] He announced that Japan would abandon the 2010 target of 53% of electricity generated with nuclear power by 2050 and "start from scratch" on a new long-term plan, with greater emphasis on renewable energy.[fq] METI nevertheless intended to return to business as usual, with METI Minister Banri Kaeda promising the host communities that safety was assured. Kan then surprised everyone in July by announcing – without consulting his cabinet – a policy of stress-testing each reactor before they could be authorised to restart.

---

29 Yes, faxing remains a common legal requirement in Japan despite long ago being discarded in most industrial countries, causing some controversy during the coronavirus pandemic.

Many reactors were already offline following the earthquake. The last one halted operation a year later, on 6th May 2012, marking the first time since May 1970 – when Japan's only two reactors at Tsuruga and Tōkai both went offline for maintenance – that the electricity grid went without atomic energy.[fr] Almost six decades of successes and failures, progress and setbacks, all led to this moment: a window for another paradigm shift in energy policy. Though the utilities were now spending an extra $50 million a day on imported natural gas and found themselves under tremendous strain, polls indicated the public wanted nuclear power gone. Some conservative politicians, career METI bureaucrats and people working in the industry, on the other hand, did not. The remaining politicians in the middle held private reservations but in public still considered its continuation inevitable – a sunk cost. The big three newspapers took their expected sides, with editorials in the *Yomiuri* arguing against denuclearisation, while the *Asahi* and *Mainichi* advocated reform.[fs] All recognised the nuclear blackout as a turning point, but after a few weeks the spectre of a summertime energy shortfall and damage to the plants' local economies forced the government to restart KEPCO's Ōi Unit 3 on 1st July while hundreds protested at its gates.[ft, fu] Unit 4 followed three weeks later, yet neither reactor met all 30 of the government's new expanded stress-test safety requirements.[fv]

This was one example of the Democratic Party of Japan (DPJ) struggling to enact a coherent energy policy since Naoto Kan stepped down. His successor, Yoshihiko Noda, walked back a momentous plan to phase out nuclear power by 2040 in September 2012 a mere five days after its announcement, under intense pressure from the businesses and host communities that relied upon it.[fw, fx] Nuclear technology was one area where Japanese companies still held a competitive edge over most of their global rivals, having lost the lead in other sectors since the 1990s. The idea of wiping that out, thus driving electricity prices up and earnings and economic growth down, was viewed as unconscionable. Broad public opposition rose as high as 90% and the country now hosted its largest-ever anti-nuclear demonstrations, but Noda's plan dissatisfied this group, too, given its allowance for plants to continue operating for 30 years and, in some cases, beyond.[fy, fz]

Optimistic voters had elected the progressive DPJ in 2009 largely as a reaction to the conservative LDP's almost unbroken reign, but the challenge proved too great a task for a party inexperienced at ruling, and those same voters punished the DPJ at a general election just three years later. People felt dissatisfied for several reasons: the DPJ's apparent incompetence (with historically low legislative productivity), its bungling of the previous year's natural and nuclear disasters, its mishandling of the September 2010 Senkaku boat collision incident with China and a scandal of illegal political donations. Control returned to the LDP – the same party which fostered the culture of ineffective and impenetrable nuclear bureaucracy to begin with – by a comfortable margin. Its pro-nuclear leader, Shinzō Abe, in November 2019 became Japan's longest-serving prime minister since the fall of the Tokugawa Shogunate. He stepped down in September 2020 for health reasons.

* * *

Monju never ran again. The JAEA admitted in June 2012 that it had failed to inspect almost 10,000 components at the facility, followed by 2,300 more.[ga] The NRA rebuked them again in November 2013, this time for inadequate security precautions at the site, which stores plutonium.[gb] After another two years of mistakes, having now racked up a lifetime cost of over ¥1.1 trillion (over $10 billion), the NRA grew tired of the JAEA and declared it "not competent to operate [Monju]," giving MEXT six months to find a replacement.[gc, gd] The ministry could not. Though workers extracted the stuck refuelling device after partial disassembly of the reactor, the government announced in December 2016 its decision to decommission Monju, almost 21 years to the day since the accident that crippled it; a sad end to a pioneering facility. Plans to use FBR technology in a demonstration reactor by 2025 and then a full-scale 1,500-MW commercial design by 2050 died along with it.

I am reminded of comments by Admiral Hyman G. Rickover, father of the US Navy's nuclear fleet, in relation to USS *Seawolf*, the world's second nuclear submarine and the only example with a sodium-cooled reactor: "Expensive to build, complex to operate, susceptible to prolonged shutdown as a result of even minor malfunctions, and difficult and time-consuming to repair." After the bright early promise of infinite energy from recycled nuclear fuel that so captured the imagination back in the 1970s, only Russia operates commercial breeder reactors today, both at the Beloyarsk plant, near the city of Yekaterinburg. It is also the only nation to ever connect FBRs to the electricity grid for any length of time, but China and India are still looking at the technology.

The NRA and local officials gave the green light to restart Onagawa in 2020, after Tōhoku Electric installed a new 29-metre-high, 800-metre-long seawall, far larger than the one that saved it in 2011. The total cost of meeting the NRA's strict new requirements is estimated to be ¥340 billion ($3.2 billion), a cost which is passed to consumers.[ge] Completion of the upgrades has been postponed several times from its original date of April 2017, and the latest delay in May 2020 currently puts the date at March 2023.[gf] Similar work continues at every other plant. At Hamaoka, the Shizuoka Prefecture plant where another "Big One" – with estimated death tolls twice as high as the 2011 quake – remains overdue, Chubu Electric built a new 18-metre, 1.6-kilometre-long seawall at a cost of $1.3 billion.[gg] Nine PWRs have completed stress tests and been restarted as of early 2021, but Onagawa's will likely be the first BWR to resume operation.

* * *

Japan's nuclear industry has, disappointingly, failed to avoid further scandal unrelated to the 2011 disaster. In September 2019, a whistle-blower revealed that 20 top KEPCO officials had been accepting bribes for contracts since at least as far back as 1987 and were under investigation by tax authorities.[30] Eiji Moriyama, former deputy mayor of the small Fukui Prefecture town of Takahama, gave almost ¥360 million (around $3.4 million) worth of cash and gifts (including gold, expensive business suits and tickets to sumo wrestling matches) in dealings related to the utility's Takahama nuclear plant and later to construction company Yoshida Kaihatsu, for whom he served as an advisor.[gh] The number of recipient KEPCO officials ballooned to 75 after the utility's own investigation.[gi] Over a third of the total value went to then executive vice president and head of the company's nuclear division Hideki Toyomatsu, and payments continued until one month before Moriyama's death in March 2019 at age 90.

---

30 This scandal is not to be confused with a similar bribery scandal involving another KEPCO, the Korea Electric Power Company, in 2015.

Moriyama had been heavily involved with approving the construction of (and quelling resistance to) Takahama Units 3 and 4 and henceforth wielded tremendous influence, according to KEPCO. Though he retired from government work in 1987, he went on to work for KEPCO subsidiary Kanden Plant Corporation, in another example of *amakudari*. Six of the original 20 executives returned their bribes in February 2018 after tax authorities launched an investigation into Yoshida Kaihatsu. The rest claimed to have made similar attempts but failed because they found the elderly politician intimidating, so instead "temporarily kept valuables in their possession," according to KEPCO president Shigeki Iwane, until the scandal broke six months after his death, at which point they hastily returned most of it to his estate.[gj]

The tax investigation uncovered numerous cases where KEPCO had contracted Yoshida Kaihatsu without the usual contractor-bidding process, yet the executives claimed the bribes had not influenced their decision and vowed not to resign. After determining that the three main recipients somehow did not break any company rules, KEPCO dealt appropriately limp punishments, including a two-month 20% pay cut to Chairman Makoto Yagi, the same one-month pay cut to Shigeki Iwane and a "stern warning" to its managing executive officer.[gk, gl] Under intense public dissatisfaction, Yagi capitulated in early October and resigned. Iwane followed soon after. Vice President Takashi Morimoto took his place in March 2020 while the former head of Keidanren, Sadayuki Sakakibara, became chairman; KEPCO promised to create new systems of oversight for its executives.[gm]

The NRA hasn't been as squeaky clean as proponents would have liked either. In January 2020, for instance, a *Mainichi Shimbun* investigation revealed a December 2018 private meeting in which NRA officials (including Fuketa) decided on countermeasures against volcanic ash for KEPCO's nuclear plants behind closed doors and without taking minutes of what was said, in contravention of the Public Records and Archives Management Act, which requires the organisation to make all decisions in public hearings.[gn] Days later, all five commissioners publicly agreed on the actions they had secretly already decided, as if discussing them for the first time. This was a minor incident in the grand scheme of things, but it raises concerns about a slippery slope.

In February 2020, the NRA accused JAPC of altering geological survey data in an apparent scheme to trick the NRA into believing the faults beneath Tsuruga Unit 2 were inactive. The authority's inspectors found that 80 areas of the original data had been deleted and rewritten in newly submitted documents.[go] JAPC apologised in June, insisting the changes were unintentional, but the matter is still unresolved.

* * *

So what does the future hold, aside from a saddening yet apparently inevitable penchant for corruption? Germany, which received around a quarter of its power from nuclear plants in 2010 despite a longstanding ambivalence towards the technology, announced its intention to phase out nuclear power within months of the Fukushima disaster (as, since, have Switzerland and South Korea). A change in 2015 to accounting methods of Japan's Electricity Business Act allowed utilities to calculate decommissioning costs in ten-year slices instead of one lump sum, spreading the cost and therefore encouraging the closure of old reactors. It had the intended effect: within days, KEPCO announced it would decommission Mihama Units 1 and 2. JAPC said the same for Tsuruga Unit 1, ending the service lives of the nation's three oldest reactors. Each had received extensions beyond their original 40-year lifespan but making expensive safety upgrades to such ancient equipment did not make financial sense. When including Monju, this made four of the six remaining reactors closed on Fukui Prefecture's nuclear peninsula in a short space of time (Fugen ceased operations in 2003). Zooming out, KEPCO announced in December 2017 that another Fukui plant, Ōi, would decommission two of its four reactors. Those closures may have been a victory for campaigners, but it was an enormous loss for the old towns of Ōi and Mihama (not to mention the wider prefecture's economy), which completely depend on those plants. As of early 2021, 21 of Japan's 54 commercial reactors have been confirmed for decommissioning since the disaster (including Fukushima Daiichi).

JAEA announced another victim of uneconomical safety upgrades in 2014: the Tōkai Reprocessing Facility.[gp] The facility had repro-cessed a total of 1,140 tons of spent fuel during its lifetime and will continue to use the materials already there until 2028, when it will be decommissioned. This is expected to take 70 years and cost ¥1 trillion ($9.2 billion).[29] Tōkai's replacement at Rokkasho – though long partially operational, with uranium enrichment, low-level waste disposal and waste storage facilities – is *still* not operating almost 30 years after construction began because of its own upgrades to withstand a 9.0 earthquake, among other things. It is now expected to be finished in mid-2022 at a cost of ¥3 trillion (almost $28 billion), four times its original price tag, with a total projected lifetime cost of ¥14 trillion ($130.4 billion).[gr] Rokkasho reached its spent fuel storage capacity of 3,000 tons in 2014.[gs] Since then, further fuel has been sent to a new dry storage site joint-owned by TEPCO and JAPC at Mutsu, which is considerably cheaper and safer than leaving it in pools.[31]

The problem of what to do with the high-level waste won't go away, however, as Rokkasho is not a permanent geological disposal site, but a location has not been found after a twenty-year search. METI produced a list of possible sites and is currently trying to convince a municipality to host one, with the government conducting preliminary surveys to construct a site in Hokkaido Prefecture beginning in November 2020, but public opposition is high. If and when they eventually succeed, the volume-reduction justification for reprocessing fuel may be questioned once again. The benefits of that reduction partially justified the high cost, not to mention the health and proliferation dangers, but a team of eight experts on the International Critical Review Committee Panel (ICRC) concluded back in 2005 that this argument makes no sense. Reducing the volume does not reduce the required space to store it, as the size of a fuel repository depends on the amount of heat produced by the fuel, not its volume. Spent reprocessed fuel produces up to almost ten times as much decay heat – and four to five times more radiation – than regular spent fuel, making the space savings almost irrelevant.[gt] The argument also ignores the volume of low- and intermediate-level long-life waste produced by the reprocessing, making it no better than ordinary spent fuel glass vitrification. The ICRC hand-delivered their report – which also argued that Japan already possessed enough plutonium to satisfy MOX requirements for decades to come – to the JAEC in 2005, to no avail.

---

31 The site is operated by the Recyclable-Fuel Storage Company, owned in a 80%/20% split. It was finished in August 2013 and offers a further 3,000 tons of storage, with plans for more.

Even when Rokkasho goes fully online, the facility will no longer have Monju for its primary source and user of plutonium, which was part of the premise for Rokkasho's existence, and with most nuclear plants currently inactive, its MOX fuel plant won't be operating anywhere near capacity (i.e. at its most economical). Toshiba and a group of French companies including Électricité de France (EDF) and Areva were collaborating on a new type of experimental fast breeder reactor called the Advanced Sodium Technological Reactor for Industrial Demonstration (ASTRID), a successor to the Superphénix, but after experiencing a 13-year delay within 10 years, the two countries cancelled the program in September 2019.[gu, gv] With these setbacks, the dream of carbon-free, recyclable and relatively cheap energy handling the bulk of Japanese requirements – still considered realistic twenty years ago – seems all but impossible now.

Or is it? Since 2011, most world leaders have finally accepted what the entire scientific community has been screaming about for decades: that our world is dying, and eliminating emissions is a key component of saving it. With this lethargic revelation comes the equally ironic realisation that, after decades of campaigns against nuclear power for environmental reasons, it is now seen in some industrialised nations as a vital tool in the fight against climate change. Renewable energy is another. Renewables have come a long way since 2011, with wind turbines in Scotland – my home country – generating close to twice the power needed for all its homes during the first half of 2019. This only covers domestic use and does not address the base load for cold, calm winter conditions, but it's nonetheless an impressive achievement, given that wind and solar only accounted for around 3% of British electricity in 2010.[32] Further afield, European renewables generated more energy than fossil fuels for the first time ever during the first half of 2020, soon before the International Energy Agency announced that the cost per megawatt to build solar plants was now cheaper than fossil fuels.[33] The technology works, then, at least for domestic and business purposes, but decades of investment will be required to handle everything in most industrial nations.

---

32 The numbers in Britain as a whole are so good, in fact, that officials are concerned about the frequency with which the country's excessive quantity of wind turbines overwhelm its national grid with all the power they're producing, forcing enormous government-funded "constraint payments" of almost £130 million in 2019 to switch many of them off.

33 This is heavily caveated and depends on generous financing but is nonetheless a remarkable achievement.

The Japan Wind Energy Association, admittedly somewhat biased, expects around 10 GW of wind capacity to be added across the islands by 2030, although vendors are hindered by the mostly remote and mountainous terrain.[gw] To circumvent this age-old problem of scant development space, the government passed a new law in 2019 to permit offshore turbines to operate for up to 30 years.[gx] Onshore, Fukushima Prefecture plays a leading role and intends to ditch coal and source 100% of its energy from renewables by 2040, putting it ahead of all other regions. Ten wind and eleven solar plants are planned for remote mountain areas and contaminated ex-farmland across the prefecture at a cost of ¥300 billion ($2.75 billion), with the power sent to Tokyo.[gy] In a perfect illustration of the capacity problems inherent to renewable energy, however, these 21 power plants will have a combined perfect-conditions output of only 600 MW – far below that of a single reactor. Japan's solar industry, which enjoyed explosive growth in 2013 with four times the 2012 figures, has seen newly installed capacity decline each year since 2015.[gz] (Japan still ranked fourth in the world for new builds in 2019.) Suitable hydroelectricity sites, meanwhile, have long been exploited to their maximum potential by over 3,000 dams.

Japan was falling far short of its obligations as a signatory of the 2016 Paris Climate Agreement until October 2020 when Shinzō Abe's successor Yoshihide Suga announced a national target of zero emissions by 2050. Just as it did in Germany, coal power has thus far replaced a large chunk of Japan's nuclear deficit to make it the world's third-largest coal importer after India and China, with plans to add a potential further 20 GW of coal plants, depending on how many reactors stay offline. [ha, hb] A 2019 study by Columbia University scientists estimated that replacing nuclear with fossil fuels in Japan had already caused up to 23,300 unnecessary deaths from air pollution by 2017.[hc] This is a depressing state of affairs, given the nation had all but quit the dirty fuel back in 1980.[34]

All this means Japan now finds itself in a difficult position. It still imports well over 90% of its fuel for electricity, despite dedicating the past 50 years to avoiding such a position.[hd] Its fleet of dormant reactors, which are stuck in limbo but capable of generating clean, almost self-sufficient electricity – not to mention an at-long-last almost complete state-of-the-art reprocessing facility to recycle

---

[34] In Japan's defence on this point, the country is promoting ultra-supercritical coal plants, which are as green as they come, but even they are still far less environmentally friendly than the alternatives.

the fuel – face overwhelming public opposition to their use. There are also ongoing concerns not just about the country's 19,000 tons of spent fuel, but also the volume of plutonium and the small but growing number of spent MOX fuel rods, which cannot be reprocessed a second time.[35] This spent recycled fuel, as we have discussed, generates several times the heat of normal used fuel and is therefore more dangerous in the event of a blackout. None of this stopped KEPCO from signing a new production contract with Orano in February 2020 for even more MOX fuel.[he]

Nor has it stopped a core group of METI officials from trying to revive the fast breeder program. Mitsubishi's Japan Sodium-cooled Fast Reactor (JSFR) – a 1,500-MW, fourth-generation design – will likely be built at some point in the future, probably with JAEA at the helm. In addition, just three months after cancelling the ASTRID program, French nuclear company Framatome and the government-funded Alternative Energies and Atomic Energy Commission signed a cooperation agreement with the JAEA and Mitsubishi.[hf] The new group plans to combine resources to produce the next generation of fast neutron reactors, which are similar to breeders but do not produce more fuel than they consume.

The utilities have, as of early 2020, applied to restart 27 reactors. The NRA has approved 15, but only nine are currently online and many of the rest may never be used again. Construction work on several reactors, which all halted in 2011, has since resumed. This includes the remote Ōma NPP at the northernmost tip of Honshū, which will be Japan's first new plant in 20 years once it is commissioned (expected after 2025), at the site where Fugen's successor was originally hoped to be built.[hg] Ōma, built by the government- and industry-owned Electric Power Development Co, uses an ABWR similar to Kashiwazaki-Kariwa Unit 6 and can accept a 100% MOX fuel load.

The overall outlook is less definitive than it was in the summer of 2012, when the industry seemed unlikely to ever recover, but it remains to be seen if the government's newest target of nuclear power providing 20–22% of the country's electricity by 2030 comes to fruition, or if any brand-new plants will ever be built.[36] METI

---

35 Currently. The Japanese government is funding research into reusing MOX, but so far there has been no luck in developing commercial solutions.

36 With the possible exception of the new units at the Higashidōri NPP, where construction had already begun before the 2011 disaster. Work will likely resume there by 2025.

has ruled out building entirely new reactors for the next decade, at least.[hh] More reactors will likely come back online during that time (although reaching 20% is doubtful) before nuclear capacity ultimately declines over the subsequent two, because public opposition is too great to overcome and renewables are gaining enormous momentum worldwide.

When combining wind and solar technologies with advancements in energy storage and the government's enquiries into the use of hydrogen power, there is hope for a clean future. There remains the matter of power density – renewables don't scale-up nearly as well as nuclear – but it's a start. And, as always, Japan has shown remarkable determination to use nuclear power, so predictions are difficult to make. One welcome change is the liberalisation of the power retail market: after a sluggish start back in 1995, April 2016 finally saw total deregulation, with small, low-voltage users (including households) able to choose their own supplier for the first time.[hi] Over a million people switched within two months. This has hopefully lowered extreme-temperature mortality caused by post-disaster electricity costing up to 40% more, estimated to be "at least an additional 1,280 deaths during 2011-2014."[hj]

For a final comparison, Japan's energy mix in 2020 – the most up-to-date numbers at the time of writing – were as follows: oil 37%, coal 28%, LNG 24.3%, biofuels 3.9%, 1.7% for hydro, and renewables such as wind and solar at 2.6%. Nuclear power sat at 2.5% (down from 4% in 2019; not so dissimilar from the energy mix back around 1970).[hk] Internationally, nuclear power generation in 2019 almost beat 2006's record output.

Rarely have disasters come without warning. Challenger, Deepwater Horizon, Bhopal, Chernobyl – just a few among countless examples where experts tried to prevent the preventable but were dismissed by authority figures, often because the looming problems would be costly and time-consuming to fix. The rise and fall of Japan's nuclear industry, as we have seen, is littered with many comparable instances of safety overlooked for the sake of money, speed and, ultimately, national security. Indeed, without discounting the Herculean efforts of the men and women at Daiichi, not to mention the remarkable resilience and high engineering standards of the technology, a worrying amount of luck with timing prevented the Fukushima disaster from being far worse than it was.

Had it occurred at night or over a national holiday, when plants operate with a skeleton crew, getting everyone back would have taken hours or even days, and most may well not have made it. Had it occurred during a storm such as Typhoon Hagibis, which battered Japan in October 2019 with winds of up to 260 kilometres per hour (160mph), venturing outside to perform any kind of recovery work would have been impossible. Had it occurred at *any other time of year* from winter 2010 to autumn 2011, a weather analysis of these periods predicted that up to 85% of radioactive particles like caesium-137 – as opposed to the 22% that played out in March 2011 – would have settled on land around the world, instead of the Arctic and Pacific Ocean where most ended up.[hl] Had it occurred several decades earlier, the shaking would have far exceeded the original design tolerances and safety systems, and the plant may well have been crippled before the tsunami reached it. Had it occurred even one year earlier, before the fire protection lines and seismic-isolation building were installed, operators would've had no way to inject water and nowhere to shelter and organise. Shimizu agreed, commenting during his June 2012 government testimony that "it is frightening to think what would have happened if TEPCO did not have [the SIB]."[hm] On the other hand, bad luck clearly played a fundamental role, with the tsunami – itself prompted by one of the largest earthquakes in recorded history – peaking at Fukushima Daiichi. Neither would have compromised nuclear plants in other, less geologically volatile countries. But who's to say that other "unforeseen" events won't happen elsewhere? After all, if regulators accounted for every possibility, nothing would get built... Maybe that's the lesson: always account for luck. Regardless, with more and more reactors receiving approval to be restarted, I hope enough people with enough influence have learned enough to make a difference.

November 2004

March 2011

March 2013

March 2020

EPILOGUE

Yasuhiro Nakasone, the Imperial Army lieutenant who first convinced the Diet to pass a nuclear power funding bill in 1954, passed away on 29th November 2019 at the remarkable age of 101, having remained influential and in the public eye for the rest of his life.

Yasuzaemon Matsunaga, the titan of early electric power, founded the Industry Planning Council in 1956 to tackle national policy issues. As president, he was involved in proposals to improve the nation's drinking water systems, the creation of several expressways and what became the Tokyo-Narita International Airport, among many other projects. He took up tea-making around age 60, eventually becoming well known as a master tea-maker and collector of antiquities. Some of his collection is still on display at the Fukuoka Art Museum. He died on 16th June 1971, aged 95, having led a remarkable life. With Matsunaga never a fan of ceremony, his family held no funeral in accordance with his wishes.

Naoto Kan lost his Diet seat in Tokyo's 18th district in the 2012 general election but retained a seat through the proportional representation system. He won back his district seat in the 2017 election and also became an advisor to Japan's Technical Committee on Renewable Energy. His party, the DPJ, collapsed in 2016, but many of its members, including Kan, are now part of the new Constitutional Democratic Party of Japan, led by Yukio Edano.

Daiichi's resourceful shift manager from the Units 1/2 control room, Ikuo Izawa, reflected on the disaster in 2014. "The impact of the tsunami was totally bigger than what we expected, trained, prepared for, or believed was possible – it was unimaginable. We must always be prepared for the possibility that something much bigger can happen."[hn] He still works for TEPCO.

Fukushima Daini's manager, Naohiro Masuda, stayed at his plant to supervise its decommissioning for several years until he took over decommissioning operations at Daiichi in April 2014. There, he helped to improve TEPCO's poor reputation for dealing with labourers – among many other things – until October 2018, when he joined Japan Nuclear Fuel Limited as an advisor. He became its president and CEO four months later.[ho]

Kazuhiko Maekawa, the tireless physician who tried to save Hisashi Ouchi and Masato Shinohara in 1999, still made a daily commute to work at Tokyo's Harada Hospital as of late 2020, aged 80. Having also attended the sarin gas attacks in 1995 and Fukushima Daiichi as an emergency radiation doctor, he remains respected as a world-leader in his field. As of early 2021 he was retired but still chaired the non-profit Disaster Humanitarian Medical Support Association (HuMa), which provides free medical care to disaster victims.

Yutaka Yokokawa, the third criticality victim, is still alive, but he refuses to talk about the accident. He gave an interview to the *Mainichi* in 2005 wherein he expressed overwhelming sorrow and loneliness over the loss of his colleagues. He had retired by this time but mowed the lawn at JCO on weekdays and still made bimonthly visits to Ouchi and Shinohara's graves. The last words he spoke to Ouchi were "Do your best," as they lay in adjacent hospital beds at NIRS.

IMAGE SOURCES

Maps & Diagrams

p. iv-v: Base Japan map by freevectormaps.com
Modified by A. Leatherbarrow.

p. vi: A. Leatherbarrow

p. vii: US Government (Public Domain). Modified by A. Leatherbarrow

p. viii-ix: US Government (Public Domain). https://commons.wikimedia.org/
wiki/File:Mark_I_Containment.svg.

p. x-xi: Government of Japan (Public Domain). Modified by A. Leatherbarrow

Images

B Cover   Source: Tokyo Electric Power Company Holdings

p. 56-57   US Government, Department of Energy.
https://www.flickr.com/photos/departmentofenergy/11824391473/

p. 68-69:   US Government, Department of Energy.
https://www.flickr.com/photos/departmentofenergy/11840039684/

p. 74:   US Government, Department of Energy.
https://www.flickr.com/photos/departmentofenergy/11840035586/
https://www.flickr.com/photos/departmentofenergy/11840035476/

p. 75 (t):   Copyright © National Land Image Information (Color Aerial
Photographs), Ministry of Land, Infrastructure, Transport and
Tourism, Government of Japan.

p. 75 (b):   International Atomic Energy Agency (Creactive Commons BY-SA
2.0). https://www.flickr.com/photos/iaea_imagebank/8388173135/

p. 76-77:   US Government, Department of Energy.
https://www.flickr.com/photos/departmentofenergy/11840034056/

p. 78:   US Government, department of Energy.
https://www.flickr.com/photos/departmentofenergy/11840034116/

p. 90-91:   US Government, department of Energy.
https://www.flickr.com/photos/departmentofenergy/11839699333/

p. 92-93:   US Government, department of Energy.
https://www.flickr.com/photos/departmentofenergy/11839700023/

p. 166-167:   International Atomic Energy (Creactive Commons BY-SA 2.0).
https://www.flickr.com/photos/iaea_imagebank/8389261826/

p. 242, 249, 257, 268, 275, 282, 283, 294, 295, 296, 297:
Source: Tokyo Electric Power Company Holdings.
https://photo.tepco.co.jp/en/index-e.html

p. 360-361: Satellite images ©2022 Google

REFERENCES AND NOTES

I wanted to include a brief note in regards to this book's "made in Japan" opening quote by NAIIC chairman Kiyoshi Kurokawa, whose remarks were met with some controversy upon their release. This was a quote written to introduce the English translation of his commission's report and was not included in the original Japanese version, though the report itself was the same. It is not difficult to imagine why this specific comment upset so many people. Many argued – not unreasonably – that he was taking the easy way out by blaming cultural stereotypes and avoiding pointing fingers at specific individuals, or that those very same sterotypes, such as rigid deference to authority, are common in other cultures. Kurokawa defended his words, however, saying that he was trying to explain to a global audience how the amalgamation of certain qualities of Japan, such as the near-single party rule and the practice of handing bureaucrats plush high-paying private jobs upon their 'retirement,' meant that the accident *was* uniquely Japanese. His critics made some good points, but in light of everything contained within this book, I find it difficult to argue that he was wrong.

The following list includes many archived links because the original pages are no longer available. Thank goodness for the Internet Archive! I tried to remember to archive as many live links as I could during my research, in the hopes that little would be lost to time. I have deliberately prioritised English-language references as much as possible so non-Japanese speakers can learn more for themselves.

PREFACE

a Moret, Leuren, "Japan's deadly game of nuclear roulette." *Japan Times*, 23 May 2004. https://www.japantimes.co.jp/life/2004/05/23/to-be-sorted/japans-deadly-game-of-nuclear-roulette/. I am aware that this article was written by a crazy conspiracy theorist (albeit a highly qualified and experienced one), but I was struck by such an article and opener being in the *Japan Times*.

b Noriko Behling, Mark C. Williams, and Shunsuke Managi, "Regulating Japan's nuclear power industry to achieve zero accidents." *Energy Policy*, vol. 127, pp. 308-319, April 2019.

c Karan, Pradyumna Prasad, *Nihon No Toshi (The Japanese City)*, p. 181. University Press of Kentucky, 2005.

d "How much fuel does it take to power the world?" *Forbes*, 20 September 2017. https://www.forbes.com/sites/startswithabang/2017/09/20/how-much-fuel-does-it-take-to-power-the-world/

CHAPTER 1

a Hawkes, Francis L. *Narrative of the Expedition of an American Squadron to The China Seas and Japan*. Washington: AOP Nicholson, Vol. 1, 1856, p. 231.

b Fujimoto, Masaru, "Black ships of 'shock and awe.'" *Japan Times*, 1 June 2003. https://www.japantimes.co.jp/community/2003/06/01/general/black-ships-of-shock-and-awe/

c   Hawkes, Francis L. *Narrative of the Expedition of an American Squadron to The China Seas and Japan*. Washington: AOP Nicholson, Vol. 1, 1856, p. 252.

d   "Letters from US president Millard Fillmore and US navy commodore Matthew C. Perry to the emperor of Japan (1852–1853)." Asia for Educators, n.d. http://afe.easia.columbia.edu/ps/japan/fillmore_perry_letters.pdf

e   Miyauchi, D. Y., "Yokoi Shōnan's response to the foreign intervention in late Tokugawa Japan, 1853–1862." *Modern Asian Studies*, vol. 4, no. 3, 1970. https://www.jstor.org/stable/311496

f   Beasley, W. G., *The Meiji Restoration*, p. 90. Stanford University Press, 1972.

g   *Narrative of the Expedition of an American Squadron to The China Seas and Japan*, p. 350.

h   Schroeder, John H., *Matthew Calbraith Perry: Antebellum Sailor and Diplomat*, p. 288. Naval Institute Press, 2001.

i   Ohno, Kenichi, "Chapter 5: The industrialization and global integration of Meiji Japan." In *East Asian Growth and Japanese Aid Strategy: Collected Essays*. National Graduate Institute for Policy Studies, 2003. http://www.grips.ac.jp/forum-e/pdf_e01/eastasia/ch5.pdf

j   Pyle, Kenneth, *Japan Rising: The Resurgence of Japanese Power and Purpose*. Hachette UK, 2009.

k   Yang, Daqing, *Technology of Empire: Telecommunications and Japanese Expansion in Asia*, p. 23. Brill, 2011.

l   Norman, Herbert E., *Japan's Emergence as a Modern State: Political and Economic Problems of the Meiji Period*. 60th anniversary edition, p. 121. UBC Press, 2011.

m   Campbell, Allen, and Noble, David S., *Japan: An Illustrated Encyclopedia*, p. 88. Kodansha International, 1993.

n   Toshiba, "The wizard of invention and the warrior of innovation," 2006. https://emea.dynabook.com/Contents/Toshiba_teg/EU/Others/Generic/visions_02_06/pdf/visions_06_02_02_en.pdf

o   Toshiba, "Chronology of history," n.d. https://www.toshiba.co.jp/worldwide/about/history_chronology.html

p   Toshiba Science Museum, "Ichisuke Fujioka: A wizard with electricity." https://toshiba-mirai-kagakukan.jp/en/learn/history/toshiba_history/spirit/ichisuke_fujioka/p2_1_3.htm

q   Yamamoto, Mitsuyoshi, and Yamaguchi, Mitsugi, "Electric power in Japan: Rapid electrification a century ago." *Institute of Electrical and Electronics Engineers (IEEE) Power & Energy* magazine, vol. 3, issue 2, March–April 2005. https://ieeexplore.ieee.org/document/1405877. (Note: This article has quite a lot of specific technical information about the early power systems in Japan.)

r   US Bureau of Foreign and Domestic Commerce, *Trade Information Bulletin*, issues 1–78, January 1922, p. 9.

s   Joint Committee on Japanese Studies, "Japanese industrialization and its social consequences," p. 302.

t   "On the front line of fire prevention technology." *Newsweek International*, 19 July 2019. https://www.pressreader.com/uk/newsweek-international/20190719/281797105547957

u   K. Okude, et al., "History of Japan's first commercial hydroelectric generation at Keage power station." *2015 ICOHTEC/IEEE International History of High-Technologies and Their Socio-Cultural Contexts Conference (HISTELCON)*, Tel Aviv, 2015, pp. 1-5. https://ethw.org/Milestones:Keage_Power_Station:_The_Japan%E2%80%99s_First_Commercial_Hydroelectric_Plant,_1890-1897

v   "Tours provide renewed interest in Japan's oldest hydro power plant." *Japan Times*, 16 November 2015. https://www.japantimes.co.jp/news/2015/11/16/national/tours-provide-renewed-interest-japans-oldest-hydro-power-plant/

w   "Japan's Birthplace of Hydroelectricity." Japan Travel, 2019. https://en.japantravel.com/miyagi/japan-s-birthplace-of-hydroelectricity/25905

"Japan's first hydro plant wins US award for 125-year history." *Japan Times*, 2016. https://www.japantimes.co.jp/news/2016/09/13/national/japans-first-hydro-plant-wins-u-s-award-125-year-history/

x   Kikkawa, Takeo, "The history of Japan's electric power companies before World War II." *Hitotsubashi Journal of Commerce and Management*, vol. 46, no. 1, pp. 1-16, October 2012. https://www.jstor.org/stable/43295037

y   Owen, Edward L. "The origins of 60-Hz as a power frequency." *IEEE Industry Applications* magazine, vol. 3, issue 6, November–December 1997. https://ieeexplore.ieee.org/document/628099

z   Low, Morris, Nakayama, Shigeru, and Yoshioka, Hitoshi, "Chapter 4: Technology versus commercial feasibility: Nuclear power and electric utilities." In *Science, Technology and Society in Contemporary Japan*. Cambridge University Press, 1999.

aa Kikkawa, "The history of Japan's electric power companies."

ab Gilbert, Richard J., and Kahn, Edward P. (eds.), *International Comparisons of Electricity Regulation*, p. 236. Cambridge University Press, 2006.

ac Owen, "Origins of 60-Hz."

ad Scalise, Paul J., "Who controls whom? Constraints, challenges, and rival policy images in Japan's post-war energy restructuring." *Critical Issues in Contemporary Japan*, 8 December 2013. https://www.academia.edu/4727471/Who_Controls_Whom_Constraints_Challenges_and_Rival_Policy_Images_in_Japans_Post-War_Energy_Restructuring

ae Minami, Ryoshin, "The introduction of electric power and its impact on the manufacturing industries: With special references to smaller scale plants." Economic Growth Center, Yale University, 1974.

af

ag "Demon of power, Yasuzaemon Matsunaga" (in Japanese). *Mita Review, Keio Journal*, June 2012. http://www.keio-up.co.jp/mita/r-shiseki/s1206_1.html

ah Kikkawa, Takeo, "The role of Matsunaga Yasuzaemon in the development of Japan's electric power industry." *Social Science Japan Journal*, vol. 9, issue 2, 18 October 2006.

ai Minami, "The introduction of electric power."

aj Federation of Power Companies, "History of electricity (Japanese electricity business and society): From Taisho to Showa (1912–1945)" (in Japanese), n.d. http://www.fepc.or.jp/enterprise/rekishi/taishou/index.html

ak Xia, Chenxiao, "Electrifying Kyoto: Business and politics in light and power, 1887–1915." *Enterprise & Society*, vol. 18, issue 4, pp. 952-970, 2017.

al Minami, "The introduction of electric power."

am Kikkawa, Takeo, "Management and regulation of the electric power industry (1923–1935)." *Japanese Yearbook on Business History*, vol. 3, pp. 82-102, 1987. https://www.jstage.jst.go.jp/article/jrbh1984/3/0/3_0_82/_article/-char/en

an Kikkawa, Takeo, "The relations between the Mitsui Bank and two electric power companies, Tokyo Electric Light Co. and Toho Electric Power Co." *Keiei Shigaku (Japan Business History Review)*, vol. 17, issue 2, pp. 23-46, 1982. https://www.jstage.jst.go.jp/article/bhsj1966/17/2/17_2_23/_article/-char/en

ao Kikkawa, "History of Japan's Electric Power Companies."

ap Kikkawa, "Role of Matsunaga Yasuzaemon."

aq Kikkawa, Takeo, "Chapter II: The government-industry relationship in Japan: What the history of the electric power industry teaches us." In Anne Holzhausen (ed.), *Can Japan Globalize? Studies on Japan's Changing Political Economy and the Process of Globalization in Honour of Sung-Jo Park*, p. 30. Springer, 2013.

ar Samuels, Richard J., *The Business of the Japanese State: Energy Markets in Comparative and Historical Perspective*. Cornell University Press, 1987.

as Nester, William R., *Japanese Industrial Targeting: The Neomercantilist Path to Economic Superpower*, p. 125. Palgrave Macmillan UK, 1991.

at Kikkawa, "History of Japan's Electric Power Companies."

au Nester, *Japanese Industrial Targeting*, p. 127.

av Samuels, *Business of the Japanese State.*

aw "Electric power nationalized in Japan." *Far Eastern Survey*, vol. 7, no. 12, pp. 139-140, 1938. www.jstor.org/stable/3022025. I've also seen numbers just under half written more recently. Either way, it was a lot.

ax Ibid.

ay Pauer, Erich, *Japan's War Economy*, p. 35. Routledge, 2002.

az Kikkawa, "Role of Matsunaga Yasuzaemon."

ba Bloch, Kurt, "Coal and power shortage in Japan." *Far Eastern Survey*, vol. 9, no. 4, pp. 39-45, 1940. www.jstor.org/stable/3022085.

bb Morgenstern, George, *Perpetual War for Perpetual Peace*. "Chapter 6: The actual road to Pearl Harbor." Caxton Printers, 1953.

bc Yoshida, Phyllis, "Japan's energy conundrum: A discussion of Japan's energy circumstances and US-Japan energy relations." Sasakawa Peace Foundation, 2018. https://spfusa.org/wp-content/uploads/2018/06/JapanEnergyConundrum_SPFUSA.pdf

bd Tomaya, Yoshihiro, "An historical overview of the electric, rail, and communication utilities in Japan." *Otemon Economic Studies*, no. 31, 1998. Xia, Chenxiao. "Business, Fascism, War: Electricity in Germany and Japan, 1931-1945."

be Low et al., *Science, Technology and Society in Contemporary Japan*, p. 68.

bf Kikkawa, "History of Japan's Electric Power Companies."

bg Luft, Gal, and Korin, Anne (eds.), *Energy Security Challenges for the 21st Century: A Reference Handbook*. ABC-CLIO, 2009.

bh Dower, John W., "The bombed: Hiroshimas and Nagasakis in Japanese memory." *Diplomatic History*, vol. 19, no. 2, pp. 275-295, 1995. www.jstor.org/stable/24912296

bi Atomic Heritage Foundation, "Survivors of Hiroshima and Nagasaki," 2017. https://www.atomicheritage.org/history/survivors-hiroshima-and-nagasaki

bj Warren, Stafford L., "Conclusions: Tests proved irresistible spread of radioactivity." *Life* magazine, p. 86, 11 August 1947.

bk Ibid., p. 88.

bl Ibid.

bm Yamamura, Kōzō, *Economic Policy in Postwar Japan: Growth versus Economic Democracy*, vol. 1, p. 227. University of California Press, 1967.

bn Hoshi, Takeo, and Kashyap, Anil, *Corporate Financing and Governance in Japan: The Road to the Future*, p. 70. MIT Press, 2004.

bo Drysdale, Peter, and Gower, Luke (eds.), *The Japanese Economy*, vol. 5, p. 58. Routledge, 1999.

bp Gilbert and Kahn (eds.), *International Comparisons of Electricity Regulation*, p. 241.

bq Ibid.

br Takao, Yasuo, *Japan's Environmental Politics and Governance: From Trading Nation to EcoNation*, p. 78. Routledge, 2016.

bs Sumiya, Mikio (ed.), *A History of Japanese Trade and Industry Policy*, p. 202 (summary volume). Oxford University Press, 2000.

bt Samuels, Richard J., *The Politics of Regional Policy in Japan: Localities Incorporated?* p. 170. Princeton University Press, 1983. I have also seen this body called the Public Works Committee.

bu Federation of Electric Power Companies, "About us," n.d. http://www.fepc.or.jp/english/about_us/index.html

bv "Will the means of transportation be revolutionized?" (in Japanese). *Asahi Shimbun*, 16 August 1945.

bw Freedman. Lawrence, *The Evolution of Nuclear Strategy*, p. 78. Macmillan, 1981.

bx Osgood, Kenneth, *Total Cold War: Eisenhower's Secret Propaganda Battle at Home and Abroad*. University Press of Kansas, 2006.

by Broinowski, Adam, *Cultural Responses to Occupation in Japan: The Performing Body during and after the Cold War*, p. 38. Bloomsbury Academic, 2016.

bz Tanaka, Yuki, "Nuclear nation: Japan's unlikely love affair with atomic energy." *Foreign Policy*, 22 March 2011. https://foreignpolicy.com/2011/03/22/nuclear-nation/

ca Taketani, Mituo, "The direction of atomic energy research" (in Japanese). *Kaizō Reconstruction*, no. 33, issue 17, pp. 70-72, November 1952.

cb Zwigenberg, Ran, "The coming of a second sun: The 1956 Atoms for Peace exhibit in Hiroshima and Japan's embrace of nuclear power." *Asia-Pacific Journal*, vol. 10, issue 6, no. 1, p. 3, 4 February 2012. https://apjjf.org/-Ran-Zwigenberg/3685/article.pdf

cc Matashichi, Ōishi, *The Day the Sun Rose in the West: Bikini, the Lucky Dragon, and I*. University of Hawaii Press, 2011.

cd Schreiber, Mark, "*Lucky Dragon*'s lethal catch." *Japan Times*, 18 March 2012. https://www.japantimes.co.jp/life/2012/03/18/general/lucky-dragons-lethal-catch/

ce Rowberry, Arianna, "Castle bravo: The largest US nuclear explosion." Brookings, February 27, 2014. https://www.brookings.edu/blog/up-front/2014/02/27/castle-bravo-the-largest-u-s-nuclear-explosion/

cf Matashichi, *The Day the Sun Rose in the West*.

cg Ibid.

ch Arnold, James R., "Effects of the recent bomb tests on human beings." *Bulletin of the Atomic Scientists*, pp. 347-348, November 1954.

ci Ropeik, David, "How the unlucky *Lucky Dragon* birthed an era of nuclear fear." *Bulletin of the Atomic Scientists*, 28 February 2018. https://thebulletin.org/how-unlucky-lucky-dragon-birthed-era-nuclear-fear11546

cj Zwigenberg, "Coming of a second sun."

ck "A memo by US assistant to secretary of defense" (22 March 1954). *Japan Press Weekly*, 3 October 2011. http://www.japan-press.co.jp/modules/feature_articles/index.php?id=301

cl Kuznick, 2011.

cm Zwigenberg, "Coming of a second sun."

cn Takekawa, Shunichi, "Drawing a line between peaceful and military uses of nuclear power: The Japanese press, 1945–1955." *Asia-Pacific Journal*, vol. 10, issue 37, no. 2, 9 September 2012. https://apjjf.org/2012/10/37/Shunichi-TAKEKAWA/3823/article.html

co Ryfle, Steve, "Godzilla's footprint." *Virginia Quarterly Review*, vol. 81, issue 1, pp. 44-68, winter

2005.

CHAPTER 2

a   Sugita, Yoneyuki (ed.), *Japan Viewed from Interdisciplinary Perspectives: History and Prospects*, p. 266. Lexington Books, 2015.

b   Declassified CIA document, vol. 1_0027, 31 December 1954. https://www.cia.gov/readingroom/docs/SHORIKI%2C%20MATSUTARO%20%20%20VOL.%201_0027.pdf

c   Declassified CIA document, vol. 2_0002, 21 April 1955. https://www.cia.gov/readingroom/docs/SHORIKI%2C%20MATSUTARO%20%20%20VOL.%202_0002.pdf

d   Ibid.

e   "Ex-PM Nakasone, influential Shoriki played key role in promoting nuclear power." *Fukushima News Online*, 2 August 2011. https://fukushimanewsresearch.wordpress.com/2011/08/02/japan-ex-pm-nakasone-influential-shoriki-played-key-role-in-promoting-nuclear-power/

f   Takao, *Japan's Environmental Politics and Governance*, p. 78.

g   Shun'ya, Yoshimi, and Loh, Shi-Lin, "Radioactive rain and the American umbrella." *Journal of Asian Studies*, vol. 71, no. 2, pp. 319-331, 2012. www.jstor.org/stable/23263422

h   Ibid.

i   Takao, *Japan's Environmental Politics and Governance*, p. 79.

j   Atomic Energy Basic Act. Act no. 186, p. 1, 19 December 1955.

k   Ibid., p. 2.

l   Shiroyama, Hideaki, "Regulatory failures of nuclear safety in Japan: The case of [the] Fukushima accident." Paper presented at *Earth System Governance Tokyo Conference: Complex Architectures, Multiple Agents, Tokyo*, 2013. http://tokyo2013.earthsystemgovernance.org/wp-content/uploads/2013/01/0202-SHIROYAMA_Hideaki-.pdf

m   The description of Shoriki's life is drawn from various sources, but much of it comes from a biography of him. The biography appears to be heavily biased in his favour and never fails to paint him in a positive light, which makes me question the finer details. As such, I have been careful about what information I take from it. Uhlan, Edward, and Thomas, Dana L, *Shoriki, Miracle Man of Japan*. Exposition Press, 1957. https://archive.org/details/shorikimiraclema001551mbp/page/n10

n   Ibid.

o   Ramseyer, J. Mark, "Privatizing police: Japanese police, the Korean massacre, and private security firms." Harvard Public Law working paper no. 19-36, 2019. https://papers.ssrn.com/sol3/papers.cfm?abstract_id=3402724

p   "Matsutaro Shoriki's character and career." Declassified CIA document vol. 1_0001, n.d. https://www.cia.gov/readingroom/docs/SHORIKI%2C%20MATSUTARO%20%20%20VOL.%201_0001.pdf

q   Partner, Simon, *Assembled in Japan: Electrical Goods and the Making of the Japanese Consumer*, pp. 76-77. University of California Press, 2000.

r   Foreign Relations of the United States, "Diplomatic papers, 1945: The British Commonwealth, the Far East," vol. VI, p. 966, 1945. https://search.library.wisc.edu/digital/AFOFYVK7LN2GSN9E/pages/ABTPYCINYFA2SS8H

s   CIA, "Matsutaro Shoriki's character and career."

t   "Memorandum for record. Subject: Matsutaro Shoriki." CIA document vol.1_0014, 27 January 1953. https://www.cia.gov/readingroom/docs/SHORIKI%2C%20MATSUTARO%20%20%20VOL.%201_0014.pdf

u   "Japan's Citizen Kane." *Economist*, 22 December 2012. https://www.economist.com/christmas-specials/2012/12/22/japans-citizen-kane

v   Arima, Tetsuo, *Genpatsu Shōriki CIA: Kimitsu bunsho de yomu Shōwa rimenshi* (in Japanese). (*Nuclear power, Shoriki, CIA: History in confidential documents*.) Shinchōsha, 2008. https://catalog.princeton.edu/catalog/5573409. Shoriki's CIA codenames were "podam" and "pojacpot-1."

w   Declassified CIA document, 31 December 1954, vol. 1_0029. https://www.cia.gov/readingroom/docs/SHORIKI%2C%20MATSUTARO%20%20%20VOL.%201_0029.pdf

x   Declassified CIA document, 31 December 1954, vol. 1_0027. https://www.cia.gov/readingroom/docs/SHORIKI%2C%20MATSUTARO%20%20%20VOL.%201_0027.pdf

y   Declassified CIA document, 9 December 1955, vol. 2_0043. https://www.cia.gov/readingroom/docs/SHORIKI%2C%20MATSUTARO%20%20%20VOL.%202_0043.pdf

z   Declassified CIA document, 20 September 1956, vol. 3_0024. https://www.cia.gov/readingroom/docs/SHORIKI%2C%20MATSUTARO%20%20%20VOL.%203_0024.pdf

aa   Declassified CIA document, 8 June 1956, vol. 3_0012. https://www.cia.gov/readingroom/docs/SHORIKI%2C%20MATSUTARO%20%20%20VOL.%203_0012.pdf

ab   Ibid.
ac   "Precis of analysis of the *Yomiuri Shimbun*." Declassified CIA document, 13 August 1956, vol.
     3_0020. https://www.cia.gov/readingroom/docs/SHORIKI%2C%20MATSUTARO%20%20%20
     VOL.%203_0020.pdf
ad   Johnston, Eric, "Key players got nuclear ball rolling." *Japan Times*, 16 July 2011. https://www.
     japantimes.co.jp/news/2011/07/16/national/key-players-got-nuclear-ball-rolling/
ae   "Letter to Eisenhower from Shoriki," 17 November 1955. Declassified CIA document, vol.
     2_0038. https://www.cia.gov/readingroom/docs/SHORIKI%2C%20MATSUTARO%20%20%20
     VOL.%202_0038.pdf
af   "Exploitation of Atoms for Peace program in Japan." Declassified CIA document, 17 January
     1955, vol. 1_0032. The document quotes Shoriki. https://www.cia.gov/readingroom/docs/
     SHORIKI%2C%20MATSUTARO%20%20%20VOL.%201_0032.pdf
ag   "Letter to Hopkins from Shoriki," 26 May 1955. Declassified CIA document, vol. 2_0007. https://
     www.cia.gov/readingroom/docs/SHORIKI%2C%20MATSUTARO%20%20%20VOL.%20
     2_0007.pdf
ah   Japan Atomic Energy Commission (JAEC), "Long-term plan for research, development and
     utilization of nuclear energy," 1956.
ai   Argonne National Laboratory, "Pyroprocessing technologies: Recycling used nuclear fuel for a
     sustainable energy future," p. 2.
aj   "Business circle ignores objection and embarks on N-reactors: US strategy influence on Japan's
     nuclear energy policy" (part 6). *Japan Press Weekly*, 5 October 2011. http://www.japan-press.co.jp/
     modules/feature_articles/index.php?id=304
ak   USIS archive video "Power for Peace."
al   Ibid., timestamp 1: 10.
am   CIA, "Letter to Eisenhower from Shoriki."
an   Tanaka, Yuki, and Kuznick, Peter J., "Japan, the atomic bomb and the 'peaceful uses of nuclear
     power.'" *Asia-Pacific Journal*, vol. 9, issue 18, no. 1, 2 May 2011. https://apjjf.org/-Peter-J--Kuz-
     nick--Yuki-Tanaka/3521/article.pdf
ao   *Chūgoku Shimbun*, 26 May 1956.
ap   Osgood, *Total Cold War*, p. 179.
aq   Low, Morris, *Japan on Display: Photography and the Emperor*, p. 116. Routledge, 2012.
ar   Nester, *Japanese Industrial Targeting*, p. 133.
as   Sovacool, Benjamin K., and Valentine, Scott V., *The National Politics of Nuclear Power: Econom-
     ics, Security, and Governance*, p. 109. Routledge, 2012.
at   Sugita, Hiroki, "USIS role revealed in Japan's tilt toward West." *Japan Times*, 21 November 2007.
     https://www.japantimes.co.jp/news/2007/11/21/news/usis-role-revealed-in-japans-tilt-toward-west/
au   Transcript of Hinton's lecture in Tokyo. https://www.cia.gov/readingroom/docs/SHORIK-
     I%2C%20MATSUTARO%20%20%20VOL.%203_0019.pdf
av   Ibid.
aw   Declassified CIA document, 6 July 1956, vol. 3_0015. https://www.cia.gov/readingroom/docs/
     SHORIKI%2C%20MATSUTARO%20%20%20VOL.%203_0015.pdf
ax   Hunter, Janet, and Sugiyama, S. *The History of Anglo-Japanese Relations 1600–2000: Volume IV:
     Economic and Business Relations*, p. 301. Springer, 2001.
ay   Kurosaki, Akira, "Nuclear energy and nuclear-weapon potential: A historical analysis of Japan
     in the 1960s." *Nonproliferation Review*, vol. 24, issue 1-2, 2017. Special section: Nuclear Asia.
     https://www.tandfonline.com/doi/pdf/10.1080/10736700.2017.1367536?needAccess=true
az   Declassified CIA document, 6 July 1956, vol. 3_0015. https://www.cia.gov/readingroom/docs/
     SHORIKI%2C%20MATSUTARO%20%20%20VOL.%203_0015.pdf
ba   "US officials are failing badly to sell 'Atoms for Peace' program." *Jasper* (Indiana) *Herald*, p. 20,
     11 July 1956. https://www.newspapers.com/newspage/32833959/
bb   Hunter and Sugiyama, *History of Anglo-Japanese Relations 1600–2000*, p. 296.
bc   Ibid., p. 299.
bd   Declassified CIA document, 5 and 10 July 1956, vol. 3_0017. https://www.cia.gov/readingroom/
     docs/SHORIKI%2C%20MATSUTARO%20%20%20VOL.%203_0017.pdf
be   Ibid.
bf   Yamashita, Kiyonobu, "History of nuclear technology development in Japan." *AIP Conference
     Proceedings 1659*, vol. 020003, p. 3, 2015. https://aip.scitation.org/doi/pdf/10.1063/1.4916842?-
     class=pdf
bg   Suttmeier, R., "The Japanese nuclear power option: Technological promise and social limitations."
     In R. Morse (ed.), *The Politics of Japan's Energy Strategy*, p. 110 (p. 124 of PDF). Institute of

East Asian Studies, University of California, 1981. https://digitalassets.lib.berkeley.edu/ieas/IEAS_03_0002.pdf

bh Federation of Electric Power Companies, "Late Showa history chronology of electricity" (in Japanese), n.d. http://www.fepc.or.jp/enterprise/rekishi/shouwa1955/index.html

bi Johnston, "Key players got nuclear ball rolling."

bj Inui, Yasuyo, "Background of creating a framework for nuclear power plant location regulation and peripheral development regulation in the early days of nuclear power development" (in Japanese). City Planning Papers, vol. 50, no. 3, pp. 968-973, 2015. https://www.jstage.jst.go.jp/article/journalcpij/50/3/50_968/_article/-char/ja/

bk Koshoji, Hiroyuki, "Japan's nuclear facilities face quake risk." UPI Asia, 12 June 2008. https://archive.is/ByMWv

bl Aldrich, Daniel P. Site Fights: Divisive Facilities and Civil Society in Japan. Cornell University Press, 2008.

bm Aldrich, Daniel P., "The limits of flexible and adaptive institutions: The Japanese government's role in nuclear power plant siting over the post-war period." In S. Lisbirel and D. Shaw (eds.), Managing Conflict in Facility Siting: An International Comparison. Edward Elgar, 2005. https://www.academia.edu/810937/The_Limits_of_Flexible_and_Adaptive_Institutions_The_Japanese_Governments_Role_In_Nuclear_Power_Plant_Siting_Over_the_Post_War_Period

bn Shimanishi, Tomoki, "Economic growth, resources and the environment: The use of fossil fuels in Japan's post-war industrialization," p. 9, 2009. https://apebhconference.files.wordpress.com/2009/09/shimanishi2.pdf

bo Nester, Japanese Industrial Targeting.

bp Yamashita, Kiyonobu, "Human resource development in the beginning phase of nuclear technology development in Japan." 11th National Conference on Nuclear Science and Technology, agenda and abstracts, 2015. https://inis.iaea.org/collection/NCLCollectionStore/_Public/47/065/47065304.pdf

bq IAEA Bulletin, vol. 1–1, April 1959.

br Low, Morris. Visualizing Nuclear Power in Japan: A Trip to the Reactor. Palgrave Macmillan, 2020.

bs Takayanagi, M., "The reduced enrichment program for JRR-4." Tokai Research Establishment, JAERI. https://inis.iaea.org/collection/NCLCollectionStore/_Public/23/087/23087458.pdf

bt "Minamata disease," n.d. Boston University. https://www.bu.edu/sustainability/minamata-disease/

bu Asahi, Schiyo, and Yakita, Akira. "SOX emissions reduction policy and economic development: A case of Yokkaichi." Modern Economy Journal, vol. 3, no. 1, p. 2, January 2012.

bv Shimanishi, "Economic Growth, Resources and the Environment," p. 7.

bw Ibid., p. 13.

bx Institute for International Cooperation and the Japan International Cooperation Agency, "Chapter 6: Environmental pollution control measures." In Japan's Experience in Public Health and Medical Systems, March 2005. https://www.jica.go.jp/jica-ri/IFIC_and_JBICI-Studies/english/publications/reports/study/topical/health/index.html

by Kokubun, Naoko, "Japan's hunger for growth: Environment as political symbolism," 2014. https://dspace.library.uvic.ca/handle/1828/5123

bz Broadbent, 1998.

ca Japan Atomic Power Company (JAPC), "Corporate history," n.d. Archive link: https://web.archive.org/web/20090101085730/http://www.japc.co.jp/english/company/history.html. (The current company history webpage has far less detail, hence the archived link.)

cb Shropshire, David E., "Lessons learned from GEN 1 carbon dioxide cooled reactors," p. 4. Twelfth International Conference on Nuclear Engineering, Idaho National Engineering and Environmental Laboratory, 2004. https://pdfs.semanticscholar.org/c33a/b21620e31670cf438121f66996e0de19c-c2c.pdf

cc Connors, Duncan, "Strategic decision making and the commercial failure of British MAGNOX nuclear reactor exports: The General Electric Company and the Tokai Nuclear Power Plant in Japan, 1955 to 1970," n.d. https://ebha.org/public/C3:paper_file:131; "Opinions expressed at public hearing on problem of safety related to improved Calder Halt type reactor," Mainichin Shimbun, 16 March 1959; "Yasukawa says reactor accident can be prevented," Yomiuri Shimbun, 14 March 1959.

cd Hicks, Raymond, An Odyssey: From Ebbw Vale to Tyneside, p. 176. Self-published, 2007. Dr Raymond Hicks was the GEC Simon-Carves principal earthquake engineer responsible for the Tokai modifications. His autobiography has an interesting and detailed account of a visit to Japan by a group delegation in September 1958.

ce Bangash, M. Y. H., *Structures for Nuclear Facilities: Analysis, Design, and Construction*, p. 27. Springer, 2011. https://www.springer.com/gp/book/9783642125591

cf "Japan's first nuclear power station." *Engineer*, March 6, 1959. https://web.archive.org/web/20150320120631/www.theengineer.co.uk/Journals/2013/03/12/m/g/b/Japan-nuclear-power-er-station.pdf

cg Ibid.

ch Muto, Kiyoshi, "Tentative response analysis of Tokai reactor structure due to earthquake ground motion on 3rd April, 1966 (structures)." *Transactions of the Architectural Institute of Japan Summaries of Technical Papers*, vol. 42, p. 206, 1967. https://www.jstage.jst.go.jp/article/ai-jsaxxe/42/0/42_KJ00005417387/_article/-char/en

ci Bangash, *Structures for Nuclear Facilities*, p. 27.

cj "Britain supplies first atomic power." *Guardian*, 29 February 1960.

ck Hashimoto, Uichi, Kihara, Hiroshi, and Ando, Yoshio, "Welding problems associated with the construction of nuclear power plants in Japan." *Journal of the Welding Society*, 1964, vol. 33, no. 4, pp. 357-363. https://www.jstage.jst.go.jp/article/qjjws1943/33/4/33_4_357/_article/-char/ja/

cl Wearne, S. H., and Bird, R. H., "UK experience of consortia engineering for nuclear power stations," p. 11, 2016.

cm Cortazzi, Hugh, *Japan Experiences: Fifty Years, One Hundred Views*, p. 476. Routledge, 2001.

cn Tromans, Stephen, *Nuclear Law: The Law Applying to Nuclear Installations and Radioactive Substances in Its Historic Context*, 2nd ed., ref. 34. Bloomsbury, March 2010.

co Yoshiyuki Nishikawa, et al., "Radiation control at Tokai Nuclear Power Station," *Japanese Journal of Health Physics*, vol. 4, issue 1, pp. 13-23, 1969. https://www.jstage.jst.go.jp/article/jhps1966/4/1/4_1_13/_article/-char/en

cp Act on Compensation for Nuclear Damage, 1961, part II: "Liability for Nuclear Damage," p. 4.

cq Ibid., p. 3.

cr Oguma, Eiji, "Japan's nuclear power and anti-nuclear movement from a socio-historical perspective." In *Towards Long-Term Sustainability: In Response to the 3/11 Earthquake and the Fukushima Nuclear Disaster* (conference), p. 3. Institute of East Asian Studies, University of California, 2012. https://web.archive.org/web/20160222025357/http://ieas.berkeley.edu/events/pdf/2012.04.20_sustainability_oguma_en.pdf

cs Tajima, Hiroshi, Ichino, Ichiro, and Gotoh, Haruo, "Aseismic design of reactor enclosure of the Japan Demonstration Reactor (JPDR) and its actual behavior for earthquake." *Bulletin of the Japan Society of Mechanical Engineers*, vol. 9, issue 34, pp. 294-299, 1966. https://www.jstage.jst.go.jp/article/jsme1958/9/34/9_34_294/_article/-char/en

ct Yamashita, Kiyonobu, "History of nuclear technology development in Japan." *AIP Conference Proceedings* vol. 1659, no. 020003, p. 1, 2015. https://aip.scitation.org/doi/pdf/10.1063/1.4916842?class=pdf

cu Eyre, John D., "Development trends in the Japanese electric power industry, 1963–68." *Professional Geographer*, vol. 22, issue 1, pp. 26-30, January 1970. https://onlinelibrary.wiley.com/doi/abs/10.1111/j.0033-0124.1970.00026.x

cv Hunter and Sugiyama, *History of Anglo-Japanese Relations 1600–2000*, p. 304.

cw International Atomic Energy Agency (IAEA), "50 years of nuclear energy." https://www.iaea.org/sites/default/files/gc/gc48inf-4-att3_en.pdf

cx Mounfield, Peter R., *World Nuclear Power*. Routledge, 2017 (1991).

cy Aldrich, *Site Fights*, pp. 126-127. Citing "Pickett 2002, 1349" for the quote.

cz Ipponmatsu, Tamaki, "Public Acceptance: A Japanese view." JAPC. https://www.iaea.org/sites/default/files/publications/magazines/bulletin/bull14-2/14210091218.pdf

CHAPTER 3

a Atomic Energy Commission (AEC), "About TEPCO's Fukushima nuclear power plant nuclear reactor installation" (in Japanese), n.d. http://www.aec.go.jp/jicst/NC/about/ugoki/geppou/V11/N11/196604V11N11.html

b Onitsuka, Hiroshi, "Hooked on nuclear power: Japanese state-local relations and the vicious cycle of nuclear dependence." *Asia-Pacific Journal*, vol. 10, issue 3, no. 1, 15 January 2012. https://apjjf.org/-Hiroshi-Onitsuka/3677/article.pdf

c Uehara, M., "Chapter 18: The long term economic value of holistic ecological planning for disaster risk." In W. Yan and W. Galloway (eds.), *Rethinking Resilience, Adaptation and Transformation in a Time of Change*. Springer, 2017. https://www.academia.edu/40345764/The_Long_Term_Economic_Value_of_Holistic_Ecological_Planning_for_Disaster_Risk

d Onitsuka, "Hooked on nuclear power."

374 • References

e   "Safety concerning installation of nuclear reactor at Fukushima Nuclear Power Station TEPCO" (in Japanese). Report by reactor safety expert review committee, 2 November 1964. http://www. aec.go.jp/jicst/NC/about/ugoki/geppou/V11/N11/196620V11N11.html

f   Shuto, Nobuo, and Fujima, Koji, "A short history of tsunami research and countermeasures in Japan." *Proceedings of the Japan Academy, Series B: Physical and Biological Sciences*, vol. 85, no. 8, pp. 267-275, October 2009. https://www.ncbi.nlm.nih.gov/pmc/articles/PMC3621565/

g   Nakao, Masayuki, "The Great Meiji Sanriku Tsunami June 15, 1896, at the Sanriku coast of the Tohoku region," n.d. Institute of Engineering Innovation, School of Engineering, University of Tokyo. https://web.archive.org/web/20081223230738/http://shippai.jst.go.jp/en/Detail?fn=2&id=-CA1000616

h   "Geotechnical, geological and earthquake engineering," 2012. Preliminary reconnaissance report of the 2011 Tohoku-Chiho Taiheiyo-Oki Earthquake. https://link.springer.com/content/pdf/bb-m%3A978-4-431-54097-7%2F1.pdf

i   Carydis, P., Pomonis, A., and Goda, K., "Fukushima Daiichi Nuclear Power Plant: A retrospective evaluation." Conference paper, *15th World Conference on Earthquake Engineering*, 2012. https://www.iitk.ac.in/nicee/wcee/article/WCEE2012_2832.pdf

j   Onishi, Norimitsu, and Glanz, James, "Japanese rules for nuclear plants relied on old science." *New York Times*, 26 March 2011. https://www.nytimes.com/2011/03/27/world/asia/27nuke.html

k   Kaburaki, Hiroshi, *Hatsuden Suiryoku* magazine, January 1969.

l   Dawson, Chester, and Hayashi, Yuka, "Fateful move exposed Japan plant." *Wall Street Journal*, 12 July 2011. Archive link: https://archive.fo/NT8Uk

m   Ibid.

n   Yoshida, Reiji, "GE plan followed with inflexibility." *Japan Times*, 14 July 2011. https://www.japantimes.co.jp/news/2011/07/14/national/ge-plan-followed-with-inflexibility/

o   Dawson and Hayashi, "Fateful move exposed Japan plant."

p   Independent Investigation Commission on the Fukushima Nuclear Accident, *The Fukushima Daiichi Nuclear Power Station Disaster: Investigating the Myth and Reality*, p. 101, footnote 9. Routledge, 2014.

q   Kikkawa, "Role of Matsunaga Yasuzaemon." Wikipedia has a nice quote on this incident, but I have not read the cited Japanese source, as I've been unable to acquire a copy. Still, I like it enough to mention as a hidden note. The note says, "Just after the earthquake, when a TV reported, with an image of a building on fire, that the Niigata thermal power station had exploded, Matsunaga, the so-called 'King of Electric Power of Japan' immediately retorted: 'That's a mistake. The power station Hirai constructed cannot be broken.' In fact, it was a case of misreporting and there was no damage; the main body of the power station only sank 20 centimeters just perpendicularly." I love the 82-year-old's faith in his protégé.

r   "Onagawa nuke plant saved from tsunami by one man's strength, determination." *Mainichi Shimbun*, 19 March 2012. Archive link: https://web.archive.org/web/20120618134231/http://mainichi.jp/english/english/perspectives/news/20120319p2a00m0na020000c.html

s   Sasagawa. Toshiro, and Hirata, Kazuo (TEPCO), "Tsunami evaluation and countermeasures at Onagawa Nuclear Power Plant." *15th World Conf. Earthquake Engineering*, Lisbon, Portugal, 24–28 September 2012. https://www.iitk.ac.in/nicee/wcee/article/WCEE2012_4545.pdf

t   *Asahi Shimbun*, 1971. Found in Loh, Shi-Lin, "Narrating Fukushima: Scales of a nuclear meltdown." *East Asian Science, Technology and Society: An International Journal*, 2013. https://www.academia.edu/5807210/Narrating_Fukushima_Scales_of_a_Nuclear_Meltdown

u   Morris, Jim, and Mehta, Aaron, "Reactors at heart of Japanese nuclear crisis raised concerns as early as 1972, memos show." Center for Public Integrity, 15 March 2011. https://www.publicintegrity.org/2011/03/15/3520/reactors-heart-japanese-nuclear-crisis-raised-concerns-early-1972-memos-show

v   Beyea, Jan, and Hippel, Frank von., "Containment of a reactor meltdown." *Bulletin of the Atomic Scientists*, pp. 53-54, September 1982.

w   Ibid.

x   Ibid.

y   NRC memo, reproduced in Beyea and Hippel, "Containment of a reactor meltdown."

z   Yamakoshi, Atsushi, "A study on Japan's reaction to the 1973 oil crisis," pp. 29-30. Waseda University, 1986. https://open.library.ubc.ca/cIRcle/collections/ubctheses/831/items/1.0076972 (Kunio Yanagida, Ohkami ga Yatte Kita Hi [Tokyo: Bungei Shunju], pp. 65-66.)

aa Lowinger, Thomas C., "Japan's nuclear energy development policies: An overview." *Journal of Energy and Development*, vol. 15, no. 2, pp. 211-230, spring 1990. https://www.jstor.org/stable/24807916

ab Aldrich, D., "The limits of flexible and adaptive institutions: The Japanese government's role in nuclear power plant siting over the post-war period." In S. Lisbirel (ed.), *Managing Conflict in Facility Siting: An International Comparison*. Edward Elgar, 2005.

ac Suzuki, Tatsujiro (Central Research Institute of the Electric Power Industry), "Energy security and the role of nuclear power in Japan," p. 4. http://oldsite.nautilus.org/archives/energy/eaef/Reg_Japan_final.PDF

ad Sumihara, Noriya, "The nuclear power visitor center as a strategic tool for hegemonic partnership in Japan." *Agora: Journal of International Center for Regional Studies*, 2003. http://www.tenri-u.ac.jp/topics/q3tncs00000fizfe-att/q3tncs00000fj089.pdf

ae Oguma, "Japan's nuclear power and anti-nuclear movement."

af Aldrich, *Site Fights*, p. 64.

ag

ah Norwood, W. D., "Fact and fiction about health and atomic power plants." *Journal of Occupational Medicine*, vol. 13, issue 4, pp. 199-204, April 1971. https://journals.lww.com/joem/Citation/1971/04000/Fact_and_Fiction_about_Health_and_Atomic_Power.7.aspx

ai "Dengen sanpo kofukin jisseki" ("Subsidies under three electricity acts"). In Japanese. Kashiwazaki City, Japan. http://www.city.kashiwazaki.lg.jp/atom/genshiryoku/kofukin/kofukin-jisseki.html

aj Ramseyer, J. Mark, "Nuclear reactors in Japan: Who asks for them, what do they do?" *European Journal of Law and Economics*, forthcoming, Harvard Law School, John M. Olin Center discussion paper no. 909, 2017; Harvard Public Law working paper no. 17-35. https://papers.ssrn.com/sol3/papers.cfm?abstract_id=2986410

ak Kingston, Jeff (ed.), *Critical Issues in Contemporary Japan*, 2nd ed. Routledge, 2019.

al Clutterbuck, David, "Rough passage for nuclear ships?" *New Scientist and Science Journal*, 13 May 1971.

am Rippon, S., "280 nuclear containerships on the high seas by 2000." *Nuclear Engineers Journal*, p. 443, June 1971.

an Sasaki, Shuichi, "General description of the first nuclear ship *Mutsu*." *Nuclear Engineering and Design*, vol. 10, issue 2, pp. 123-125, June 1969. https://www.sciencedirect.com/science/article/pii/0029549369900351

ao Bauman, James Robert, "Analysis of past, present and future applications of nuclear power for propulsion of marine vehicles." Massachusetts Institute of Technology, May 1972. https://core.ac.uk/download/pdf/36709227.pdf This document contains extensive technical information on *Mutsu* and other nuclear ships of the time.

ap Fujii-e, Yoichi, "Control systems and transient analysis." *Nuclear Engineering and Design*, vol. 10, issue 2, pp. 231-242, 1969.

aq Freire, L., and Andrade, D., "Historic survey on nuclear merchant ships." *Nuclear Engineering and Design*, vol. 293, pp. 176-186, 2015.

ar Ōi, Hiroshi, and Tanigaki, Kazuo, "The ship design of the first nuclear ship in Japan." *Nuclear Engineering and Design*, vol. 10, issue 2, pp. 211-219, June 1969.

as Akita, Yoshio, and Kitamura, Katsuhide, "A study on collision by an elastic stem to a side structure of [nuclear] ships." *Journal of the Society of Naval Architects of Japan*, issue 131, pp. 307-317, 1972. https://www.jstage.jst.go.jp/article/jjasnaoe1968/1972/131/1972_131_307/_pdf

at "Radiation leakage investigation report" (in Japanese). *Mutsu* Radiation Leakage Problem Investigation Committee, May 1975. http://www.aec.go.jp/jicst/NC/about/ugoki/geppou/V20/N05/197524V20N05.html

au Kobayashi, Iwao, and Yamazaki, Hiroshi, "Measurements of streaming neutrons on nuclear ship *Mutsu* by a two-detector-method." *Journal of Nuclear Science and Technology*, vol. 13, no. 11, pp. 663-675, 2012. https://www.tandfonline.com/doi/abs/10.1080/18811248.1976.9734087

av McKean, Margaret A., *Environmental Protest and Citizen Politics in Japan*. University of California Press, 1992. Reference cited: "JT, 12 August 1974, 16 October 1974."

aw "Japan's trouble-plagued nuclear-powered ship *Mutsu* put to sea Tuesday." UPI, 31 August 1982. https://www.upi.com/Archives/1982/08/31/Japans-trouble-plagued-nuclear-powered-ship-Mutsu-put-to-sea-Tuesday/4756399614400/

ax "Radiation leak 'Mutsu' after 40 years. What do Sasebo citizens think now?" (in Japanese) *Nagasaki Shimbun*, 17 October 2018. https://this.kiji.is/425110314843636833

ay "Radiation leakage investigation report" (in Japanese). *Mutsu* Radiation Leakage Problem Investigation Committee, May 1975. http://www.aec.go.jp/jicst/NC/about/ugoki/geppou/V20/N05/197524V20N05.html

az Freire, Luciano Ondir, and Andrade, Delvonei Alves de, "Historic survey on nuclear merchant ships." *Nuclear Engineering and Design*, vol. 293, pp. 176-186, 2015.

ba Nakao, Masayuki, "Radiation leaks from nuclear power ship *Mutsu*," n.d. Association for the Study of Failure. http://www.shippai.org/fkd/en/cfen/CA1000615.html

bb Kishimoto, A., "Chapter 12: Public attitudes and institutional changes in Japan following nuclear accidents." In E. Balleisen, et al. (eds.), *Policy Shock: Recalibrating Risk and Regulation after Oil Spills, Nuclear Accidents and Financial Crises*, p. 316. Cambridge University Press, 2017.

bc Kuroda, Yasumasa, "Protest movements in Japan: A new politics." *Asian Survey*, vol. 12, no. 11, pp. 947-952, November 1972. https://www.jstor.org/stable/2643115

bd Kurosaki, "Nuclear energy and nuclear-weapon potential."

be Statement on nuclear policy, President Gerald Ford. http://www.presidency.ucsb.edu/ws/?pid=6561

bf Morris and Mehta, "Reactors at heart of Japanese nuclear crisis."

bg Mosk, Matthew, "Fukushima: Mark 1 nuclear reactor design caused GE scientist to quit in protest." *ABC News*, 15 March 2011. https://abcnews.go.com/Blotter/fukushima-mark-nuclear-reactor-design-caused-ge-scientist/story?id=13141287

bh Lowinger, "Japan's nuclear energy development policies."

bi National Diet of Japan Fukushima Nuclear Accident Independent Investigation Commission (NAIIC), *Final Report*. National Diet of Japan, 2012. https://warp.da.ndl.go.jp/info:ndljp/pid/3856371/naiic.go.jp/en/report/

bj Matsuo, Yuhji, and Nei, Hisanori, "Assessing the historical trend of nuclear power plant construction costs in Japan." Institute of Energy Economics, Japan (IEEJ), June 2018. https://eneken.ieej.or.jp/data/7922.pdf

bk IAEA, "Annex I: Historical development of the governmental, legal and regulatory framework for nuclear safety in Japan." *The Fukushima Daiichi Accident, Technical Volume 2/5: Safety Assessment*, 2015.

bl Atomic Energy Commission, "Section 6: Safety design examination guidelines for light-water nuclear reactor facilities for power generation" (in Japanese), 14 June 1977. http://www.aec.go.jp/jicst/NC/about/hakusho/wp1977/ss1010106.htm

bm Gale, R., "Chapter 4: Tokyo Electric Power Company: Its role in shaping Japan's coal and LNG policy. In R. Morse (ed.), *The Politics of Japan's Energy Strategy*. Institute of East Asian Studies, University of California, Berkeley, 1981. http://digitalassets.lib.berkeley.edu/ieas/IEAS_03_0002.pdf

bn Independent Investigation on the Fukushima Nuclear Accident (henceforth Independent Fukushima Investigation), *The Fukushima Daiichi Nuclear Power Station Disaster: Investigating the Myth and Reality*, p. 45. Routledge, 2014.

bo Ibid.

bp Shadrina, Elena, "Fukushima fallout: Gauging the change in Japanese nuclear energy policy." *International Journal of Disaster Risk Science*, vol. 3, pp. 69-83, 2012. https://link.springer.com/content/pdf/10.1007/s13753-012-0008-0.pdf

CHAPTER 4

a Nagae, Shusaku, "Information of the Japan Public Relations Society," Japan Society for Corporate Communication Studies (JSCCS), n.d. http://jsccs.jp/about/index.html

b Samuels, *Business of the Japanese State*.

c Soukup, James R., "Japan." *Journal of Politics*, vol. 25, no. 4, pp. 737-756, 1963. www.jstor.org/stable/2127429. Accessed 27 Jan. 2021.

d Wang, Qiang, and Chen, Xi. "Regulatory failures for nuclear safety – The bad example of Japan – Implication for the rest of world." *Renewable and Sustainable Energy Reviews*, vol. 16, pp. 2610-2617, 2012. http://web.mit.edu/12.000/www/m2018/pdfs/japan/regulatory-failures.pdf

e Stanwick, Peter A., and Stanwick, Sarah D., *Understanding Business Ethics*. Sage, 2015.

f Shadrina, "Fukushima fallout."

g "Japan ex-official to quit TEPCO post." *Nuclear Power Daily*, 19 April 2011. https://www.nuclearpowerdaily.com/reports/Japan_ex-official_to_quit_TEPCO_post_999.html

h Clenfield, Jason, and Sato, Shigeru, "Japan nuclear energy drive compromised by conflicts of interest." Bloomberg, 12 December 2007. http://www.bloomberg.com/apps/news?pid=newsarchive&sid=awR8KsLlAcSo. Archived link: https://web.archive.org/web/20150924190250/http://www.bloomberg.com/apps/news?pid=newsarchive&sid=awR8KsLlAcSo

i Colignon, Richard A., *Amakudari: The Hidden Fabric of Japan's Economy*. ILR Press Books, 2003.

j "The problem with Amakudari." *Diplomat*, 23 May 2011. https://thediplomat.com/2011/05/the-problem-with-amakudari/

k Meyer, Cordula, "Atomic industry too close to government for comfort." *Der Spiegel*, 27 May

2011. http://www.spiegel.de/international/world/japan-s-nuclear-cartel-atomic-industry-too-close-to-government-for-comfort-a-764907-2.html

l   Onishi, Norimitsu, and Belson, Ken, "Culture of complicity tied to stricken nuclear plant." *New York Times*, 26 April 2011. https://www.nytimes.com/2011/04/27/world/asia/27collusion.html

m Jones, Colin P. A., "Amakudari and Japanese law." *Michigan State International Law Review*, vol. 22, issue 3, 2003. https://digitalcommons.law.msu.edu/ilr/vol22/iss3/9/

n   Abe, Yasuhito, "Risk assessment of nuclear power by Japanese newspapers following the Chernobyl nuclear disaster." *International Journal of Communication*, vol. 7, 2013.

o   Suzuki, Tatsuru, *New Challenge for the Power Industry: Beyond a Turbulent Decade*. Nihon Kogyo Shinbunsha, 1983.

p   Ibid.

q   Meyer, "Atomic industry too close to government for comfort."

r   Independent Fukushima Investigation, *Fukushima Daiichi Nuclear Power Station Disaster*, p. 57, footnote 10.

s   Karan, Pradyumna P., and Suganuma, Unryu. *Japan after 3/11: Global Perspectives on the Earthquake, Tsunami, and Fukushima Meltdown*. University Press of Kentucky, 2016, p. 220.

t   Independent Fukushima Investigation, *Fukushima Daiichi Nuclear Power Station Disaster*, pp. 42-43.

u   Carter, Jimmy. "Nuclear Power Policy Statement on Decisions Reached Following a Review." 7 April 1977. https://www.presidency.ucsb.edu/documents/nuclear-power-policy-statement-decisions-reached-following-review

v   US State Department action memorandum, "Options paper to the president on the Japanese nuclear reprocessing facility," 30 July 1977. https://nsarchive2.gwu.edu//dc.html?doc=3859730-Document-05-Memorandum-from-Ambassador-at-Large

w United States of America and Japan, "Joint determination for reprocessing of special nuclear material of United States origin (with joint communiqué)." Signed at Washington, DC, 12 September 1977. https://treaties.un.org/doc/Publication/UNTS/Volume%201084/volume-1084-I-16582-English.pdf

x   Malcolm, Andrew H., "US and Japan agree on Tokyo's opening of atom fuel plant." *New York Times*, 2 September 1977 (citing US officials). https://www.nytimes.com/1977/09/02/archives/us-and-japan-agree-on-tokyos-opening-of-atom-fuel-plant-2year.html

y   Dahl, Per F., *From Nuclear Transmutation to Nuclear Fission, 1932–1939*. Routledge, 2002.

z   Yoshioka, Hitoshi, *A Social History of Nuclear Power: Its Development in Japan* (in Japanese). Asahi Shimbun Press, 2011.

aa "Outline of the power plant invitation [September 1967–November 28, 1986]." In Japanese. Kashiwazaki City, Japan. https://www.city.kashiwazaki.lg.jp/genshiryoku/hatsudenshoyuchikara-koremadenokeii/hatsudenshoyuchikaranokeikagaiyoshiryo/11449.html

ab Aldrich, *Site Fights*, pp. 129-130.

ac Suzuki, Akira, "Japanese labour unions and nuclear energy: A historical analysis of their ideologies and worldviews." *Journal of Contemporary Asia*, vol. 46, issue 4, pp. 591-613, 2016. https://www.tandfonline.com/doi/full/10.1080/00472336.2016.1178321

ad Juraku, Kohta, Suzuki, Tatsujiro, and Sakura, Osamu, "Social decision-making processes in local contexts: An STS case study on nuclear power plant siting in Japan." *East Asian Science, Technology and Society: An International Journal*, vol. 1, issue 1, pp. 53-75, December 2007. https://link.springer.com/article/10.1007/s12280-007-9002-9

ae Hirabayashi, Yuko, "Collective recognition and shared identity: Factors behind the emergence and mobilization process in a referendum movement." *East Asian Social Movements: Power, Protest, and Change in a Dynamic Region*, pp. 81-97. Springer, 2010. https://link.springer.com/chapter/10.1007%2F978-0-387-09626-1_4

af "Abandoned plans for Suzu and Maki Nuclear Power Plants: The beginning of the end?" *Nuke Info Tokyo*, issue 98, 2003. http://www.cnic.jp/english/?p=621

ag Dawson, Chester, "A nuclear-free town fades away in Japan." *Wall Street Journal*, 3 January 2012. https://www.wsj.com/articles/SB10001424052970203707504577009373547835612

ah Onishi, Norimitsu, and Fackler, Martin. "Japanese officials ignored or concealed dangers." *New York Times*, 16 May 2011. https://www.nytimes.com/2011/05/17/world/asia/17japan.html

ai US State Department telegram no. 05853. https://wikileaks.org/plusd/cables/1979TOKYO-05853_e.html; "International aspects of the Three Mile Island incident II: International reaction." National Archives, 19 June 2018. https://text-message.blogs.archives.gov/2018/06/19/international-aspects-of-the-three-mile-island-incident-ii-international-reaction/

aj Kishimoto, Atsuo, "Impacts of five nuclear accidents on public attitudes and national policies in

Japan." Research Institute of Science for Safety and Sustainability (RISS), National Institute of Advanced Industrial Science and Technology (AIST), p. 16.

ak Yoshioka, Hitoshi, *A Social History of Nuclear Power.*

al Cook, R., and Woods, D., "Chapter 20: Distancing through differencing: An obstacle to organizational learning following accidents." In D. Woods and E. Hollnagel (eds.), *Resilience Engineering: Concepts and Precepts.* CRC Press, 2006. https://www.researchgate.net/publication/292504703_Distancing_Through_Differencing_An_Obstacle_to_Organizational_Learning_Following_Accidents

am The President's Commission on the Accident at Three Mile Island (the Kemeny Commission), John G. Kemeny, chair, "The accident at Three Mile Island, part 3," p. 9 (p. 14 of PDF), 1980. http://large.stanford.edu/courses/2012/ph241/tran1/docs/188.pdf

an Gale, "Tokyo Electric Power Company."

ao Federation of Electric Power Companies, "Graphical flip-chart of nuclear and energy related topics," 2014. https://www.fepc.or.jp/english/library/graphical_flip-chart/

ap Powell, John W., "Nuclear power in Japan." *Bulletin of the Atomic Scientists,* May 1983.

aq Nester, *Japanese Industrial Targeting,* p. 134.

ar "Thanks to nuclear power, the city of Tsuruga has ..." UPI, 3 May 1981. https://www.upi.com/Archives/1981/05/03/Thanks-to-nuclear-power-the-city-of-Tsuruga-has/9488357710400/

as history.com, "This Day in History," "Japanese power plant leaks radioactive waste." https://www.history.com/this-day-in-history/japanese-power-plant-leaks-radioactive-waste. I have also seen the date listed as 9 March.

at "Japanese fear 56 exposed in nuclear cleanup." *Pacific Stars and Stripes* (US newspaper for government/military personnel stationed in Japan), p. 1, 23 April 1981. https://newspaperarchive.com/pacific-stars-and-stripes-apr-23-1981-p-1/

au Japan Atomic Energy Agency, "Accidents of radioactive waste liquid leakage from radioactive waste treatment facility at Tsuruga Power Plant" (in Japanese), 1998. https://atomica.jaea.go.jp/data/detail/dat_detail_02-07-02-12.html

av "45 Japanese workers are reported exposed to nuclear radiation." *New York Times.* UPI, 26 April 1981. https://www.nytimes.com/1981/04/26/world/45-japanese-workers-are-reported-exposed-to-nuclear-radiation.html

aw "More exposed to radiation in nuclear accident." UPI, 27 April 1981. https://www.upi.com/Archives/1981/04/27/More-exposed-to-radiation-in-nuclear-accident/4310357192000/

ax "Japan Nuclear Power Co. disclosed today that six company ..." UPI, 27 April 1981. https://www.upi.com/Archives/1981/04/27/Japan-Nuclear-Power-Co-disclosed-today-that-six-company/1378357192000/

ay International Atomic Energy Agency, "Radioactivity leak accidents of Tsuruga Nuclear Power Station, 2." https://inis.iaea.org/search/search.aspx?orig_q=RN:13714953

az "Japanese nuclear plant admits to leaking wastes." *Tyrone Daily Herald,* 29 April 1981. https://www.newspapers.com/clip/1964267/nuclear_power_plant_admits_to_leaking/

ba UPI, "Thanks to nuclear power."

bb Powell, "Nuclear power in Japan."

bc Kishimoto, "Public attitudes and institutional changes."

bd Powell, "Nuclear power in Japan."

be Aldrich, Daniel P., "Japan's nuclear power plant siting: Quelling resistance." *Asia-Pacific Journal,* vol. 3, issue 6, 10 June 2005. https://apjjf.org/-Daniel-P--Aldrich/2047/article.pdf

bf The following is a very thorough account of how it all came about: Nishiguchi, Toshihiro, and Brookfield, Jonathan, "The evolution of Japanese subcontracting." *MIT Sloan Management Review,* 15 October 1997. https://sloanreview.mit.edu/article/the-evolution-of-japanese-subcontracting/

bg Hong, Emily, "State of the world's minorities and indigenous peoples 2013 – Case study: Japan's Burakumin minority hired to clean up after Fukushima." Refworld, the United Nations High Commissioner for Refugees (UNHCR), 24 September 2013. https://www.refworld.org/docid/526f-b70e5.html

bh Nee, Brett, "Sanya: Japan's internal colony." *Bulletin of Concerned Asian Scholars,* vol. 6, no. 3, p. 13, September–October 1974.

bi Tanaka, Y., "Chapter 7: Nuclear power and the labour movement." In G. McCormack and Y. Sugimoto (eds.), *The Japanese Trajectory: Modernization and Beyond.* Cambridge University Press, 2009.

bj Suzuki, Tomohiko, *Yakuza and the nuclear industry: Diary of An undercover reporter working at the Fukushima plant.* In Japanese. Bungei Shunju, 2011.

bk "Toshiba reports teens worked at nuke plant." UPI, 4 June 2008. http://www.upi.com/NewsTrack/ Top_News/2008/06/04/toshiba_reports_teens_worked_at_nuke_plant/9483/. Archived link: https:// web.archive.org/web/20080605132813/http://www.upi.com/NewsTrack/Top_News/2008/06/04/ toshiba_reports_teens_worked_at_nuke_plant/9483/

bl McCurry, Justin, "Beyond the bright lights, Japan's biggest slum is nation's dark secret." *Guardian*, 22 August 2008. https://www.theguardian.com/business/2008/aug/22/japan.socialexclusion

bm Tanaka, Yuki, "Nuclear power plant gypsies in high-tech society." *Bulletin of Concerned Asian Scholars*, vol. 18, no. 1, January–March 1986. http://criticalasianstudies.org/assets/files/bcas/ v18n01.pdf

bn Ibid.

bo Adelstein, Jake, and Nakajima, Stephanie, "TEPCO: Will someone turn off the lights?" *Atlantic*, 28 June 2011. https://www.theatlantic.com/international/archive/2011/06/tepco-will-someone-turn-lights/352260/

bp Efron, Sonni, "System of disposable laborers." *Los Angeles Times*, 30 December 1999. https:// www.latimes.com/archives/la-xpm-1999-dec-30-mn-49042-story.html

bq Glionna, John M., "Japan's 'nuclear gypsies' face radioactive peril at power plants." *Los Angeles Times*, 4 December 2011. http://articles.latimes.com/2011/dec/04/world/la-fg-japan-nuclear-gypsies-20111204

br Tanaka, "Nuclear power plant gypsies."

bs Efron, "System of disposable laborers."

bt "Japanese nuclear 'slaves' at risk." BBC, 29 October 1999. http://news2.thls.bbc.co.uk/hi/ english/world/asia-pacific/newsid_493000/493133.stm. Archived link: https://web.archive. org/web/20000525220822/http://news2.thls.bbc.co.uk/hi/english/world/asia-pacific/news-id_493000/493133.stm

bu Ramseyer, J. Mark, "Nuclear power and the mob: Extortion and social capital in Japan." 31 March 2015. Social Science Research Network (SSRN). https://papers.ssrn.com/sol3/papers.cfm?abstract_id=2614087

bv Suzuki, "Yakuza and the nuclear industry."

bw This and other quotes on Reagan lifting the reprocessing ban are taken from his speech "Statement Announcing a Series of Policy Initiatives on Nuclear Energy," 8 October 1981. http://www.presidency.ucsb.edu/ws/?pid=44353

bx National Security Decision Directive (NSDD) 39, "United States policy on foreign reprocessing and use of plutonium subject to United States control." https://catalog.archives.gov/id/6879641

by *Yomiuri Shimbun*, p. 3, 26 May 1986.

bz Abe, Yasuhito, "Risk assessment of nuclear power by Japanese newspapers following the Chernobyl nuclear disaster." *International Journal of Communication*, vol. 7, 2013.

ca Atomic Energy Commission, "White paper on nuclear energy," 31 October 1986.

cb *Asahi Shimbun*, 1986.

cc Takao, *Japan's Environmental Politics and Governance*, p. 90.

cd Nuclear Safety Commission, "Nuclear Safety Commission Soviet Union nuclear power plant accident investigation special committee report" (in Japanese), May 1987. http://www.aec.go.jp/ jicst/NC/about/ugoki/geppou/V32/N05/198704V32N05.html

ce Ibid.

cf *Inside NRC*, vol. 8, no. 12, 9 June 1986.

cg Chernobyl hearing transcription. http://s3-euw1-ap-pe-ws4-cws-documents.ri-prod.s3.amazonaws. com/9781138917644/other/Doc_4E.pdf

ch Park, Y. J., and Hofmayer, C. H., "NUREG/CR-6241: Technical guidelines for aseismic design of nuclear power plants (translation of JEAG 4601-1987)." Japan Electric Association, Brookhaven National Laboratory, 1987. A full English translation (almost 1,000 pages long) is available here: https://www.nrc.gov/docs/ML1307/ML13079A392.pdf https://www.nrc.gov/docs/ML1713/ ML17131A127.pdf

ci Park and Hofmayer, "NUREG/CR-6241," p. 65 (p. 107 of the linked PDF).

cj Japan Atomic Energy Agency, "Reactor recirculation pump for Fukushima Daini Nuclear Power Station Unit 3 damage event" (in Japanese), May 1998. https://atomica.jaea.go.jp/data/detail/ dat_detail_02-07-02-08.html

ck "Pump rupture at Fukushima II arouses serious safety concern." *Nuke Info Tokyo*, March–April 1989. https://www.cnic.jp/english/old/newsletter/pdffiles/nit10_.pdf

cl "Fukushima II-3 re-started despite referendum." *Nuke Info Tokyo*, November–December 1990. https://www.cnic.jp/english/old/newsletter/pdffiles/nit20_.pdf

cm Lowinger, "Japan's nuclear energy development policies."

cn Howles, L. R., "Annual review of 9 power plant performance trends." *Nuclear Engineering International*, p. 19, June 1988.

co Sanger, David E., "Ideas and trends; A crack in Japan's nuclear sangfroid." *New York Times*, 2 February 1991. https://www.nytimes.com/1991/02/17/weekinreview/ideas-trends-a-crack-in-japans-nuclear-sangfroid.html

CHAPTER 5

a US Nuclear Regulatory Commission (USNRC), "Information notice no. 91-43: Recent incidents involving rapid increases in primary-to-secondary leak rate," 5 July 1991. https://www.nrc.gov/reading-rm/doc-collections/gen-comm/info-notices/1991/in91043.html

b Sanger, David E., "Japan now tells of radiation release." *New York Times*, 12 February 1991. https://www.nytimes.com/1991/02/12/world/japan-now-tells-of-radiation-release.html

c World Information Service on Energy (WISE) Amsterdam, "Update on Mihama accident/Japan." *Nuclear Monitor*, issue 351, 26 April 1991. https://www.wiseinternational.org/nuclear-monitor/351/update-mihama-accidentjapan

d Reid, T. R., "Japan concludes mistakes caused nuclear accident." *Washington Post*, 14 March 1991. https://www.washingtonpost.com/archive/politics/1991/03/14/japan-concludes-mistakes-caused-nuclear-incident/efc01c50-2039-4bcf-ad67-b591d2e2fa92/

e Japan Atomic Energy Agency, "Mihama Power Plant 2 steam generator heat transfer tube damage overview," May 1998 (updated). https://atomica.jaea.go.jp/data/detail/dat_detail_02-07-02-04.html

f Sanger, David E., "Japanese are hardly reassured on nuclear peril." *New York Times*, 5 March 1991. https://www.nytimes.com/1991/03/05/world/japanese-are-hardly-reassured-on-nuclear-peril.html

g Japanese government comments in the Diet, 27 March, 1991.

h Institute of Nuclear Safety System, Inc. http://www.inss.co.jp/english/institute.html

i Nuclear Safety Commission, "About accident management as severe accident countermeasure in light water nuclear reactor facilities for power generation," 28 May 1992. http://www.mext.go.jp/b_menu/hakusho/nc/t19920528001/t19920528001.html

j National Diet of Japan Fukushima Nuclear Accident Independent Investigation Commission (henceforth National Diet NAIIC), *Final Report*. National Diet of Japan, 2012. https://warp.da.ndl.go.jp/info:ndljp/pid/3856371/naiic.go.jp/en/report/

k Clenfield, Jason, "Nuclear regulator dismissed seismologist on Japan quake threat." Bloomberg, 21 November 2011. https://www.bloomberg.com/news/articles/2011-11-21/nuclear-regulator-dismissed-seismologist-on-japan-quake-threat

l Kimura, Makoto, "Damage statistics." *Special Issue of Soils and Foundations, Japanese Geotechnical Society*, 1-5 January 1996. https://www.jstage.jst.go.jp/article/sandf1995/36/Special/36_1/_pdf

m Headquarters for Earthquake Research Promotion, "The basic objectives and roles of the Headquarters for Earthquake Research Promotion," n.d. https://www.jishin.go.jp/main/w_shokai-e.htm

n "Report of the Japanese government to the IAEA Ministerial Conference on Nuclear Safety: The accident at TEPCO's Fukushima Nuclear Power Stations." Nuclear Emergency Response Headquarters, Government of Japan, June 2011. http://flexrisk.boku.ac.at/zitate/full_report.pdf. Apparently, this was a "long-standing" prediction.

o Sanger, David E., "Japan's nuclear fiasco." *New York Times*, 20 December 1992. https://www.nytimes.com/1992/12/20/business/japan-s-nuclear-fiasco.html

p Skolnikoff, Eugene, Suzuki, Tatsujiro, and Oye, Kenneth, "International responses to Japanese plutonium programs." Center for International Studies, Massachusetts Institute of Technology, August 1995. https://archive.org/stream/InternationalResponsesToJapanesePlutoniumPrograms/1995-08IntlResponseToJpnsPlutoniumProgram_djvu.txt. This paper is incredibly interesting and well worth a read; it articulates the criticism of Japan's breeder and recycling programs far better and with more detail than I can.

q Federation of Electric Power Companies, "Graphical flip-chart."

r Shimazaki, Masaki, "Liberalization of the Japanese electricity market." *Journal of Energy and Development*, vol. 20, no. 1, pp. 79-96, autumn 1994. https://www.jstor.org/stable/24808929

s Scalise, Paul, "Whatever happened to Japan's energy deregulation?" Research Institute of Economy, Trade and Industry, 24 June 2009. https://www.rieti.go.jp/en/events/bbl/09062401.html

t Fast Breeder Reactor Research and Development Centre, JAEC, "Basic specifications: Monju Plant." http://www.jaea.go.jp/04/monju/EnglishSite/contents02-1.html. Archived link: https://web.archive.org/web/20130123183123/http://www.jaea.go.jp/04/monju/EnglishSite/contents02-1.html. This page has a huge amount of technical information on the plant.

u   Mikami, H., Shono, A., and Hiroi, H., "Sodium leak at Monju: Cause and consequences," n.d. Reactor and System Engineering Section, Monju Construction Office, Power Reactor and Nuclear Fuel Development Corp. https://inis.iaea.org/collection/NCLCollectionStore/_Public/31/044/31044840.pdf

v   Japan Atomic Energy Agency, "Current status of investigation" (in Japanese), https://web.archive.org/web/20070222035737/http://www.jaea.go.jp/04/monju/category05/mj_accirep/mj_accirep05.html

w   Mikami et al., "Sodium leak at Monju."

x   *Daily News*, 9 December 1995.

y   "Delay worsened coolant leak of Japanese reactor." *Orlando Sentinel*, 19 December 1995. http://articles.orlandosentinel.com/1995-12-19/news/9512180765_1_reactor-leak-coolant

z   *Daily News*, 22 December 1995.

aa   Kaito, Yuichi, "I want to reveal the truth of my husband's death!" Litigation notes, 2012. (This is a lengthy blog post written in Japanese by the lawyer in the Shigeo Nishimura suicide appeal case.) http://www.news-pj.net/npj/2007/dounen-20071020.html

ab   "After cover-up, shame and suicide." *International Herald Tribune*, 15 January 1996. (English edition of a French daily newspaper.) https://archive.org/stream/InternationalHeraldTribune-1996FranceEnglish/Jan+15+1996%2C+International+Herald+Tribune%2C+%2335108%2C+-France+%28en%29_djvu.txt

ac   Kaito, "I want to reveal the truth of my husband's death!"

ad   Pollack, Andrew, "Reactor accident in Japan imperils energy program." *New York Times*, 24 February 1996. https://www.nytimes.com/1996/02/24/world/reactor-accident-in-japan-imperils-energy-program.html

ae   "Court rejects lawsuit over Monju coverup suicide." *Japan Times*, 15 May 2007. https://www.japantimes.co.jp/news/2007/05/15/national/court-rejects-lawsuit-over-monju-coverup-suicide/

af   This and a lot of other information on the Monju leak comes from Miyakawa, A., et al., "Sodium leakage experience at the prototype FBR Monju," 2004. Monju Construction Office, Japan Nuclear Cycle Development Institute. https://pdfs.semanticscholar.org/a04a/45f4851c8623a6c7405701fa-80c69681543a.pdf

ag   *Nuke Info Tokyo*, no. 54. July–August 1996. http://www.cnic.jp/english/newsletter/pdffiles/nit54.pdf

ah   Board of Audit of Japan, fiscal 1995 settlement inspection report, "About the sodium leakage accident of fast breeder prototype reactor Monju" (in Japanese). https://report.jbaudit.go.jp/org/h07/1995-h07-0433-0.htm

ai   Pangi, Robyn, "Consequence management in the 1995 sarin attacks on the Japanese subway system," discussion paper 2002-4, Belfer Center for Science and International Affairs, February 2002. https://www.belfercenter.org/sites/default/files/legacy/files/consequence_management_in_the_1995_sarin_attacks_on_the_japanese_subway_system.pdf

aj   Itagaki, Haruhiko, Kobayashi, Mitsuo, and Tamura, Shozo, "Fire explosion accident in PNC asphalt solidification processing facility" (in Japanese), 1997. Failure Knowledge Database, Japan Science and Technology Agency. http://www.jst.go.jp/fnr/info/info1/db.pdf

ak   JAEC, "Part 1, section 2.1: Asphalt solidification treatment facility fire explosion accident and subsequent improper response" (in Japanese), white paper on nuclear energy, 1998. http://www.aec.go.jp/jicst/NC/about/hakusho/hakusho10/siryo1021.htm

al   This source has a more detailed description of what caused the explosion: JAEA, "Fire explosion accident at Tokai reprocessing plant" (in Japanese), 2000. https://atomica.jaea.go.jp/data/detail/dat_detail_04-10-02-01.html

am   Pollack, Andrew, "After accident, Japan rethinks its nuclear hopes." *New York Times*, 25 March 1997. https://www.nytimes.com/1997/03/25/world/after-accident-japan-rethinks-its-nuclear-hopes.html

an   JAEC, "Power Reactor and Nuclear Fuel Development Corporation (PNC): Problems and organization reform," n.d. http://www.aec.go.jp/jicst/NC/about/hakusho/hakusho98/siryo12.htm

ao   Clenfield, "Nuclear regulator dismissed seismologist."

ap   USNRC, "Severe accident risks: An assessment for five US nuclear power plants – Final summary report (NUREG-1150, vol. 1), part II," December 1990. https://www.nrc.gov/reading-rm/doc-collections/nuregs/staff/sr1150/v1/

aq   Tsuchiya, S., et al., "An analysis of Tokaimura nuclear criticality accident: A systems approach." Chiba Institute of Technology, 2001. https://web.archive.org/web/20191101075412/http://www.dinamica-de-sistemas.com/paper/08_04.pdf

ar   "JCO ends effort to reprocess nuke fuel." *Asahi Shimbun*, 19 April 2003. https://web.archive.org/

web/20030421232914/http://www.asahi.com/english/national/K2003041900174.html

as Not directly related to Tokai, but this source includes descriptions of how the process works: Nuclear Materials Disposition Program Office Defense Programs, "Conversion and blending facility highly enriched uranium to low enriched uranium as uranyl nitrate hexahydrate," 1995. https://inis.iaea.org/collection/NCLCollectionStore/_Public/27/031/27031561.pdf

at USNRC, "NRC review of the Tokai-mura criticality accident," April 2000, p. 4. https://www.nrc.gov/reading-rm/doc-collections/commission/secys/2000/secy2000-0085/attachment1.pdf

au "JCO used electric stove to process nuclear fuel." *Yomiuri Shimbun*, 5 November 1999. http://www.yomiuri.co.jp/newse/1105cr08.htm

av IAEA, "Report on the preliminary fact finding mission following the accident at the nuclear fuel processing facility in Tokaimura, Japan," 1999. https://www-pub.iaea.org/MTCD/publications/PDF/TOAC_web.pdf

aw Judges of the Criminal Affairs Department of the Mito District Court, Judge Hideyuki Suzuki, Judge Kenji Shimozu, Judge Kazunobu Eguchi, "Court hearing sentencing document" (in Japanese), 3 March 2003. https://www.courts.go.jp/hanrei/pdf/9A577AEACFA7154A49256CFE-002AE108.pdf. Archive link: https://web.archive.org/web/20111116103938/https://www.courts.go.jp/hanrei/pdf/9A577AEACFA7154A49256CFE002AE108.pdf. This document is very thorough in detailing how the lack of safety procedures at JCO allowed for the accident to happen.

ax "Nuclear workers knew little about critical reaction." Agence France-Presse, 11 October 1999. Archive link: https://web.archive.org/web/20000505090750/http://www.smh.com.au/news/9910/11/text/world2.html

ay "Police to quiz JCO employee over negligence in N-accident." *Yomiuri Shimbun*, 25 October 1999. http://www.yomiuri.co.jp/newse/1025so05.htm

az Chiba, S., et al., "Transplantation for accidental acute high-dose total body neutron and gamma-radiation exposure." *Bone Marrow Transplantation* vol. 29, pp. 935-939, 2002.

ba Maekawa, K., "Overview of medical care for highly exposed victims of Tokaimura accident." In R. Ricks and M. Berger (eds.), *The Medical Basis for Radiation-Accident Preparedness: The Clinical Care of Victims*. CRC Press, 2002.

bb Akashi, Makoto, et al., "Initial symptoms of acute radiation syndrome in the JCO criticality accident in Tokai-mura." *Journal of Radiation Research*, September 2001. http://paperity.org/p/34424931/initial-symptoms-of-acute-radiation-syndrome-in-the-jco-criticality-accident-in-tokai

bc Akashi, Makoto, "Radiation emergency medical preparedness in Japan and a criticality accident at Tokai-mura." In H. Tsushima (ed.), *Proceedings of the 1st Hirosaki University International Symposium on Radiation Emergency Medicine*. Hirosaki University Press, 2009. https://www.hs.hirosaki-u.ac.jp/hibaku/img/pdf/210801symposium-report.pdf

bd Akashi et al., "Initial symptoms of acute radiation syndrome."

be "Radiation: How much is considered safe for humans?" *MIT News*, Massachusetts Institute of Technology, 5 January 1994. http://news.mit.edu/1994/safe-0105

bf Moss, William, and Eckhardt, Roger, "The human plutonium injection experiments." Los Alamos Science. Radiation Protection and the Human Radiation Experiments, 1995.

bg Welsome, Eileen, *The Plutonium Files: America's Secret Medical Experiments in the Cold War*. Dell Publishing, 1993.

bh Japan Broadcasting Corporation (Nippon Hōsō Kyōkai [NHK]), Vertical Inc., *A Slow Death: 83 Days of Radiation Sickness*, 2008. This is an excellent book based on work done for an NHK documentary of the same name, charting Ouchi's time in the hospital. The documentary won the Gold Nymph Award (the highest honour) at the 42nd Monte Carlo Television Festival in 2002. The book is very difficult to read owing to its painful descriptions and interviews with the hospital staff, but it's a vital resource, and I'd encourage anyone to check it out. Many details of Ouchi's time in hospital are referenced from this book.

bi Ogawa, Seishi, and Takeuchi, Kengo, "Transplantation for accidental acute high-dose total body neutron- and y-radiation exposure." *Bone Marrow Transplantation*, 2002. https://www.researchgate.net/publication/11295492_Transplantation_for_accidental_acute_high-dose_total_body_neutron-_and_-radiation_exposure. Note: the first case of such a transplant in general use on leukaemia was by pioneering scientist and Nobel laureate E. Donnall Thomas: Brenner, Benjamin, "Hematopoietic stem cell transplantation: 50 years of evolution and future perspectives." https://www.ncbi.nlm.nih.gov/pmc/articles/PMC4222417/

bj *A Slow Death*, p. 6. Note: this is not a quote from Maekawa but a quote from the book describing his reaction, based on interviews.

bk "NHK special: Tokai-mura criticality accident" (in Japanese). NHK, 2001. https://youtu.be/ZWomuWd7-to

bl Watts, Jonathan, "The day they blew the roof off corners that led to critical mass." *Guardian*, 2 October 1999. https://www.theguardian.com/world/1999/oct/02/jonathanwatts

bm USNRC, "NRC review of the Tokai-Mura criticality accident," p. 3, April 2000. https://www.nrc.gov/reading-rm/doc-collections/commission/secys/2000/secy2000-0085/attachment1.pdf

bn Lamar, Joe, "Japan's worst nuclear accident leave [sic] two fighting for life." *British Medical Journal*, vol. 319, issue 7215, p. 942, 9 October 1999. https://www.ncbi.nlm.nih.gov/pmc/articles/PMC1174623/

bo "Nuclear accident shakes Japan." BBC News, 30 September 1999. http://news.bbc.co.uk/1/hi/world/asia-pacific/461446.stm

bp "Saved by the rain: How Japan muddled through its latest nuclear accident." *Economist*, 7 October 1999. https://www.economist.com/asia/1999/10/07/saved-by-the-rain

bq "Japan N-plant has 15,000 times normal radiation." Reuters, 30 September 1999. The following link includes the chronology and press reports of the Tokaimura criticality, Institute for Science and International Security: http://isis-online.org/isis-reports/detail/chronology-and-press-reports-of-the-tokaimura-criticality/37

br IAEA, "Report on the preliminary fact finding mission following the accident at the nuclear fuel processing facility in Tokaimura, Japan." International Atomic Energy Agency (IAEA), 1999. https://www-pub.iaea.org/MTCD/publications/PDF/TOAC_web.pdf

bs Ryan, Michael E. "The Tokaimura Accident." Department of Chemical Engineering, University at Buffalo, State University of New York. https://web.archive.org/web/20150227195925/http://library.buffalo.edu:80/libraries/projects/cases/tokaimura/tokaimura_2.html

bt Tanaka, Shun-ichi. "Summary of the JCO criticality accident in Tokai-mura and a dose assessment." *Journal of Radiation Research*, 2001, Volumevol. 42, suppl., pp. S1-S9, 28 October 2001. https://www.jstage.jst.go.jp/article/jrr/42/SUPPL/42_SUPPL_S1/_article/-char/en

bu Science and Technology Agency, 15 October 1999.

bv Nuclear Safety Commission, "A summary of the report of the Criticality Accident Investigation Committee," 24 December 1999. http://www-bcf.usc.edu/~meshkati/tefall99/NSC.pdf

bw Tanaka, "Summary of the JCO criticality accident."

bx Watts, "The day they blew the roof off.";

by Farley, Maggie, "Nuclear workers appeared unaware of dangers." *Los Angeles Times*, 7 October 1999. http://articles.latimes.com/1999/oct/07/news/mn-19738

bz "Japan nuclear accident." Associated Press, 8 October 1999. "Toshiki Takagi, president of the Metal Mining Agency and Moriki Aoyagi, president of Sumitomo Metal Mining Company, have both resigned. Takagi served as president of JCO Company from June 1995 until June this year, Kyodo News reported."

ca Chiba, et al., "Transplantation for accidental acute high-dose."

cb "US expert arrives to help treat nuke accident victims." *Kyodo News*, 8 October 1999.

cc Ishii, Takeshi, "Brief note and evaluation of acute-radiation syndrome and treatment of a Tokai-mura criticality accident patient." *Journal of Radiation Research*, September 2001. http://paperity.org/p/34453684/brief-note-and-evaluation-of-acute-radiation-syndrome-and-treatment-of-a-tokai-mura

cd "NHK special: Tokai-mura criticality accident" (in Japanese). NHK, 2001.

ce Maekawa, "Overview of medical care."

cf "NHK special: Tokai-mura criticality accident" (in Japanese). NHK, 2001.

cg Hirano, Katsuya, "Fukushima and the crisis of democracy: Interview with Murakami Tatsuya." *Asia-Pacific Journal*, 25 May 2015. https://apjjf.org/Katsuya-Hirano/4320.html

ch *Mainichi Shimbun*, 4 October 1999. Referenced in Farley, Maggie, "Nuclear reliance, unease grow in Japan." *Los Angeles Times*, 6 October 1999. http://articles.latimes.com/1999/oct/06/news/mn-19273

ci Hindell, Juliet, "Japanese nuclear 'slaves' at risk." BBC, 29 October 1999. http://news2.thls.bbc.co.uk/hi/english/world/asia-pacific/newsid_493000/493133.stm. Archived link: https://web.archive.org/web/20001027063054/http://news2.thls.bbc.co.uk/hi/english/world/asia-pacific/newsid_493000/493133.stm

cj McCurry, Justin, "Nuclear accident claims new death." *Guardian*, 28 April 2000. https://www.theguardian.com/world/2000/apr/28/justinmccurry

ck Furata, Kazuo (JCO Accident Investigation Special Working Group), "View on JCO criticality accident" (in Japanese). Human Machine System Research Group, Atomic Energy Society of Japan. http://www.aesj.or.jp/~hms/link/joc-j.html

cl "Japan tightens nuclear safety." BBC, 13 December 1999. http://news2.thls.bbc.co.uk/hi/

english/world/asia-pacific/newsid_562000/562260.stm. Archived link: https://web.archive.org/web/20000521041118/http://news2.thls.bbc.co.uk/hi/english/world/asia-pacific/news-id_562000/562260.stm

cm Nuclear Safety Commission, "A summary of the report."

cn Ibid.

co World Nuclear Association, "Tokaimura criticality accident 1999," 2013. https://www.world-nuclear.org/information-library/safety-and-security/safety-of-plants/tokaimura-criticality-accident.aspx

cp Organisation for Economic Co-Operation and Development (OECD), "Tokai-mura accident, Japan: Third-party liability and compensation aspects." *Nuclear Law Bulletin*, no. 66, December 2000. https://www.oecd-nea.org/law/nlb/nlb66.pdf

cq "System to monitor radiation went unused." *Yomiuri Shimbun*, 5 October 1999.

cr "Tokai accident leaves lasting scars." *Japan Times*, 23 September 2000. https://www.japantimes.co.jp/news/2000/09/23/national/tokai-accident-leaves-lasting-scars/

cs "Substance showing DNA damage detected in 8 people." Associated Press, 7 November 1999.

ct "Six JCO staff arrested over atomic accident." *Japan Times*, 12 October 2000. https://www.japantimes.co.jp/news/2000/10/12/national/six-jco-staff-arrested-over-atomic-accident/

cu "Japanese nuclear firm raided." BBC, 16 December 1999. http://news2.thls.bbc.co.uk/hi/english/world/asia-pacific/newsid_567000/567376.stm. Archived link: https://web.archive.org/web/20001027150048/http://news2.thls.bbc.co.uk/hi/english/world/asia-pacific/news-id_567000/567376.stm

cv "Six JCO employees sentenced over fatal '99 nuclear accident." *Japan Times*, 4 March 2003. https://www.japantimes.co.jp/news/2003/03/04/national/six-jco-employees-sentenced-over-fatal-99-nuclear-accident/

cw "Judgement on JCO criticality accident." *Nuke Info Tokyo*, March–April 2003. http://www.cnic.jp/english/newsletter/pdffiles/nit94.pdf

cx "JCO ends effort to reprocess nuke fuel." *Asahi Shimbun*, 19 April 2003. https://web.archive.org/web/20030421232914/http://www.asahi.com/english/national/K2003041900174.html; "JCO calls off plan to work with uranium." *Japan Times*, 20 April 2003. https://www.japantimes.co.jp/news/2003/04/20/national/jco-calls-off-plan-to-work-with-uranium/

cy "Inspectors sent in as Sellafield admits to serious safety lapses." *Independent*, 14 September 1999. Archived link: https://web.archive.org/web/20010223130043/http://www.state.nv.us/nucwaste/news/nn10174.htm. (Why the old State of Nevada government website had this stuff saved, I have no idea.)

cz The following government investigation report contains ample information for anyone who's interested: "An investigation into the falsification of pellet diameter data in the MOX demonstration facility at the BNFL Sellafield site and the effect of this on the safety of MOX fuel in use," Her Majesty's Nuclear Installations Inspectorate, 18 February 2000. Archive link: https://web.archive.org/web/20060924030658/http://www.hse.gov.uk/nuclear/mox/mox1.htm

da Ibid.

db Johnston, Eric, "KEPCO may cut off MOX supplier." *Japan Times*, 17 December 1999. https://www.japantimes.co.jp/news/1999/12/17/national/kepco-may-cut-off-mox-supplier/

dc Watts, Jonathan, and Brown, Paul, "Sellafield shipment rejected by Japan." *Guardian*, 17 December 1999. https://www.theguardian.com/world/1999/dec/17/nuclear.uk

dd "Sellafield body parts families given government apology." BBC, 16 November 2010. https://www.bbc.co.uk/news/uk-england-cumbria-11768944

de "The Redfern Enquiry into human tissue analysis in UK nuclear facilities, vol. 1: Report," p. 320. UK Government, 2010. https://assets.publishing.service.gov.uk/government/uploads/system/uploads/attachment_data/file/229155/0571_i.pdf

CHAPTER 6

a "Japan tightens nuclear safety," BBC.

b Shiroyama, H., "Chapter 14: Nuclear safety regulation in Japan and impacts of the Fukushima Daiichi accident." In J. Ahn, et al. (eds.), *Reflections on the Fukushima Daiichi Nuclear Accident: Toward Social-Scientific Literacy and Engineering Resilience*. Springer, 2015.

c The relevant act is a 1999 revision of the Act for Establishment of the Atomic Energy Commission and the Nuclear Safety Commission: Shoriyama, Hideaki, "Regulatory failures of nuclear safety in Japan: The case of [the] Fukushima accident." Paper presented at *Earth System Governance Tokyo Conference: Complex Architectures, Multiple Agents*, 2013. Archive link: https://web.archive.org/web/20200620072312/https://nanopdf.com/downloadFile/regulatory-failures-of-nuclear-safety-in-japan-the-case-of-hideaki-shiroyama-the_pdf d   IAEA, "Historical development."

e   "Act on Special Measures Concerning Nuclear Emergency Preparedness" (act no. 156 of 17 December 1999). http://www.cas.go.jp/jp/seisaku/hourei/data/ASMCNEP.pdf

f   OECD / Nuclear Energy Agency (NEA), *Nuclear Law Bulletin*, issue 66, December 2000, p. 43. http://www.oecd-nea.org/law/nlb/nlb66.pdf

g   Electric Power Research Institute, "BWRVIP-1 81 NP: BWR vessel and internals project: Steam dryer repair design criteria," November 2007. https://www.nrc.gov/docs/ML0735/ML073551147. pdf

h   "福島原発陰謀 東電トラブル隠し事件 Kei Sugaoka GE / TEPCO whistleblower" (in Japanese, but Sugaoka's remarks are in English). *JNN News*, 2003. https://www.youtube.com/ watch?v=fBjiLaVOsl4

i   Minoura, K., et al., "The 869 Jogan tsunami deposit and recurrence interval of large-scale tsunami on the Pacific coast of northeast Japan." *Journal of Natural Disaster Science*, vol. 23, no. 2, pp. 83-88, 2001. https://www.jsnds.org/jnds/23_2_3.pdf

j   Pritchard, Justin, and Kageyama, Yuri, "Fukushima tsunami safety plan: A single page." NBC News, 27 June 2011. http://www.nbcnews.com/id/43193695/ns/world_news-asia_pacific/t/fukushima-tsunami-safety-plan-single-page/

k   Nuclear and Industrial Safety Agency, "Investigation into false recordings of licensee's self-imposed inspection works at nuclear power plants," p. 7, 29 August 2002. Archive link: https://web. archive.org/web/20030826133608/http://www.nisa.meti.go.jp:80/text/kokusai/020829.pdf

l   "Tokai-area earthquake could claim 8,100 lives, cost 345 billion yen a day." *Japan Times*, 30 August 2002. https://www.japantimes.co.jp/news/2002/08/30/national/tokai-area-earthquake-could-claim-8100-lives-cost-345-billion-yen-a-day/

m   "Radiation leak shuts down TEPCO reactor." *Japan Times*, 4 September 2002. https://www.japantimes.co.jp/news/2002/09/04/national/radiation-leak-shuts-down-tepco-reactor/

n   McNeill, David, "Warnings of nuclear disaster not heeded, claims former governor." *Independent*, 23 April 2011. https://www.independent.co.uk/news/world/asia/warnings-of-nuclear-disaster-not-heeded-claims-former-governor-2273764.html

o   "Nuclear hosts demand investigation into TEPCO." *Japan Times*, 3 September 2002. https://www.japantimes.co.jp/news/2002/09/03/national/nuclear-hosts-demand-investigation-into-tepco/

p   "Fukushima to scrap pluthermal project." *Japan Today*, 25 September 2002. http://japantoday.com/e/?content=news&cat=4&id=231593. Archived link: https://web.archive.org/ web/20021014152750/http://japantoday.com/e/?content=news&cat=4&id=231593

q   "Hiranuma sorry for delay in probe of TEPCO scandal." *Japan Times*, 4 September 2002. https:// www.japantimes.co.jp/news/2002/09/04/national/hiranuma-sorry-for-delay-in-probe-of-tepco-scandal/

r   Leyse, Mark, and Paine, Christopher, "Preventing hydrogen explosions in severe nuclear accidents: Unresolved safety issues involving hydrogen generation and mitigation." Natural Resources Defense Council (NRDC), March 2014. https://www.nrdc.org/sites/default/files/hydrogen-generation-safety-report.pdf

s   "Players identified in TEPCO cover-up." *Asahi Shimbun*, 18 November 2002. https://web.archive. org/web/20030213004558/http://www.asahi.com/english/national/K2002111800299.html; "TEPCO punishes 9 for fake leak rates." *Asahi Shimbun*, 12 December 2002. https://web.archive.org/ web/20030212144334/http://www.asahi.com/english/national/K2002121200477.html

t   "TEPCO tried to rush nuke reactor checks." *Asahi Shimbun*, 28 November 2002. https://web. archive.org/web/20021215210540/http://www.asahi.com/english/business/K2002112800231.html

u   Krolicki, Kevin, and Fujioka, Chisa, "Special report: Japan's 'throwaway' nuclear workers." Reuters, 24 June 2011. https://www.reuters.com/article/idINIndia-57893220110624

v   "TEPCO in-house probe reveals division chiefs' coverup role." *Japan Times*, 6 September 2002. https://www.japantimes.co.jp/news/2002/09/06/national/tepco-in-house-probe-reveals-division-chiefs-coverup-role/

w   "30 TEPCO head office officials may be involved in cover-ups." *Japan Today*, 8 September 2002. https://web.archive.org/web/20020920053554/http://www.japantoday.com/e/?content=news&cat=2&id=229490

x   "TEPCO official suspected of giving damage coverup order." *Japan Times*, 10 September 2002. https://www.japantimes.co.jp/news/2002/09/10/national/tepco-official-suspected-of-giving-damage-coverup-order/

y   "TEPCO chairman, president announce resignations over nuclear coverups." *Japan Times*, 3 September 2002. https://www.japantimes.co.jp/news/2002/09/03/national/tepco-chairman-president-announce-resignations-over-nuclear-coverups/

z   Nuclear and Industrial Safety Agency (NISA), Ministry of Economy, Trade and Industry (METI), "Investigation results of falsified records issue at nuclear power plants and prevention of recurrence," 7 October 2002. https://web.archive.org/web/20030322170754/http://www.nisa.meti.go.jp:80/text/kokusai/TEPCO021007.pdf

aa Edwards, Rob, "Japan's nuclear safety 'dangerously weak.'" *New Scientist*, 1 October 2002. https://www.newscientist.com/article/dn2859-japans-nuclear-safety-dangerously-weak/

ab "Japan Atomic Power hid cracks." *Japan Times*, 26 September 2002. https://www.japantimes.co.jp/news/2002/09/26/national/japan-atomic-power-hid-cracks/

ac "Shimane offline following faulty inspections." *World Nuclear News*, 31 March 2010. http://www.world-nuclear-news.org/RS-Shimane_offline_following_faulty_inspections-3103105.html

ad "New TEPCO head vows change." *Asahi Shimbun*, 16 October 2002. https://web.archive.org/web/20021018193045/http://www.asahi.com/english/business/K2002101600352.html

ae TEPCO, Mr Katsumata speech BGM 2003: "Reconstruction after misconduct: The pursuit of excellence," 10 March 2003.

af More details as to what steps it took to regain public trust are available in this speech from a TEPCO manager: Kuroda, Hiroyuki, "Lesson learned from TEPCO nuclear power scandal." Tokyo Electric Power Company, 25 March 2004. http://www.tepco.co.jp/en/news/presen/pdf-1/040325-s-e.pdf

ag A summary of things that were changed can be found here: Nuclear and Industrial Safety Agency, "Status report: Countermeasures against falsification related to inspections at nuclear power stations," 10 December 2002. https://web.archive.org/web/20030826123118/http://www.nisa.meti.go.jp:80/text/kokusai/TEPCO021210.pdf

ah Kishimoto, "Public attitudes and institutional changes."

ai Kitada, Atsuko, "Impact of the TEPCO incident on the public's attitude to nuclear power generation: Periodic survey no. 3" (in Japanese). Institute of Nuclear Safety System, Inc., 2003. http://www.inss.co.jp/wp-content/uploads/2017/03/2003_10J044_062.pdf

aj "Power firms struggling to generate public trust." *Yomiuri Shimbun*, 30 December 2006. Archived link: https://web.archive.org/web/20070106225931/http://www.yomiuri.co.jp/dy/editorial/20061230TDY04005.htm. "Power firms hid 10,000 problems." *Yomiuri Shimbun*, 6 April 2007. Archived link: https://web.archive.org/web/20070410040440/http://www.yomiuri.co.jp/dy/national/20070406TDY02006.htm

ak "TEPCO must probe 199 plant check coverups." *Japan Times*, 2 February 2007. https://www.japantimes.co.jp/news/2007/02/02/national/tepco-must-probe-199-plant-check-coverups/

al "TEPCO owns up to past nuclear plant accidents." *Japan Times*, 31 March 2007. https://web.archive.org/web/20070813134225/http://search.japantimes.co.jp/cgi-bin/nn20070331a4.html

am "'98 reactor emergency in Miyagi covered up." *Japan Times*, 13 March 2007. https://www.japantimes.co.jp/news/2007/03/13/national/98-reactor-emergency-in-miyagi-covered-up/

an "Outrage over Japan Nuclear Reactor coverup." Reuters, 15 March 2007. https://www.thestar.com.my/news/world/2007/03/15/outrage-over-japan-nuclear-reactor-coverup/ (I have no idea why the *Star* has a working link to this but Reuters doesn't).

ao "Three nuclear testing facilities in Tokai ordered to shut down." *Japan Times*, 2 September 2007. https://www.japantimes.co.jp/news/2007/09/02/national/three-nuclear-testing-facilities-in-tokai-ordered-to-shut-down/

ap Nuclear and Industrial Safety Agency, "Analytical evaluation results for the interim reports on the comprehensive inspection results related to appropriateness of self-imposed inspection of nuclear facilities," 24 December 2002. https://web.archive.org/web/20030322190519/http://www.nisa.meti.go.jp:80/text/kokusai/TEPCO021224soutenken.pdf

aq "Japan orders stricter checks at reactors." *World Nuclear News*, 20 April 2007. http://www.world-nuclear-news.org/regulationSafety/200407-Japan_orders_stricter_checks_at_reactors.shtml

ar "Seven nuke utilities told to conduct new checks." *Japan Times*, 21 April 2007. https://web.archive.org/web/20070813134251/http://search.japantimes.co.jp/cgi-bin/nn20070421a1.html

as Wise International. "Japan's nuclear industry plagued by scandals." 20 April 2007. https://www.wiseinternational.org/nuclear-monitor/654/japans-nuclear-industry-plagued-scandals, Citing: *Niigata Nippo*, 2 February 2007.

at "Two recirculation pumps fail simultaneously at Hamaoka Unit 1." *Nuke Info Tokyo*, March–April 1988. https://www.cnic.jp/english/old/newsletter/pdffiles/nit04_.pdf

au Kansai Electric Power Company (KEPCO), "A report on the conditions of the victims of the event at Mihama Unit 3 and on investigations of the automatic reactor trip." Press release, 10 August 2004. Archived link: https://web.archive.org/web/20041010051743/http://www.kepco.co.jp/english/press/2004/e20040810_1.html

av   Nuclear and Industry Safety Agency, "Automatic shut-down of Unit 3, Mihama Nuclear Power Plant (Kansai Electric Power Co., Inc.), 3rd report," 10 August 2004. Archived link: https://web. archive.org/web/20041109070336/http://www.meti.go.jp/english/newtopics/data/mihama3e.html

aw USNRC, "Secondary piping rupture at the Mihama Power Station in Japan," 16 March 2006. https://www.nrc.gov/reading-rm/doc-collections/gen-comm/info-notices/2006/in200608.pdf

ax Brooke, James, "Blown pipe in Japan nuclear plant accident had been used, but not checked, since 1976." *New York Times*, 11 August 2004. https://www.nytimes.com/2004/08/11/world/blown-pipe-japan-nuclear-plant-accident-had-been-used-but-not-checked-since-1976.html

ay USNRC, "Erosion/corrosion-induced pipe wall thinning in US nuclear power plants. 2. Major incidents of pipe wall thinning and rupture in feedwater systems," p. 2/3 (p. 14 of PDF), 1 April 1989. https://www.osti.gov/servlets/purl/6152848

az Markey, Edward J., "Action needed to ensure that utilities monitor and repair pipe damage." US General Accounting Office, 18 March 1988. https://www.gao.gov/assets/150/146211.pdf

ba "Criminal charges sought in fatal nuclear plant accident." *Asahi Shimbun*, 13 August 2004. Archived link: https://web.archive.org/web/20041018000740/http://www.asahi.com/english/nation/TKY200408130176.html

bb "MHI used notebooks to keep track of reactor pipes." *Japan Times*, 22 August 2004. https://www.japantimes.co.jp/news/2004/08/22/national/mhi-used-notebooks-to-keep-track-of-reactor-pipes/

bc Comment by Yonezo Tsujikura, "Officials grill operator over nuke mishap." *New York Times* (Associated Press), 20 August 2004. Archive link: https://web.archive.org/web/20060508045259/http://www.csitechnologies.com/Mihama/The%20New%20York%20Times%20%20AP%20%20International%20%20Officials%20Grill%20Operator%20Over%20Nuke%20Mishap.htm

bd Brooke, "Blown pipe in Japan nuclear plant accident."

be In total, there were 66 pipes thinner than allowed by the regulations, all of which were replaced. "KEPCO reactor pipes fail multiple checks: Scores of corrosion dangers." *Japan Times*, 23 February 2007. https://www.japantimes.co.jp/news/2007/02/23/national/kepco-reactor-pipes-fail-multiple-checks/

bf Japan Atomic Energy Agency, "Damage to the secondary system piping at Unit 3 of the Kanden Mihama Power Plant," February 2007. https://atomica.jaea.go.jp/data/detail/dat_detail_02-07-02-23.html

bg "Japan's shaky nuclear record." BBC, 24 March 2006. http://newsvote.bbc.co.uk/1/hi/world/asia-pacific/3548192.stm

bh "Nuclear plant accident splits Japan press." BBC, 10 August 2004. http://news.bbc.co.uk/1/hi/world/asia-pacific/3551902.stm

bi Brooke, "Blown pipe in Japan nuclear plant accident."

bj "Editorial: Nuclear reactor accident." *Asahi Shimbun*, 3 September 2004. Archived link: https://web.archive.org/web/20040903085011/http://www.asahi.com/english/opinion/TKY200409030116.html

bk "2 more N-plants report problems." *Yomiuri Shimbun*, 21 March 2007. Archived link: https://web.archive.org/web/20070323223054/http://www.yomiuri.co.jp/dy/national/20070321TDY02009.htm

bl "Monju, modified." *Nuclear Engineering International*, 4 March 2011. https://archive.is/20130130012503/http://www.neimagazine.com/story.asp?storyCode=2059044

bm National Diet NAIIC, "Chapter 5: Organizational issues of the parties involved in the accident, p. 17. *Final Report*. National Diet of Japan, 2012. https://warp.da.ndl.go.jp/info:ndljp/pid/3856371/naiic.go.jp/en/report; http://warp.da.ndl.go.jp/info:ndljp/pid/3856371/naiic.go.jp/en/report/

bn Clenfield and Sat, "Japan nuclear energy drive."

bo Birch, Douglas, Adelstein, Jake, and Smith, R. Jeffry, "Unarmed guards, bogus terror drills, and 96 tons of plutonium." *Foreign Policy*, 10 March 2014. https://foreignpolicy.com/2014/03/10/unarmed-guards-bogus-terror-drills-and-96-tons-of-plutonium/

bp National Diet NAIIC, "Chapter 5," p. 14 *Final Report*.

bq "No tsunami effect on nuclear plant." rediff.com (India), 7 January 2005. https://www.rediff.com/news/2005/jan/07inter1.htm

br Okuyama, Toshihiro, "Weaknesses discovered by eight TEPCO engineers." *Asahi Shimbun*, 22 November 2019. https://www.asahi.com/articles/ASMBZ1DVXMBYULZU017.html

bs National Diet NAIIC, "Chapter 1: Was the accident preventable?" p. 26. *Final Report*. National Diet of Japan, 2012. https://warp.da.ndl.go.jp/info:ndljp/pid/3856371/naiic.go.jp/en/report

bt National Diet NAIIC, "Chapter 5," p. 7. *Final Report*.

bu Ishibashi, Katsuhiko, "About Nuclear Safety Commission / Seismic Guideline Study Subcommittee," 19 September 2006. https://historical.seismology.jp/ishibashi/opinion/bunkakai060919.html

bv Cryanoski, David, "Quake shuts world's largest nuclear plant." *Nature – International Journal of*

Science, 26 July 2007. https://www.nature.com/articles/448392a

bw Clenfield, "Nuclear regulator dismissed seismologist."

bx Toshiba, "Toshiba completes Westinghouse acquisition," 27 October 2006. https://www.toshiba. co.jp/about/press/2006_10/pr1702.htm. Westinghouse by this point was already owned by British Nuclear Fuels; Toshiba bought the company from them.

by "Areva and MHI in nuclear plant plan." *Financial Times*, 20 October 2006. https://www.ft.com/content/85d370b0-5fcb-11db-a011-0000779e2340

bz USNRC, "Seismic design standards and calculational methods in the United States and Japan, p. 16 (p. 44 of PDF), February 2013. https://www.nrc.gov/docs/ML1713/ML17131A127.pdf

ca Nuclear Safety Commission of Japan decision of 19 September 2006, "About revision of safety design regulatory guides for seismic safety including the 'NSC, 2006-59, revision of regulatory guide for reviewing seismic design of nuclear power reactor facilities.'"

cb Comment by Hitoshi Sato, deputy director general for nuclear safety, Nuclear and Industrial Safety Agency. "Nuclear doubts spread in wake of Niigata." *Japan Times*, 1 September 2007. https://www.japantimes.co.jp/news/2007/09/01/national/nuclear-doubts-spread-in-wake-of-niigata/

cc Shiroyama, "Regulatory failures of nuclear safety in Japan."

cd Nuclear Safety Commission of Japan decision of 19 September 2006, "About revision of safety design."

ce National Diet NAIIC, "Introduction," p. 40. *Final Report*. National Diet of Japan, 2012. https://warp.da.ndl.go.jp/info:ndljp/pid/3856371/naiic.go.jp/en/report

cf "Former governor of Sato and Fukushima arrested for bribery charges Tokyo District public prosecutor" (in Japanese). *Asahi Shimbun*, 23 October 2006. http://www.asahi.com/special/060905/TKY200610230336.html

cg "Fukushima rigging corruption composition" (in Japanese). NHK, 11 October 2006. http://www.nhk.or.jp/gendai/articles/2307/index.html

ch Tanaka, Yoshinori, "What does the movie 'Truth of the Governor' tell?" (in Japanese). *Yahoo! Japan*, 14 October 2016. https://news.yahoo.co.jp/byline/tanakayoshitsugu/20161014-00063223/

ci "Former governor of Fukushima, second instance reduced sentence. Only 'intangible bribe' is recognized" (in Japanese). *Asahi Shimbun*, 14 October 2009. Archive link: https://web.archive.org/web/20091017071604/https://www.asahi.com/national/update/1014/TKY200910140253.html

cj "Supreme Court dismisses appeal of Sato and former Fukushima prefectural governor" (in Japanese). *Asahi Shimbun*, 16 October 2012. Archive link: https://web.archive.org/web/20121017131100/https://www.asahi.com/national/update/1016/TKY201210160382.html

ck Earthquake Engineering Research Institute (EERI), "Preliminary observations on the Niigata-Chuetsu Oki, Japan, earthquake of July 16, 2007," September 2007. EERI special earthquake report. https://www.eeri.org/lfe/pdf/japan_niigata_chuetsu_oki_eeri_preliminary_report.pdf. (Different organisations gave different estimates for the earthquake depth, hence "eight to ten.")

cl Ibid.

cm Gatti, F., et al., "On the effect of the 3-D regional geology on the seismic design of critical structures: the case of the Kashiwazaki-Kariwa Nuclear Power Plant." *Geophysical Journal International*, vol. 213, issue 2, pp. 1073-1092, 1 May 2018. https://doi.org/10.1093/gji/ggy027

cn OECD/NEA, "Status report on seismic re-evaluation," p. 16, 1998. http://www.oecd-nea.org/nsd/docs/1998/csni-r98-5.pdf

co Tokyo Electric Power Company (TEPCO), "Impact of the Niigata Chuetsu-oki earthquake on the TEPCO Kashiwazaki-Kariwa NPP and countermeasures," p. 19, September 2007. http://www.tepco.co.jp/en/news/presen/pdf-1/0709-e.pdf

cp Ibid.

cq "Damaged firefighting pipe kept plant workers at bay." *Yomiuri Shimbun*, 22 July 2007. Archived link: https://web.archive.org/web/20070809080427/http://www.yomiuri.co.jp/dy/national/20070722TDY02005.htm

cr "Japan accepts IAEA inspectors after quake troubles." Reuters, 22 July 2007. https://www.reuters.com/article/us-quake-japan-nuclear/japan-accepts-iaea-inspectors-after-quake-troubles-idUST20337520070722

cs Hogg, Chris, "Japan post-quake power shake-up." BBC, 20 July 2007. http://news.bbc.co.uk/1/hi/world/asia-pacific/6908177.stm

ct "N-plant checks must be rigorous to regain trust." *Yomiuri Shimbun*, 22 August 2007. Archived link: https://web.archive.org/web/20070909055211/http://www.yomiuri.co.jp/dy/editorial/20070822TDY04006.htm

cu Nakajima, Tatsuo, "N-safety reports inadequate / Gap between public, official views of incidents must be addressed." *Daily Yomiuri*, 16 November 2007. http://www.yomiuri.co.jp/dy/national/20071116TDY04303.htm

cv "No radiation leak after fire at Japan's TEPCO plant." Reuters, 12 April 2009. https://www.reuters. com/article/tepco-nuclear/no-radiation-leak-after-fire-at-japans-tepco-plant-idUST3545120090412

cw "Flammables ban lifted for TEPCO nuclear plant." Reuters, 27 March 2009. https://www.reuters. com/article/tepco-nuclear/flammables-ban-lifted-for-tepco-nuclear-plant-idUST7669420090327

cx Nishiyama, George, "Japan quake stirs nuclear fears." Reuters, 17 July 2007. https://www. reuters.com/article/us-quake-japan/japan-quake-stirs-nuclear-fears-idUST14801720070717?s-rc=071707_1022_DOUBLEFEATURE_peace_push_in_doubt

cy Fackler, Martin, "Japan's quake-prone atomic plant prompts wider worry." New York Times, 25 July 2007. https://www.nytimes.com/2007/07/25/world/asia/25japan.html

cz "Nuclear plant hit by earthquake closed indefinitely in Japan." International Herald Tribune (Associated Press), 18 July 2007. Archived link: https://web.archive.org/web/20070919145221/http://www.iht.com/articles/2007/07/18/asia/japan.php

da "Japan admits more nuclear leaks." BBC, 17 July 2007. http://news.bbc.co.uk/1/hi/world/asia-pa-cific/6902142.stm

db "Towels used to mop up nuke spill." Asahi Shimbun, 26 July 2007. Archived link: https://web.ar-chive.org/web/20070929130258/http://www.asahi.com/english/Herald-asahi/TKY200707250500. html

dc Fackler, "Japan's quake-prone atomic plant."

dd Soble, Jonathan, "Safety key for Japan to buy back into nuclear." Financial Times, 26 October 2007. Archive link: https://web.archive.org/web/20071105150806/http://www.ft.com/cms/s/0/1df-be196-834e-11dc-b042-0000779fd2ac.html

de "Point of view / Katsuhiko Ishibashi: Nuclear plants at grave risk of quake damage." Asahi Shim-bun, 11 August 2007. Archived link: https://web.archive.org/web/20071130094257/http://www. asahi.com/english/Herald-asahi/TKY200708110090.html

df "Japan court nixes nuke plant suspension." Guardian, 26 October 2007. Archived link: https://web. archive.org/web/20071026192213/http://www.guardian.co.uk/worldlatest/story/0,,-7027238,00. html

dg "Update 1: Chubu to replace nuclear units, sees 08/09 net loss." Reuters, 22 December 2008. https://www.reuters.com/article/nuclear-chubu-elec/update-1-chubu-to-replace-nuclear-units-sees-08-09-net-loss-idUST3036752008122

dh Freeman, Laurie-Anne, "Ties that bind: Press, state, and society in contemporary Japan." Program on US-Japan Relations, Harvard University, 1995. Excerpt found here: Freeman, Laurie-Anne, "Japan's press clubs as information cartels." Japan Policy Research Institute, working paper no. 18, April 1996. http://www.jpri.org/publications/workingpapers/wp18.html

di 3-4 December Nuclear Safety and Security Group meeting. US Embassy telegram, 17 December 2008. https://wikileaks.org/plusd/cables/08TOKYO3432_a.html

dj Krolicki, Kevin, DiSavino, Scott, and Fuse, Taro, "Special report: Japan engineers knew tsunami could overrun plant." Reuters, 29 March 2011. https://www.reuters.com/article/us-japa-nuclear-risks/special-report-japan-engineers-knew-tsunami-could-overrun-plant-idUS-TRE72S2UA20110329

dk "TEPCO faces rare red ink after nuke plant's halt." Japan Times (Kyodo News), 1 November 2007. https://www.japantimes.co.jp/news/2007/11/01/business/tepco-faces-rare-red-ink-after-nuke-plants-halt/

dl Hayashi, Yuka, "In Japan, resistance rises to nuclear-power plans." Wall Street Journal, 9 July 2008. https://www.wsj.com/articles/SB121554213823836415

dm Katori, Keisuke, and Suzuki, Ayako, "Atomic power safety questions still unanswered." Asahi Shimbun, 5 October 2009. Archived link: https://web.archive.org/web/20091009050020/http:// www.asahi.com/english/Herald-asahi/TKY200910050059.html

dn TEPCO, "Safety measures of nuclear power station." http://www.tepco.co.jp/en/challenge/csr/ nuclear/stations-e.html#content-head-wrapper. Interestingly, this page seems to be a holdover from before the Fukushima disaster.

do This information is from TEPCO, but this interesting paper from an engineering journal has extensive references on the topic: Warn, Gordon, and Ryan, Keri, "A review of seismic isolation for buildings: Historical development and research needs." Buildings, vol. 2, issue 4, pp. 300-325, 2012. https://www.researchgate.net/publication/274544499_A_Review_of_Seismic_Isolation_for_ Buildings_Historical_Development_and_Research_Needs

dp Johnston, Eric, "Nuclear plants rural Japan's economic fix." Japan Times, 4 September 2007. https://www.japantimes.co.jp/news/2007/09/04/national/nuclear-plants-rural-japans-economic-fix/

dq Averaged statistics from 20 credible sources, with variances viewable in the report: World Nuclear

Association, "Comparison of lifecycle greenhouse gas emissions of various electricity generation sources," July 2011. http://www.world-nuclear.org/uploadedFiles/org/WNA/Publications/Working_Group_Reports/comparison_of_lifecycle.pdf

dr "Japan must lead way on nuclear energy, say advisors." Agence France-Presse, 21 March 2008. http://afp.google.com/article/ALeqM5gEHaZZKFDHEoZA8hyy2uA7z8qFyw; archived link: https://web.archive.org/web/20080325064840/http://afp.google.com/article/ALeqM5gEHaZZKF-DHEoZA8hyy2uA7z8qFyw

ds Heiser, Stephen, "Japan intends to build over 14 new nuclear power plants by 2030." NuclearStreet.com, 22 April 2010. http://nuclearstreet.com/nuclear_power_industry_news/b/nuclear_power_news/archive/2010/04/22/japan-intends-to-build-over-14-new-nuclear-power-plants-by-2030-04225

dt "Active fault exists near Monju." Yomiuri Shimbun, 1 April 2008. Archived link: https://web.archive.org/web/20080417024629/http://www.yomiuri.co.jp/dy/national/20080401TDY01302.htm

du "Monju reactor's future uncertain / Spate of problems likely to doom Oct. target for restart of operations." Yomiuri Shimbun, 11 April 2008. https://web.archive.org/web/20080417025324/http://www.yomiuri.co.jp/dy/national/20080411TDY04303.htm

dv "Monju N-reactor key to energy security." Yomiuri Shimbun, 23 April 2009. Archived link: https://web.archive.org/web/20090430174733/http://www.yomiuri.co.jp/dy/editorial/20090423T-DY04305.htm

dw "Nuclear fuel cycle delay is intolerable." Yomiuri Shimbun, 16 October 2010. Archived link: https://web.archive.org/web/20101024002245/http://www.yomiuri.co.jp/dy/editorial/T101016001479.htm

dx Sanger, David E., "Japan's nuclear fiasco." New York Times, 20 December 1992. https://www.nytimes.com/1992/12/20/business/japan-s-nuclear-fiasco.html

dy JAEC, Technical Subcommittee on Nuclear Power, Nuclear Fuel Cycle, etc., "Estimation of nuclear fuel cycle cost," 10 November 2011. Data sheet 1. http://www.aec.go.jp/jicst/NC/about/kettei/seimei/111110_1_e.pdf

dz "Japan delays MOX nuclear fuel goal by 5 years." Reuters, 12 June 2009. https://www.reuters.com/article/nuclear-japan-mox/japan-delays-mox-nuclear-fuel-goal-by-5-years-idUST34756220090612

ea Ofuji, Kenta, "Fukushima's non-nuclear power plants: Their history, damage by disasters, and prospects for the future." Energy & Environment Journal, vol. 24, no. 5, pp. 711-725, 2013. https://www.jstor.org/stable/43735193

eb Yoshida, Reiji, "TEPCO refused safety agency's proposal to simulate Fukushima tsunami nine years before meltdown disaster." Japan Times, 30 January 2018. https://www.japantimes.co.jp/news/2018/01/30/national/japan-scrapped-proposed-fukushima-tsunami-simulation-nine-years-disaster/

ec "IAEA leadership team transition and US influence in the agency." Telegram from the UN embassy in Vienna to the Department of Energy National Security Council, 7 July 2009. https://wikileaks.org/plusd/cables/09UNVIEVIENNA322_a.html

ed Takao, M., "Tsunami assessment for nuclear power plants in Japan." Tokyo Electric Power Company, 2010. In First Kashiwazaki International Symposium on Seismic Safety of Nuclear Installations (JNES), 24-26 November 2010, Kashiwazaki, Japan: Niigata Institute of Technology.

ee "Fisheries took ¥6 billion hit from Chile tsunami." Japan Times, 28 March 2010. https://www.japantimes.co.jp/news/2010/03/28/national/fisheries-took-6-billion-hit-from-chile-tsunami/

ef Okal, Emily A., "The quest for wisdom: Lessons from 17 tsunamis, 2004–2014." Philosophical Transactions of the Royal Society A – Mathematical, Physical and Engineering Sciences, vol. 373, issue 2053, 2015. https://royalsocietypublishing.org/doi/10.1098/rsta.2014.0370

eg Funabashi, Yoichi, and Kitazawa, Kay, "Fukushima in review: A complex disaster, a disastrous response." Bulletin of the Atomic Scientists, vol. 68, issue 2, p. 14, 2012. https://www.tandfonline.com/doi/full/10.1177/0096340212440359

eh TEPCO, "TEPCO's tsunami countermeasure preparation and tsunami prediction positioning," p. 4 of PDF, n.d. http://www.tepco.co.jp/en/nu/fukushima-np/interim/images/111202_01-e.pdf

ei National Diet NAIIC, "Chapter 1," p. 26. Final Report.

ej "Reactors continue through earthquake." World Nuclear News, 9 March 2011. https://www.world-nuclear-news.org/Articles/Reactors-continue-through-earthquake

CHAPTER 7

a Ishigaki, Akemi, et al., "The Great East-Japan Earthquake and devastating tsunami: An update and lessons from the past great earthquakes in Japan since 1923." Tohoku Journal of Experimental

Medicine, vol. 229, issue 4, 2013. https://www.jstage.jst.go.jp/article/tjem/229/4/229_287/_html/-char/ja

b Takahashi, Hideki, "The unsung hero of the meltdowns." *Japan Times*, 9 October 2014. https://www.japantimes.co.jp/news/2014/10/09/national/remembering-fukushima-plant-chief-helped-prevent-catastrophe/

c Tokyo Electric Power Company, "Fukushima nuclear accident analysis report," p. 130 (p. 146 of PDF), June 2012. https://www.tepco.co.jp/en/press/corp-com/release/betu12_e/images/120620e0104.pdf

d Masuda, Naohiro (former site superintendent at Fukushima Daini NPP), "East Japan Earthquake on March 11, 2011 and emergency response at Fukushima Daini Nuclear Power Plant." Tokyo Electric Power Company, March 2014. https://www-pub.iaea.org/iaeameetings/cn233p/Opening-Session/6Masuda.pdf

e IAEA, "IAEA mission to Onagawa Nuclear Power Station to examine the performance of systems, structures and components following the Great East Japanese Earthquake and Tsunami (report)," 2012. https://www.iaea.org/sites/default/files/iaeamissionOnagawa.pdf

f Atomic Energy Society of Japan (AESJ), "Chapter 4: Overview of events occurring at power stations other than the Fukushima Daiichi Nuclear Power Station." *The Fukushima Daiichi Nuclear Accident: Final Report of the AESJ Investigation Committee*. Springer, 2015.

g "Japanese nuclear plant 'remarkably undamaged' in earthquake – UN atomic agency." *UN News*, 12 August 2012. https://news.un.org/en/story/2012/08/417412-japanese-nuclear-plant-remarkably-undamaged-earthquake-un-atomic-agency#.UV5n_crpyJM

h Kushida, Kenji E., "Japan's Fukushima nuclear disaster: Narrative, analysis, and recommendations." Shorenstein Asia-Pacific Research Center, 2012. https://www.academia.edu/3611514/Japan_s_Fukushima_Nuclear_Disaster_Narrative_Analysis_and_Recommendations. Note: Twenty tons per hour is for Unit 1; Units 2 and 3 require 30 tons, then 10.

i USNRC, "Reactor core isolation cooling reliability study," 2005. https://nrcoe.inl.gov/resultsdb/publicdocs/SystemStudies/rcic-system-description.pdf

j Tateiwa, Kenji (Tokyo Electric Power Co.), "Fukushima nuclear accident: Personal perspective on the human impact." National Radiological Emergency Preparedness Conference notes, 8 April 2013. http://www.nationalrep.org/2013Presentations/Session%207_SLIDES_Human%20Impact%20from%20Fukushima_Tateiwa.pdf

k Yeh, Harry, Barbosa, Andre R., and Mason, Ben, "Tsunamis [sic] effects in man-made environment." *Encyclopedia of Complexity and Systems Science*, Springer Berlin Heidelberg, pp. 1-27, 2015. https://www.researchgate.net/publication/282949701_Tsunamis_Effects_in_Man-Made_Environment

l IAEA, "IAEA mission to Onagawa Nuclear Power Station to examine performance of systems, structures and components following the Great East Japanese Earthquake and Tsunami," 2012. https://www.iaea.org/sites/default/files/iaeamissionOnagawa.pdf

m Tojima, Wako, "How Onagawa responded at the time? [sic] The nuclear power station that withstood the Great East Japan Earthquake." *Hiroba*, no. 431, June 2014. https://web.archive.org/web/20190419043716/https://www.t-enecon.com/cms/wp-content/uploads/2015/ebooks/OnagawaE.pdf

n IAEA, *The Fukushima Daiichi Accident, Technical Volume 1/5: Description and Context of the Accident*, p. 13, 2015. https://www.iaea.org/publications/10962/the-fukushima-daiichi-accident

o "Japan: Fukushima nuclear workers' bodies found." BBC, 3 April 2011. https://www.bbc.co.uk/news/world-asia-pacific-12949783

p Investigation Committee on the Accident at the Fukushima Nuclear Power Stations of Tokyo Electric Power Company (henceforth Fukushima Investigation Committee), "Chapter 4: Accident response at TEPCO's Fukushima Dai-ichi NPS," p. 225. *Interim Report*, December 2011. https://www.cas.go.jp/jp/seisaku/icanps/eng/interim-report.html

q TEPCO, "Employees of TEPCO who were missing at Fukushima Daiichi Nuclear Power Station," 3 April 2011. Press release. http://www.tepco.co.jp/en/press/corp-com/release/11040302-e.html

r Kadota, Ryusho, *On the Brink: The Inside Story of Fukushima Daiichi*. Kurodahan Press, 2014.

s Tateiwa, "Fukushima nuclear accident."

t Tokyo Electric Power Company, "Fukushima nuclear accident analysis report," p. 12.

u Ministry of Economy, Trade and Industry (METI), "Receipt of a report regarding a legally reportable event that occurred at the Tokai-Daini Power Station, owned by Japan Atomic Power Company," 2 September 2011. Archive link: https://web.archive.org/web/20130213201927/http://www.atomdb.jnes.go.jp/content/000119422.pdf

v JAPC, "Status of the Tokai Daini Power Station at the time of the earthquake" (in Japanese).

Archive link: https://web.archive.org/web/20130403041306/http://www.japc.co.jp/tohoku/tokai/tsunami_to.html

w   Independent Fukushima Investigation, *Fukushima Daiichi Nuclear Power Station Disaster*, p. 9.

x   Fukushima Investigation Committee, "Chapter 3: Emergency responses required and taken by governments and other bodies," p. 88, footnote 31. *Interim Report*. Government of Japan, December 2011. https://www.cas.go.jp/jp/seisaku/icanps/eng/interim-report.html

y   Gavett, Gretchen, "TEPCO's Akio Komori: 'The options we had available ... were rather limited.'" PBS, 28 February 2012. https://www.pbs.org/wgbh/frontline/article/tepcos-akio-komori-the-options-we-had-available-were-rather-limited/

z   There's a video of the relevant comments here: https://www.youtube.com/watch?v=vLmEgkpBnak

aa   Takahara, Kanako, "Tight-lipped TEPCO lays bare exclusivity of press clubs." *Japan Times*, 3 May 2011. https://www.japantimes.co.jp/news/2011/05/03/national/tight-lipped-tepco-lays-bare-exclusivity-of-press-clubs/

ab   Fukushima Investigation Committee, "Chapter 3: Emergency Responses Required and Taken by Governments and Other Bodies," p. 86. *Interim Report*. Government of Japan, December 2011. https://www.cas.go.jp/jp/seisaku/icanps/eng/interim-report.html

ac   Kushida, "Japan's Fukushima nuclear disaster."

ad   AESJ, "Chapter 3: Overview of the accident at the Fukushima Daiichi Nuclear Power Station," p. 21. *The Fukushima Daiichi Nuclear Accident: Final Report of the AESJ Investigation Committee*. Springer, 2015.

ae   Comments during Yoshida's government testimony.

af   Yoshida's feelings at this moment vary from source to source, from calm to worry to panic. I felt it was better for him to speak for himself, so this quote comes straight from his own government testimony (as translated by the *Yomiuri Shimbun*): "Yoshida interviews / Strong words on Fukushima N-crisis from TEPCO's manager on the ground." *Japan News by Yomiuri Shimbun*, 10 September 2014. Archive link: https://web.archive.org/web/20140915210028/http://the-japan-news.com/news/article/0001553303

ag   The following is a translation from Yoshida's government testimony by the *Asahi Shimbun*: "The Yoshida report, chapter 3: Hubris of wisdom." *Asahi Shimbun*. http://www.asahi.com/special/yoshida_report/en/3-2.html

ah   IAEA, *Fukushima Daiichi Accident, Technical Volume 1/5*, p. 17.

ai   "Yoshida interviews," *Japan News by Yomiuri Shimbun*.

aj   Gulati, Ranjay, Casto, Charles, and Krontiris, Charlotte, "How the other Fukushima plant survived." *Harvard Business Review*, July–August 2014. https://hbr.org/2014/07/how-the-other-fukushima-plant-survived

ak   TEPCO, "TEPCO plant status of Fukushima Daini Nuclear Power Station." Press release, 21 August 2011. https://www.tepco.co.jp/en/press/corp-com/release/betu11_e/images/110821e5.pdf

al   National Diet NAIIC, "Chapter 3," *Final Report*.

am   "Top court upholds HIV-scandal sentence." *Japan Times*, 5 March 2008. https://www.japantimes.co.jp/news/2008/03/05/national/top-court-upholds-hiv-scandal-sentence/

an   OECD/NEA, "Benchmark Study of the Accident at the Fukushima Daiichi Nuclear Power Plant (BSAF Project): Phase 1 summary report," p. 23. March 2015. https://www.oecd-nea.org/upload/docs/application/pdf/2020-01/csni-r2015-18.pdf

ao   Fukushima Investigation Committee, "Attachment II-1-1: Investigation results concerning the damage to major systems and facilities," p. 24, footnote 40. *Final Report*. Government of Japan, July 2012. https://www.cas.go.jp/jp/seisaku/icanps/eng/02Attachment1.pdf

ap   Harutaka Hoshi and Masashi Hirano, "Severe accident analyses of Fukushima-Daiich Units 1 to 3." Japan Nuclear Energy Safety Organization (JNES), 17 September 2012. http://www.simplyinfo.org/wp-content/uploads/2013/01/JNES_accidentprogression_Fukushima_P-4.pdf

aq   Tanabe, Fumiya, "Analyses of core melt and re-melt in the Fukushima Daiichi nuclear reactors." *Journal of Nuclear Science and Technology*, vol. 49, issue 1, 2012. https://www.tandfonline.com/doi/full/10.1080/18811248.2011.636537. Note: Zirconium oxide's melting point is actually slightly below this, at 2,715°C, but the figures are close.

ar   Fukushima Investigation Committee, "Attachment II-1-1," p. 24, footnote 40.

as   Kan, Naoto, *My Nuclear Nightmare: Leading Japan through the Fukushima Disaster to a Nuclear-Free Future*, p. 44. Cornell University Press, 2018.

at   Ibid.

au   Independent Fukushima Investigation, *Fukushima Daiichi Nuclear Power Station Disaster*, p. 9.

av   "TEPCO boss wanted use of military plane." *Yomiuri Shimbun*, 27 April 2011. https://web.archive.org/web/20171030040957/https://www.asiaone.com/News/Latest%2BNews/Asia/Story/A1Sto-

ry20110427-275888.html
aw Dvorak, Phred, "Japan utility president stranded as crisis began." *Wall Street Journal*, 27 April 2011. Archive link: https://web.archive.org/web/20150615225130/https://www.wsj.com/articles/SB10001424052748704729304576286741965541546
ax Ibid.
ay Fukushima Investigation Committee, "Chapter 4: Accident response," p. 185. *Interim Report*.
az Hatamura, Yotaro, et al., *The 2011 Fukushima Nuclear Power Plant Accident: How and Why It Happened*, p. 42. Woodhead Publishing, 2014.
ba National Diet NAIIC, "Introduction," p. 40. *Final Report*.
bb Repeta, L., "Chapter 6: The nuclear future: New questions – Could the meltdown have been avoided?" In J. Kingston (ed.), *Tsunami: Japan's Post-Fukushima Future*. Foreign Policy, 2011.
bc Kan, *My Nuclear Nightmare*, p. 47.
bd Independent Fukushima Investigation, *Fukushima Daiichi Nuclear Power Station Disaster*, prologue, p. xlviii.
be Hatamura et al., *2011 Fukushima Nuclear Power Plant Accident*, p. 45.
bf IAEA, *The Fukushima Daiichi Accident, Technical Volume 3/5: Emergency Preparedness and Response*, p. 15, 2015. https://www.iaea.org/publications/10962/the-fukushima-daiichi-accident
bg IAEA, *Fukushima Daiichi Accident, Technical Volume 1/5*, p. 18.
bh Kadota, *On the Brink*, pp. 84-85.
bi Japan Nuclear Technology Institute (JANTI), Examination Committee on Accident at Fukushima Daiichi Nuclear Power Station (henceforth JANTI Examination Committee), "Examination of accident at Tokyo Electric Power Co., Inc.'s Fukushima Daiichi Nuclear Power Station and proposal of countermeasures," pp. 2-24, October 2011. http://www.gengikyo.jp/english/shokai/Tohoku_Jishin/report.pdf. (Note: JANTI became the Nuclear Safety Promotion Association in November 2012.)
bj TEPCO, "Report on the investigation and study of unconfirmed/unclear matters in the Fukushima Nuclear accident." *Progress Report No. 2*, p. 29, 6 August 2014. https://www.tepco.co.jp/en/press/corp-com/release/betu14_e/images/140806e0101.pdf
bk IAEA, *Fukushima Daiichi Accident, Technical Volume 1/5*, p. 20.
bl Dvorak, Phred, and Landers, Peter, "Japanese plant had barebones risk plan." *Wall Street Journal*, 31 March 2011. https://www.wsj.com/articles/SB10001424052748703712504576232961004646464
bm Comment by Yoshida in Kadota, *On the Brink*, p. 96.
bn National Diet NAIIC, "Chapter 4: Overview of the damage and how it spread," p. 7. *Final Report*. National Diet of Japan, 2012. https://warp.da.ndl.go.jp/info:ndljp/pid/3856371/naiic.go.jp/en/report/
bo Kan, *My Nuclear Nightmare*, p. 50.
bp Kadota, *On the Brink*, p. 110. Kan stated to Kodata that he learned this at the plant.
bq TEPCO, "Report ... on unconfirmed/unclear matters," p. 10.
br TEPCO, "Report on initial responses to the accident at Tokyo Electric Power Co.'s Fukushima Daiichi Nuclear Power Plant," p. 30, 22 December 2011. http://www.tepco.co.jp/en/press/corp-com/release/betu11_e/images/111222e17.pdf
bs IAEA, *Fukushima Daiichi Accident, Technical Volume 1/5*, p. 98.
bt Institute of Nuclear Power Operators (INPO), "Special report on the nuclear accident at the Fukushima Daiichi Nuclear Power Station," p. 79, November 2011. https://www.nrc.gov/docs/ML1134/ML11347A454.pdf
bu IAEA, *Fukushima Daiichi Accident, Technical Volume 1/5*, p. 99. https://www.iaea.org/publications/10962/the-fukushima-daiichi-accident
bv Maeda, Yukiko, and Shinohara, Yuya, "Yoshida's call on seawater kept reactor cool as Tokyo dithered." *Japan Times*, 14 September 2014. https://www.japantimes.co.jp/news/2014/09/14/national/yoshidas-call-seawater-kept-reactor-cool-tokyo-dithered/
bw This is a comment Yoshida gave in a video letter recorded for Hideaki Yabuhara in August 2012.
bx Fukushima Investigation Committee, "Chapter 4: Accident response," p. 246. *Interim Report*.
by Fukushima Investigation Committee, "Chapter 2: The damage and accident responses at the Fukushima Daiichi NPS and the Fukushima Daini NPS," p. 70, footnotes 133, 134. *Final Report*. Government of Japan, July 2012. https://www.cas.go.jp/jp/seisaku/icanps/eng/final-report.html
bz Kan, *My Nuclear Nightmare*, pp. 56-57.
ca Fukushima Investigation Committee, "Chapter 4: Emergency response measures primarily implemented outside the Fukushima Dai-ichi Nuclear Power Station in response to the accident," p. 328. *Final Report*. Government of Japan, July 2012. https://www.cas.go.jp/jp/seisaku/icanps/eng/

final-report.html
cb Fukushima Investigation Committee, "Chapter 4: Accident response," p. 193. *Interim Report*.
cc JANTI Examination Committee, "Examination of accident," pp. 2-42.
cd Maeda and Shinohara, "Yoshida's call on seawater." The exact wording and translation depend on the source, though all agree on the basic gist of the brief discussion. I have kept this and Yoshida's later comment on the decision confined to the same source for consistency on the incident.
ce Comments made during Yoshida's government testimony, translation by the *Yomiuri Shimbun*: "Yoshida interviews."
cf Maeda and Shinohara, "Yoshida's call on seawater."
cg National Diet NAIIC, "Chapter 4." *Final Report*.
ch National Diet NAIIC, "Chapter 3: Emergency responses required and taken by governments and other bodies," p. 89. *Interim Report*. Government of Japan, December 2011. https://www.cas.go.jp/jp/seisaku/icanps/eng/interim-report.html
ci National Diet NAIIC, "Chapter 4," p. 39, *Final Report*.
cj Ishikawa, Michio, *A Study of the Fukushima Daiichi Nuclear Accident Process: What Caused the Core Melt and Hydrogen Explosion?* Springer, August 2015. Please note again that this water was not actually cold, just colder than the reactor water.
ck TEPCO, "Report on … unconfirmed/unclear matters," p. 20.
cl Ibid., p. 21.
cm Tateiwa, "Fukushima nuclear accident."
cn Maeda, Yukiko, "Friend's death etched in memory of young Fukushima No. 1 worker." *Japan Times* (*Kyodo News*), 9 October 2014. https://www.japantimes.co.jp/news/2014/10/09/national/friends-death-etched-in-memory-of-young-fukushima-no-1-worker/
co Takahashi, Hideki, "A melted shoe and a farewell letter in the dark." *Japan Times*, 14 September 2014. https://www.japantimes.co.jp/news/2014/09/14/national/melted-shoe-farewell-letter-dark-fukushima-1/ Note: the IAEA report states that he "checked the valve indication locally in the torus [i.e. pressure suppression] room, but the valve indicated [that it was] closed." The *Japan Times* article, where they have obviously interviewed him, states that Hiyashizaki "gave up on checking on the valve and returned to the control room." It's unclear which is correct.
cp Fukushima Investigation Committee, "Chapter 4: Accident response," p. 206. *Interim Report*.
cq TEPCO, "Fukushima nuclear accident investigation report," p. 54. *Interim Report – Supplementary Volume*. December 2, 2011. https://www.tepco.co.jp/en/press/corp-com/release/betu11_e/images/111202e16.pdf
cr *Lessons Learned from the Fukushima Nuclear Accident for Improving Safety of US Nuclear Plants*, p. 149. National Academies Press, 2014. https://www.nap.edu/catalog/18294/lessons-learned-from-the-fukushima-nuclear-accident-for-improving-safety-of-us-nuclear-plants
cs Hatamura et al., *2011 Fukushima Nuclear Power Plant Accident*, p. 49.
ct IAEA, "Annex III: Unit 3 sequence of events," p. 9. *The Fukushima Daiichi Accident, Technical Volume 1/5: Description and Context of the Accident*, 2015. https://www.iaea.org/publications/10962/the-fukushima-daiichi-accident
cu IAEA, "Annex II: Unit 2 sequence of events." *The Fukushima Daiichi Accident, Technical Volume 1/5: Description and Context of the Accident*, 2015. https://www.iaea.org/publications/10962/the-fukushima-daiichi-accident
cv "The Yoshida testimony: Chapter 2." *Asahi Shimbun*, 2014. http://www.asahi.com/special/yoshida_report/en/2-1.html. This is the *Asahi Shimbun*'s English translation of Yoshida's original testimony.
cw IAEA, "Storage of water reactor spent fuel in water pools," p. 12. Technical reports series no. 218, 1982. http://large.stanford.edu/courses/2018/ph241/saffari2/docs/sti-doc-10-218.pdf
cx Kadota, *On the Brink*, p. 164.
cy Japan Nuclear Safety Institute (JANSI), "TEPCO Fukushima Daini Nuclear Power Station: Research on the status of response to the Tohoku-Pacific Ocean Earthquake and Tsunami and lessons learned therefrom," October 2012. http://www.genanshin.jp/report/data/F2jiko_Report.pdf. (This document, in my opinion, is the single best and most comprehensive English-language report on the events at Daini. The countless others I've seen provide only cursory information, as Daiichi is almost always the focus, even when the stated function is a comparison between the two.)
cz Gulati et al., "How the other Fukushima plant survived."
da INPO, "Special report," p. 42.
db Tateiwa, Kenji, "Fukushima nuclear accident: A TEPCO nuclear engineer's perspective." TEPCO. Presentation to the Nuclear Energy Institute, Washington, DC, 9 July 2012. http://docplayer.net/144085628-Fukushima-nuclear-accident.html

dc Video released by TEPCO of the moment Unit 3 exploded: https://www.youtube.com/watch?-time_continue=78&v=OzzLtjfRWQ8

dd "Yoshida interviews." *Yomiuri Shimbun*, 2014.

de Maeda, Yukiko, "Responders cowed by explosion at Reactor 3 building of Fukushima No. 1." *Japan Times*, 14 September 2014. https://www.japantimes.co.jp/news/2014/09/14/national/responders-cowered-by-explosion-at-no-3-reactor-building/

df Recorded on the video feed: Tabuchi, Hiroko, "Videos shed light on chaos at Fukushima as a nuclear crisis unfolded." *New York Times*, 9 August 2012. https://www.nytimes.com/2012/08/10/world/asia/fukushima-videos-shed-light-on-chaos-in-nuclear-crisis.html

dg "The Yoshida testimony: In went 'suicide squads.'" *Asahi Shimbun*, 2014. http://www.asahi.com/special/yoshida_report/en/3-1.html. This is the *Asahi Shimbun*'s English translation of Yoshida's original testimony.

dh Fukushima Investigation Committee, "Chapter 2: The damage and accident responses," p. 206, *Final Report*.

di Ishikawa, *A Study of the Fukushima Daiichi Nuclear Accident Process*, p. 53.

dj OECD/NEA, "Benchmark Study," p. 29.

dk AESJ, "Chapter 6: Accident analysis and issues," pp. 338-339. *The Fukushima Daiichi Nuclear Accident: Final Report of the AESJ Investigation Committee*. Springer, 2015.

dl Kitazawa, Kay, "Bungled crisis management: Fukushima's painful lessons." *Global Asia*, vol. 9, no. 1, 16 March 2014. Rebuild Japan Initiative Foundation. http://www.globalasia.org/v9no1/focus/bungled-crisis-management-fukushimas-painful-lessons_kay-kitazawa

dm Takahashi, Hideki, "Helplessness as Reactor 2 lost cooling." *Japan Times*, 23 September 2014. https://www.japantimes.co.jp/news/2014/09/23/national/helplessness-reactor-2-lost-cooling/

dn National Diet NAIIC, "Chapter 2: Escalation of the accident," p. 29. *Final Report*. National Diet of Japan, 2012. https://warp.da.ndl.go.jp/info:ndljp/pid/3856371/naiic.go.jp/en/report

do Ishikawa, *A Study of the Fukushima Daiichi Nuclear Accident Process*, p. 56.

dp IAEA, "Annex II," p. 10. *Fukushima Daiichi Accident, Technical Volume 1/5*.

dq Independent Fukushima Investigation, *Fukushima Daiichi Nuclear Power Station Disaster*, p. 28.

dr Comments by Yukio Edano in the Rebuild Japan Initiative Foundation's 2012 report.

ds Comments made to Kadota in Kadota, *On the Brink*, p. 189.

dt This quote is from *Shi no Fuchi o Mita Otoko ("The Man Who Saw the Edge of Death")*, the Japanese version of Ryusho Kodata's book, which is named *On the Brink* in English. The translation of this quote is more powerful than the one that appears in *On the Brink*. Seen in Soble, Jonathan, "Hero who strove to contain a nuclear calamity – Masao Yoshida, atomic engineer 1955–2013." *Financial Times*, 12 July 2013. https://www.ft.com/content/43af89dc-ea23-11e2-b2f4-00144feab-dc0

du Shozugawa, Katsumi, Nogawa, Norio, and Matsuoa, Motoyuki, "Deposition of fission and activation products after the Fukushima Dai-ichi Nuclear Power Plant accident." *Environmental Pollution*, vol. 163, pp. 243-247, April 2012. https://www.sciencedirect.com/science/article/pii/S0269749112000024

dv Ota, Hisashi, "TEPCO plea to evacuate enraged Kan." *Japan Times*, 23 September 2014. https://www.japantimes.co.jp/news/2014/09/23/national/tepco-plea-evacuate-enraged-kan/

dw These numbers were announced by Edano at a press conference but are repeated here: Japan Atomic Industrial Forum Inc., "Status of nuclear power plants in Fukushima as of 13:00 March 15," 15 March 2011. https://web.archive.org/web/20110409114645/http://www.jaif.or.jp/english/news_images/pdf/ENGNEWS01_1300168169P.pdf

dx Yoshida's comments were recorded by the video-conference system, which is the source of the quote, translated here: Takahashi, Hideki, "As radiation levels soared at Fukushima No. 1, plant chief Yoshida rescinded evacuation order." *Japan Times*, 28 September 2014.https://www.japantimes.co.jp/news/2014/09/28/national/radiation-levels-soared-fukushima-1-plant-chief-yoshida-rescinded-evacuation-order/

dy Bradsher, Keith, and Tabuchi, Hiroko, "Last defense at troubled reactors: 50 Japanese workers." *New York Times*, 15 March 2011. https://www.nytimes.com/2011/03/16/world/asia/16workers.html

dz This is as relayed in Kadota, *On the Brink*, based on interviews the author conducted with Yoshida and other people in the room, but I have seen subtly different versions of this moment that were also based on interviews with several of the same people. Everyone inevitably recalls moments like this differently as memory is unreliable, so take it with a small grain of salt.

ea "Kan berates TEPCO for tardy response." *Yomiuri Shimbun*, 16 March 2011. Archive link: https://web.archive.org/web/20110318021803/https://www.yomiuri.co.jp/dy/national/T110315004235.htm

eb "Inspection sheds light on Fukushima torus." *World Nuclear News*, 24 April 2012. https://www. world-nuclear-news.org/RS_Inspection_sheds_light_on_Fukushima_torus_2404121.html

ec Fukushima Investigation Committee, "Chapter 2: The damage and accident responses, p. 73 (p. 67 of PDF). *Final Report.*

ed Ibid., pp. 73-74.

ee INPO, "Special report," p. 33-34.

ef National Diet NAIIC, "Chapter 4," p. 22. *Final Report.*

eg "SPEEDI report deepens suspicions." *Japan Times*, 11 August 2012. https://www.japantimes.co.jp/ opinion/2012/08/11/editorials/speedi-report-deepens-suspicions/

eh National Diet NAIIC, "Chapter 4," p. 39. *Final Report.* This makes you wonder how many countries *are* prepared for an accident of this scale...

CHAPTER 8

a 112th US Congress, "Transcript of the joint hearing before the Subcommittee on Energy and Power and the Subcommittee on Environment and the Economy of the Committee on Energy and Commerce – House of Representatives," 16 March 2011. https://www.gpo.gov/fdsys/pkg/CHRG-112hhrg68480/pdf/CHRG-112hhrg68480.pdf

b Gilligan, Andrew, and Mendick, Robert, "Japan tsunami: Fukushima Fifty, the first interview." *Telegraph*, 27 March 2011. https://www.telegraph.co.uk/news/worldnews/asia/japan/8408863/ Japan-tsunami-Fukushima-Fifty-the-first-interview.html

c Phase 2, chapter 2: "Lessons learned for spent fuel storage." In *Lessons Learned from the Fukushima Nuclear Accident for Improving Safety and Security of US Nuclear Plants.* National Academies Press, 2016. https://www.nap.edu/catalog/21874/lessons-learned-from-the-fukushima-nuclear-accident-for-improving-safety-and-security-of-us-nuclear-plants

d NHK special: "The truth revealed by the fateful Unit 1 '30,000 conversations'" (in Japanese), n.d. NHK. https://www3.nhk.or.jp/news/special/shinsai6genpatsu/

e "Where is Japan's nuclear power CEO?" Reuters, 20 March 2011. https://www.reuters.com/article/ japan-quake-absent-ceo/rpt-where-is-japans-nuclear-power-ceo-idUSL3E7EK0IR20110320

f "TEPCO president admitted to hospital on Tuesday evening." Nikkei, 30 March 2011. Archive link: https://web.archive.org/web/20110404053744/http://e.nikkei.com/e/fr/tnks/Nni-20110330D30JF440.htm

g This quote appeared in several news media articles at the time, though the translation was sometimes subtly different. Di-Natale, Dominic, "Japan's nuclear rescuers: 'Inevitable some of them may die within weeks.'" Fox News, 31 March 2011. https://www.foxnews.com/world/japans-nuclear-rescuers-inevitable-some-of-them-may-die-within-weeks

h "Plant water highly radioactive / 3 emergency workers suffered exposure levels near 1-year limit." *Yomiuri Shimbun*, 26 March 2011. Archive link: https://web.archive.org/web/20110326215713/ http://www.yomiuri.co.jp/dy/national/T110325004911.htm

i Goldenberg, Suzanne, "Fukushima crisis: Radiation fears grow for low-paid heroes battling disaster." *Guardian*, 27 March 2011. https://www.theguardian.com/world/2011/mar/27/fukushima-crisis-workers-radiation-fears

j Sasaki, Manabu, et al., "Heroes and realists found among the brave 'Fukushima 700.'" *Asahi Shimbun*, 12 April 2011. Archive link: https://web.archive.org/web/20110412220100/http://www. asahi.com/english/TKY201104110137.html

k Krolicki, Kevin, "Japanese retirees ready to risk Fukushima front line." Reuters, 6 June 2011. https://af.reuters.com/article/energyOilNews/idAFL3E7H60ZD20110606

l The original report can be viewed here: Kondo, Shunsuke, "Fukushima Daiichi Nuclear Power Station – Sketch of contingency scenario" (in Japanese), 25 March 2011. http://www.asahi-net. or.jp/~pn8r-fjsk/saiakusinario.pdf

m "Cabinet kept alarming nuke report secret." *Japan Times*, 22 January 2012. https://www.japan-times.co.jp/news/2012/01/22/news/cabinet-kept-alarming-nuke-report-secret/

n Clenfield, Jason, "Fukushima engineer says he covered up flaw at shut Reactor No. 4." Bloomberg, 23 March 2011. https://www.bloomberg.com/news/articles/2011-03-23/fukushima-engineer-says-he-covered-up-flaw-at-shut-reactor

o IEEJ, "Impacts of East Japan Great Earthquake power supply," 22 March 2011. https://eneken.ieej. or.jp/data/3752.pdf

p Edahiro, Junko, "Japan's power shortages and countermeasures after the Tohoku earthquake, tsunami and Fukushima nuclear crisis." *Japan for Sustainability*, no. 104, April 2011. https://www. japanfs.org/en/news/archives/news_id030904.html

q IEEJ, "Short-term energy supply/demand outlook: Analysis on scenario through FY2012," 22

December 2011. https://eneken.ieej.or.jp/data/4231.pdf

r   Okumura, Tetsu, and Tokuno, Shinichi, "Case study of medical evacuation before and after the Fukushima Daiichi nuclear power plant accident in the great East Japan Earthquake." *Disaster and Military Medicine*, vol. 1, article no. 19, 2015. https://disastermilitarymedicine.biomedcentral.com/articles/10.1186/s40696-015-0009-9

s   "Families want answers after 45 people die following evacuation from Fukushima hospital." *Mainichi Daily News*, 26 April 2011. Archive link: https://web.archive.org/web/20110930063423/http://mdn.mainichi.jp/features/archive/news/2011/04/20110426p2a00m0na006000c.html

t   "Government probe: Many Fukushima hospital patients died during botched rescue operation." *Asahi Shimbun*, 24 July 2012. Archived link: https://web.archive.org/web/20120801064235/http://ajw.asahi.com/article/0311disaster/life_and_death/AJ201207240092

u   Tanigawa, Koichi, et al., "Loss of life after evacuation: Lessons learned from the Fukushima accident." *Lancet*, vol. 379, issue 9819, pp. 889-891, 10 March 2012. https://www.thelancet.com/journals/lancet/article/PIIS0140-6736(12)60384-5/fulltext

v   Ibid.

w   National Diet NAIIC, "Chapter 4," p. 33. *Final Report.*

x   Bloxham, Andy, "Japan nuclear evacuation kills 14 elderly hospital patients." *Telegraph*, 18 March 2011. https://www.telegraph.co.uk/news/worldnews/asia/japan/8388383/Japan-nuclear-evacua-tion-kills-14-elderly-hospital-patients.html

y   Koyama, Atsushi, et al., "Medical relief activities, medical resourcing, and inpatient evacuation conducted by Nippon Medical School due to the Fukushima Daiichi Nuclear Power Plant accident following the Great East Japan Earthquake 2011." *Journal of Nippon Medical School*, vol. 78, issue 6, pp. 393-396, 2011. https://www.jstage.jst.go.jp/article/jnms/78/6/78_6_393/_article

z   Hasegawa, A., et al., "Emergency responses and health consequences after the Fukushima accident: Evacuation and relocation." *Clinical Oncology*, vol. 28, issue 4, pp. 237-244, 12 February 2016. https://www.clinicaloncologyonline.net/article/S0936-6555(16)00005-4/fulltext

aa   Okumura and Tokuno, "Case study of medical evacuation."

ab   Fuller, Patrick, "Japan: Second disaster looms for elderly survivors." *International Federation of Red Cross*, 18 March 2011. https://www.ifrc.org/en/news-and-media/news-stories/asia-pacific/japan/japan-second-disaster-looms-for-elderly-survivors/

ac   Kubota, Yoko, "High radiation levels at Japanese plant raise new worry." Reuters, 25 March 2011. https://www.reuters.com/article/us-japan-quake-idUSTRE72A0SS20110325

ad   National Diet NAIIC, "Chapter 4," p. 22. *Final Report.*

ae   EX-SKF (blog), Translation by Dr Shunichi Yamashita, radiation advisor to Fukushima: "Fuku-sima will be world-famous! It's just great!" 21 June 2011. http://ex-skf.blogspot.com/2011/06/dr-shunichi-yamashita-radiation-advisor.html

af   "People are suffering from radiophobia." *Der Spiegel*, 19 August 2011. https://www.spiegel.de/international/world/studying-the-fukushima-aftermath-people-are-suffering-from-radiopho-bia-a-780810.html

ag   Brumfiel, Geoff, "Fukushima health-survey chief to quit post." *Nature*, 20 February 2013. https://www.nature.com/news/fukushima-health-survey-chief-to-quit-post-1.12463

ah   Yabe, Hirooki, et al., "Psychological distress after the Great East Japan Earthquake and Fukushima Daiichi Nuclear Power Plant accident: Results of a mental health and lifestyle survey through the Fukushima Health Management Survey in FY2011 and FY2012." *Fukushima Journal of Medical Science*, vol. 60, issue 1, pp. 57-67, 2014. https://www.jstage.jst.go.jp/article/fms/60/1/60_2014-1/_article

ai   National Diet NAIIC, "Chapter 4," p. 16. *Final Report.*

aj   Hasegawa et al., "Emergency responses and health consequences," pp. 237-244.

ak   Broinowski, A., "Chapter 6: Informal labour, local citizens and the Tokyo Electric Fukushima Daiichi nuclear crisis – Responses to neoliberal disaster management." In T. Morris-Suzuki and E. Jeong Soh (eds.), *New Worlds from Below: Informal Life Politics and Grassroots Action in Twen-ty-First Century Northeast Asia*. Australian National University Press, 2017. https://www.jstor.org/stable/j.ctt1pwtd47

al   Knight, Sophie, "Nine years on, Fukushima's mental health fallout lingers." *Wired*, 23 June 2020. https://www.wired.co.uk/article/fukushima-evacuation-mental-health

am   Dolak, Kevin, et al., "Officials unable to plug radioactive leak found at Japan's nuclear plant." ABC News, 2 April 2011. https://abcnews.go.com/International/tokyo-electric-unable-plug-radio-active-leak/story?id=13281691

an   National Diet NAIIC, "Chapter 4," p. 4. *Final Report.*

ao   Ibid., pp. 290-293.

ap   Tabuchi, Hiroko, "Japanese workers braved radiation for a temp job." *New York Times*, 9 April 2011. https://www.nytimes.com/2011/04/10/world/asia/10workers.html. Note: though this article is ostensibly about day labourers at Fukushima after the 2011 disaster, it's implied that this sort of thing had been happening since before the accident, and it ties in with other, decades-old reports I've seen.

aq   "The health effects of Fukushima." *World Nuclear News*, 28 August 2012. https://www.world-nu-clear-news.org/RS_The_health_effects_of_Fukushima_2808121.html

ar   Slodkowski, Antoni, and Saito, Mari, "Special report: Help wanted in Fukushima: Low pay, high risks and gangsters." Reuters, 25 October 2013. https://www.reuters.com/article/us-fukushi-ma-workers-specialreport/special-report-help-wanted-in-fukushima-low-pay-high-risks-and-gang-sters-idUSBRE99O04320131025

as   "TEPCO subcontractor used lead to fake dosimeter readings at Fukushima plant." *Asahi Shimbun*, 21 July 2012. Archived link: https://web.archive.org/web/20120723234142/http://ajw.asahi.com/article/0311disaster/fukushima/AJ201207210069

at   Saito, Mari, and Slodkowski, Antoni, "Special report: Japan's homeless recruited for murky Fukushima clean-up." Reuters, 30 December 2013. https://www.reuters.com/article/us-fukushi-ma-workers/special-report-japans-homeless-recruited-for-murky-fukushima-clean-up-idUS-BRE9BT00520131230

au   TEPCO, "Fukushima Daiichi Nuclear Power Station status (daily report)," 2 August 2015. http://www.tepco.co.jp/nu-news/2015/1256674_6869.html

av   TEPCO, "Fukushima Daiichi Nuclear Power Station status (daily report)," 30 July 2016. http://www.tepco.co.jp/press/report/2016/1314410_8693.html

aw   Tabuchi, Hiroko, "Unskilled and destitute are hiring targets for Fukushima cleanup." *New York Times*, 16 March 2014. https://www.nytimes.com/2014/03/17/world/asia/unskilled-and-destitute-are-hiring-targets-for-fukushima-cleanup.html

ax   Adelstein, Jake, "The Yakuza and the nuclear mafia: Nationalization looms for TEPCO." *Atlantic*, 30 December 2011. https://www.theatlantic.com/international/archive/2011/12/yakuza-and-nucle-ar-mafia-nationalization-looms-tepco/333697/ Note: quote likely originally from Suzuki, Tomohi-ko, *Yakuza and the Nuclear Industry: Diary of an Undercover Reporter Working at the Fukushima Plant*. Bungei Shunju, 2011.

ay   Adelstein, Jake, "The Yakuza and the nuclear mafia."

az   Adelstein, Jake, "Mobsters on a mission: How Japan's mafia launched an aid effort." *Independent*, 9 April 2011. https://www.independent.co.uk/news/world/asia/mobsters-on-a-mission-how-japans-mafia-launched-an-aid-effort-2264031.html

ba   Adelstein, Jake, "Goodbye to the Yakuza." *Atlantic*, 1 October 2011. https://www.theatlantic.com/international/archive/2011/10/goodbye-yakuza/337236/

bb   Adelstein, "The Yakuza and the nuclear mafia."

bc   Kazumi, Takaki, "Listen to their silent cry: The devastated lives of Japanese nuclear power plant workers employed by subcontractors or labour–brokering companies." *Bulletin of Social Medicine*, vol. 31, issue 1, 2014. http://jssm.umin.jp/report/english/31-1-02Takagi_2014.pdf

bd   Okwu, Michael, "Gangsters and 'slaves': The people cleaning up Fukushima." *Al Jazeera*, 7 January 2014. http://america.aljazeera.com/watch/shows/america-tonight/america-to-night-blog/2014/1/7/fukushima-cleanupworkerssubcontractors.html

be   "Embattled TEPCO now facing a harsh public backlash." *Asahi Shimbun*, 3 April 2011. Archive link: https://web.archive.org/web/20110403215750/http://www.asahi.com/english/TKY201104020226.html

bf   "Fukushima workers quit Japan utility in droves over stigma, pay cuts." *Al Jazeera*, 10 July 2014. http://america.aljazeera.com/articles/2014/7/10/fukushima-japan-workers.html

bg   "Outgoing nuclear agency chief was aware of possible meltdown at Fukushima plant." *Mainichi Shimbun*, 11 August 2011. Archive link: https://web.archive.org/web/20110811225216/http://mdn.mainichi.jp/mdnnews/news/20110811p2a00m0na005000c.html

bh   "Agency admits 'melting' of N-fuel." *Yomiuri Shimbun*, 20 April 2011. Archive link: https://web.archive.org/web/20130321160753/http://www.yomiuri.co.jp/dy/national/T110419004267.htm

bi   "Japan utility: Delay in declaring 'meltdown' was a cover-up." *Los Angeles Times* (Associated Press), 21 June 2016. https://www.latimes.com/world/la-fg-japan-nuclear-meltdown-20160621-snap-story.html

bj   "Concealment of core meltdown, denial of instruction at government residence – Verification committee of TEPCO and Niigata" (in Japanese). Nikkei, 26 December 2017. https://www.nikkei.com/article/DGXMZO25077350W7A221C1000000/

bk   "TEPCO discovers after 5 years that it could have quickly declared Fukushima plant meltdown." *Asahi Shimbun*, 25 February 2016. http://ajw.asahi.com/article/0311disaster/fukushima/AJ201602250043

bl Boyd, John, "Japan's nuclear energy comeback takes a tumble." *IEEE Spectrum*, 28 March 2016. https://spectrum.ieee.org/energywise/energy/nuclear/japans-nuclear-energy-comeback-takes-a-tumble

bm Poll by *Kyodo News*, 25 July 2011. Hirokawa, Takashi, Sakamaki, Sachiko, and Ikeda, Akiko, "Japan has 'slim' chance of splitting utilities as blackouts loom." Bloomberg, 31 July 2011. https://www.bloomberg.com/news/articles/2011-07-31/japan-has-slim-chance-of-splitting-power-utilities-as-blackouts-loom

bn "74% in favor of 'ending reliance on nuclear power' in the future" (in Japanese). *Asahi Shimbun*, 13 June 2011. Archive link: https://web.archive.org/web/20110617135547/http://www.asahi.com/national/update/0613/TKY201106130401.html

bo "TEPCO shares plunge to record low on radiation crisis." BBC, 5 April 2011. https://www.bbc.co.uk/news/business-12968812

bp Layne, Nathan, and Uranaka, Taiga, "TEPCO chief quits after $15 billion loss on nuclear crisis." Reuters, 23 May 2011. https://www.reuters.com/article/us-tepco/tepco-chief-quits-after-15-billion-loss-on-nuclear-crisis-idUSTRE74I89G20110523

bq Kubota, Yoko, "Japan to take over TEPCO after Fukushima disaster." Reuters, 9 May 2012. https://www.reuters.com/article/us-tepco/japan-to-take-over-tepco-after-fukushima-disaster-idUS-BRE8480Z220120509

br Tabuchi, Hiroko, "Japan to nationalize Fukushima utility." *New York Times*, 9 May 2012. https://www.nytimes.com/2012/05/10/business/global/japan-to-nationalize-fukushima-utility.html

bs Yamaguchi, Mari, "Japan audit: Millions of dollars wasted in Fukushima cleanup." Associated Press, 24 March 2015. Archive link: https://web.archive.org/web/20150719134503/http://bigstory.ap.org/article/75dd3b31041949b7bbd4de14a2d5b287/japan-audit-millions-dollars-wasted-fukushima-cleanup

bt "TEPCO ordered to cough up after it refused deal on compensation." *Asahi Shimbun*, 20 February 2020. http://www.asahi.com/ajw/articles/13144481

bu McNeill, David, "The Fukushima nuclear crisis and the fight for compensation." *Asia-Pacific Journal*, vol. 10, issue 10, no. 6, 27 February 2012. https://apjjf.org/-David-McNeill/3707/article.pdf

bv Jobin, Paul, "The Fukushima nuclear disaster and civil actions as a social movement." *Asia-Pacific Journal*, vol. 18, issue 9, no. 1, 1 May 2020. https://apjjf.org/2020/9/Jobin.html

bw "TEPCO to pay damages in Fukushima suicide case." BBC, 26 August 2014. https://www.bbc.co.uk/news/world-asia-28933726

bx McCurry, Justin, "Fukushima evacuee to tell UN that Japan violated human rights." *Guardian*, 11 October 2017. https://www.theguardian.com/environment/2017/oct/11/fukushima-evacuee-un-japan-human-rights

by McCurry, Justin, "Japanese government held liable for first time for negligence in Fukushima." *Guardian*, 17 March 2017. https://www.theguardian.com/world/2017/mar/17/japanese-government-liable-negligence-fukushima-daiichi-nuclear-disaster

bz Rich, Motoko, "Japanese government and utility are found negligent in nuclear disaster." *New York Times*, 17 March 2017. https://www.nytimes.com/2017/03/17/world/asia/japan-fukushima-nuclear-disaster-tepco-ruling.html

ca Iwata, Tomohiro, "TEPCO: Radioactive substances belong to landowners, not us." *Asahi Shimbun*, 24 November 2011. Archive link: https://web.archive.org/web/20111126223905/http://ajw.asahi.com/article/behind_news/social_affairs/AJ201111240030

cb "Former president of TEPCO repeats 'no memory,' memo put by prosecutor" (in Japanese). *Asahi Shimbun*, 26 November 2019. https://www.asahi.com/articles/ASMC27V3HMC2ULZU005.html

cc "Ex-TEPCO exec denies being told of need for tsunami steps." *Asahi Shimbun*, 16 October 2018. http://www.asahi.com/ajw/articles/AJ201810160055.html

cd World Bank, "The recent earthquake and tsunami in Japan: Implications for East Asia," March 2011. https://siteresources.worldbank.org/INTEAPHALFYEARLYUPDATE/Resources/550192-1300567391916/EAP_Update_March2011_japan.pdf

ce "Japan estimates post-3/11 reconstruction costs at ¥1.5 trillion for fiscal 2021–25." *Japan Times*, 10 December 2019. https://www.japantimes.co.jp/news/2019/12/10/national/311-reconstruction-1-5-trillion/

cf "Why make the Onagawa Nuclear Power Plant an evacuation destination? 240 people living [have] 'nowhere to go' [that is] 'sturdy and safe'" (in Japanese). MSN, 26 March 2011. Archive link: https://web.archive.org/web/20111111145009/http://sankei.jp.msn.com/affairs/news/110326/

dst11032622340091-n1.htm

cg   "Onagawa: Japanese tsunami town where nuclear plant is the safest place." *Guardian* (Associated Press), 30 March 2011. https://www.theguardian.com/world/2011/mar/30/Onagawa-tsunami-refugees-nuclear-plant

ch   "Evacuation centre complex feels like 'Fukushima' at Onagawa Nuclear Power Plant" (in Japanese). *Asahi Shimbun*, 2 June 2011. Archive link: link:https://web.archive.org/web/20110605080119/http://mytown.asahi.com/miyagi/news.php?k_id=04000001106020004

ci   Maeda, Risa, "Japanese nuclear plant survived tsunami, offers clues." Reuters, 11 October 2011. https://www.reuters.com/article/us-japan-nuclear-tsunami/japanese-nuclear-plant-survived-tsunami-offers-clues-idUSTRE79J0B420111020

cj   Ishikawa, Michio, "Onagawa Nuclear Power Station, Tohoku Electric Power Company." *Denki Shinbun* (*Electric Daily News*), 18 November 2011. http://www.gengikyo.jp/english/shokai/Tohoku_Jishin/article_20111118.html

ck   Kushida, Kenji E., "The Fukushima nuclear disaster and the DPJ: Leadership, structures, and information challenges during the crisis." *Japanese Political Economy*, vol. 40, no. 1, pp. 29-68, spring 2014. https://papers.ssrn.com/sol3/papers.cfm?abstract_id=2334523

cl   TEPCO, "Fukushima nuclear accident investigation report." p. 31, *Interim Report – Supplementary Volume*. https://www.tepco.co.jp/en/press/corp-com/release/betu11_e/images/111202e16.pdf

cm   National Diet NAIIC, "Chapter 2," p. 21. *Final Report*.

cn   National Diet NAIIC, "Chapter 3," p. 50. *Final Report*.

co   Ibid., pp. 50 and 73.

cp   Tabuchi, Hiroko, "Japan ignored nuclear risks, official says." *New York Times*, 15 February 2012. https://www.nytimes.com/2012/02/16/world/asia/japanese-official-says-nations-atomic-rules-are-flawed.html

cq   Ito, Masami, "Nuclear safety boss faults agency, utilities." *Japan Times*, 16 February 2012. https://www.japantimes.co.jp/news/2012/02/16/national/nuclear-safety-boss-faults-agency-utilities/

cr   Okumura, Nobuyiki, et al., "Japan's media fails its watchdog role: Lessons learned and unlearned from the 2011 earthquake and the Fukushima disaster." Disaster and Media Research Group, 2019. https://www.academia.edu/41372220/Japans_media_fails_its_watchdog_role_Lessons_learned_and_unlearned_from_the_2011_earthquake_and_the_Fukushima_disaster?email_work_card=title

cs   Takahashi, Hideki, "The unsung hero of the meltdowns." *Japan Times*, 9 October 2014. https://www.japantimes.co.jp/news/2014/10/09/national/remembering-fukushima-plant-chief-helped-prevent-catastrophe/

ct   "Fukushima Daiichi Nuclear Power Plant: Former director Yoshida full remarks on video" (in Japanese). *Mainichi Shimbun*, 11 August 2012. Archive link: https://web.archive.org/web/20120812004526/http://mainichi.jp/select/news/20120812k0000m040027000c.html

cu   Imanaka, Tetsuji, "Chapter 20: Comparison of radioactivity release and contamination from the Fukushima and Chernobyl Nuclear Power Plant accidents." In M. Fukumoto (ed.), *Low-Dose Radiation Effects on Animals and Ecosystems: Long-Term Study on the Fukushima Nuclear Accident*. Springer, 2020.

cv   "Japan cites radiation in milk, spinach near plant." Associated Press, 19 March 2001. https://web.archive.org/web/20110322033200/http://www.google.com/hostednews/ap/article/ALeqM5gI6DCsmEuRwI3Y1Z1Otl4akQ-2GA?docId=c3827c1a078c46d9ab4451ed41d8641e

cw   Beech, Hannah, "Radioactive water: Japan watches what it drinks." *Time*, 24 March 2011. http://content.time.com/time/world/article/0,8599,2061183,00.html

cx   "Rainwater banned at water plants." *Japan Times*, 29 March 2011. https://www.japantimes.co.jp/news/2011/03/29/national/rainwater-banned-at-water-plants/

cy   Buesseler, Ken, Aoyama, Michio, and Fukasawa, Masao, "Impacts of the Fukushima Nuclear Power Plants on marine radioactivity." *Environmental Science & Technology* (American Chemical Society), vol. 45, no. 23, pp. 9931-9935, 2011. https://pubs.acs.org/doi/10.1021/es202816c

cz   Buesseler, Ken, "Fishing for answers off Fukushima." *Science*, vol. 338, issue 6106, pp. 480-482, 26 October 2012. https://science.sciencemag.org/content/338/6106/480.summary

da   Morita, Takami, et al., 2020. "Chapter 3: Impacts of the Fukushima nuclear accident on fishery products and fishing industry." In M. Fukomoto (ed.), *Low-Dose Radiation Effects on Animals and Ecosystems: Long-Term Study on the Fukushima Nuclear Accident*. Springer, 2020.

db   "Losses of farm, fishery industries top 2 trillion yen." *Mainichi Shimbun*, 2 July 2011. Archive link: https://web.archive.org/web/20110703115957/http://mdn.mainichi.jp/mdnnews/business/news/20110702p2g00m0bu035000c.html

dc   "Fishermen unload at Chiba port to avoid Fukushima label." *Mainichi Shimbun*, 22 June 2011. Archive link: https://web.archive.org/web/20110622222157/http://mdn.mainichi.jp/mdnnews/

news/20110622p2a00m0na005000c.html

dd Umeda, Sayuri, "Japan: New permissible levels of radioactivity in foods." Library of Congress, 9 April 2012. http://www.loc.gov/law/foreign-news/article/japan-new-permissible-levels-of-radioactivity-in-foods/

de Department of Food Safety Pharmaceutical & Food Safety Bureau, Ministry of Health, Labour and Welfare, "New standard limits for radionuclides in foods," 2011. https://www.mhlw.go.jp/english/topics/2011eq/dl/new_standard.pdf

df European Commission, "Questions and answers: Safety of food products imported from Japan," 1 April 2011. https://ec.europa.eu/commission/presscorner/detail/en/MEMO_11_215

dg Strickland, Eliza, "Fukushima fish still radioactive." *IEEE Spectrum*, 25 October 2012. https://spectrum.ieee.org/tech-talk/energy/nuclear/fukushima-fish-still-radioactive

dh Ashe, Katy, "Tuna caught off California carry radiation from the Japanese disaster, Stanford scientist finds." *Stanford News*, 30 May 2012. https://news.stanford.edu/news/2012/may/tuna-radioactive-materials-053012.html

di Nakata, Kaoru, and Sugisaka, Hiroya (eds.), *Impacts of the Fukushima Nuclear Accident on Fish and Fishing Grounds*. Springer Japan, 2015.

dj Ibid., section 12.2: "Radiocesium contamination of fat greenlings off the coast of Fukushima." Springer Japan, 2015.

dk Tateyama, Ryo, and Goto, Shunsuke, "Fukushima's fishing industry claws back from the brink." NHK, 15 October 2020. https://www3.nhk.or.jp/nhkworld/en/news/backstories/1325/

dl "Worker from nuclear power plant is certified as the first worker to have leukemia." NHK, 20 October 2015. Archive link: https://web.archive.org/web/20151020221214/http://www3.nhk.or.jp/news/html/20151020/k10010276091000.html

dm Fukushima Prefectural Government, "Damage caused by the 2011 earthquake and tsunami," 27 March 2019. http://www.pref.fukushima.lg.jp/site/portal-english/en03-01.html

dn "Number of deaths related to nuclear power plant increased to 1,368." *Tokyo Shimbun*, 6 March 2016. Archive link: https://web.archive.org/web/20160306060111/http://www.tokyo-np.co.jp/article/national/list/201603/CK2016030602000127.html

do McCurry, Justin, "Japan whistleblowers face crackdown under proposed state secrets law." *Guardian*, 5 December 2013. https://www.theguardian.com/world/2013/dec/05/whistleblowers-japan-crackdown-state-secrets

dp Knight, Sophie, "Nine years on, Fukushima's mental health fallout lingers." *Wired*, 23 June 2020. https://www.wired.co.uk/article/fukushima-evacuation-mental-health

dq Nuclear Damage Compensation and Decommissioning Facilitation Corporation, "Technical strategic plan 2019 for the decommissioning of the Fukushima Daiichi Nuclear Power Station of Tokyo Electric Power Company Holdings, Inc.," 9 September 2019. http://www.dd.ndf.go.jp/en/strategic-plan/book/20191101_SP2019eFT.pdf

dr "Absorbent yet to soak up radioactive water at Fukushima plant." *Mainichi Shimbun*, 3 April 2011. Archive link: https://web.archive.org/web/20110405201139/http://mdn.mainichi.jp/mdnnews/news/20110403p2g00m0dm025000c.html

ds Westall, Sylvia, and Dahl, Fredrik, "Japan to dump 11,500 metric tons of radioactive water." Reuters, 4 April 2011. https://www.reuters.com/article/us-japan-nuclear-water/japan-to-dump-11500-metric-tons-of-radioactive-water-idUSTRE7336L720110404. (Note: 1,500 tons was uncontaminated water.)

dt Fackler, Martin, "Flow of tainted water is latest crisis at Japan nuclear plant." *New York Times*, 29 April 2013. https://www.nytimes.com/2013/04/30/world/asia/radioactive-water-imperils-fukushima-plant.html

du McCurry, Justin, "Fukushima operator reveals leak of 300 tonnes of highly contaminated water." *Guardian*, 20 August 2013. https://www.theguardian.com/world/2013/aug/20/fukushima-leak-nuclear-pacific

dv Sheldrick, Aaron, and Saito, Mari, "Record radiation readings near Fukushima contaminated water tanks." Reuters, 4 September 2013. https://www.reuters.com/article/us-japan-fukushima-tanks/record-radiation-readings-near-fukushima-contaminated-water-tanks-idUSBRE98301020130904

dw Nuclear and Industrial Safety Agency (NISA), Ministry of Economy, Trade and Industry (METI), "Seismic damage information (110th release)," p. 23, 23 April 2011. https://web.archive.org/web/20120201081408/http://www.nisa.meti.go.jp/english/files/en20110423-6-1.pdf

dx "Japan MP Yasuhiro Sonoda drinks Fukushima water." BBC, 1 November 2011. https://www.bbc.co.uk/news/world-asia-pacific-15533018

dy Health Physics Society, "Tritium," 2011 (revised 2020). https://hps.org/documents/tritium_fact_sheet.pdf

dz Looney, Brain, et al., "Independent technical support for the frozen soil barrier installation and operation at the Fukushima Daiichi Nuclear Power Station (F1 site)." Savannah River National Laboratory, Pacific Northwest National Laboratory, SRNL-STI-2015-00215, February 2015. https://sti.srs.gov/fulltext/SRNL-STI-2015-00215.pdf

ea Strickland, Eliza, "Construction of ice wall begins at Fukushima Daiichi." *IEEE Spectrum*, 3 June 2014. https://spectrum.ieee.org/energywise/energy/nuclear/construction-of-ice-wall-begins-at-fukushima-daiichi

eb "Four coolant leaks found in Fukushima nuke plant 'ice wall' pipes." *Mainichi Shimbun*, 17 January 2020. https://mainichi.jp/english/articles/20200117/p2a/00m/0na/022000c

ec Nuclear Damage Compensation and Decommissioning Facilitation Corporation, "Technical strategic plan 2019."

ed Fackler, Martin, "Six years after Fukushima, robots finally find reactors' melted uranium fuel." *New York Times*, 19 November 2017. https://www.nytimes.com/2017/11/19/science/japan-fukushima-nuclear-meltdown-fuel.html

ee McCurry, Justin, "Fukushima robot stranded after stalling inside reactor." *Guardian*, 13 April 2015. https://www.theguardian.com/environment/2015/apr/13/fukushima-robot-stalls-reactor-abandoned

ef "Significantly corrected the radiation dose in the containment container of Unit 2" (in Japanese). NHK, 28 July 2017. Archive link: https://web.archive.org/web/20170728010236/http://www3.nhk.or.jp/news/html/20170728/k10011077611000.html

eg "Fukushima robot finds potential fuel debris hanging like icicles in reactor 3." *Japan Times*, 21 July 2017. https://www.japantimes.co.jp/news/2017/07/21/national/fukushima-robot-finds-potential-fuel-debris-hanging-like-icicles-reactor-3/

eh Beiser, Vince, "The robot assault on Fukushima." *Wired*, 26 April 2018. https://www.wired.com/story/fukushima-robot-cleanup/

ei Agency for Natural Resources and Energy (ANRE), Ministry of Economy, Trade and Industry (METI), "Article no. 3: The frontier of technological development," 28 February 2020. *The Challenge of Retrieving Fuel Debris.* https://www.enecho.meti.go.jp/en/category/special/article/detail_155.html

ej Ministry of the Environment, "Environmental remediation in Japan," March 2018. http://josen.env.go.jp/en/pdf/progressseet_progress_on_cleanup_efforts.pdf

ek Comments by Jiro Hiratsuka, environment ministry official: McCurry, Justin, "Fukushima grapples with toxic soil that no one wants." *Guardian*, 11 March 2019. https://www.theguardian.com/world/2019/mar/11/fukushima-toxic-soil-disaster-radioactive

el Nakanishi, Tomoko, O'Brian, Martin, and Tanoi, Keitaro (eds.), *Agricultural Implications of the Fukushima Nuclear Accident (III): After 7 Years.* Springer, 2019. https://link.springer.com/book/10.1007%2F978-981-13-3218-0

em Evrard, Olivier, Laceby, J. Patrick, and Nakao, Atsushi, "Effectiveness of landscape decontamination following the Fukushima nuclear accident: A review." *SOIL Journal*, vol. 5, pp. 333-350, 2019. https://doi.org/10.5194/soil-5-333-2019

en Nagata, Kazuaki, "Utility says NISA sought 'plants' to talk up MOX bid." *Japan Times*, 30 July 2011. https://www.japantimes.co.jp/news/2011/07/30/national/utility-says-nisa-sought-plants-to-talk-up-mox-bid/

eo Boyd, John, "Japanese nuclear agency and utilities tried to manipulate public opinion." *IEEE Spectrum*, 2 August 2011. https://spectrum.ieee.org/tech-talk/energy/nuclear/japanese-nuclear-agency-and-utilities-tried-to-manipulate-public-opinion

ep "NISA linked to other faked support for nuclear power." *Japan Times*, 3 August 2011. https://www.japantimes.co.jp/news/2011/08/03/national/nisa-linked-to-other-faked-support-for-nuclear-power/

eq Ryall, Julian, "Japanese Nuclear City reveals huge plan to clean every building and road of radiation." *Telegraph*, 14 July 2011. https://www.telegraph.co.uk/news/worldnews/asia/japan/8636925/Japanese-nuclear-city-reveals-huge-plan-to-clean-every-building-and-road-of-radiation.html

er Konaori, Y. "Seismic risk assessment of an active fault system: The example of the Tsurugawan–Isewan tectonic line." *Engineering Geology*, vol. 56, pp. 109-123, April 2000. https://www.sciencedirect.com/science/article/pii/S0013795299001374

es Chapman, N., et al., "Active faults and nuclear power plants." *Eos, Transactions, American Geophysical Union*, vol. 95, no. 4, pp. 33-40, 28 January 2014. https://agupubs.onlinelibrary.wiley.com/doi/pdf/10.1002/2014EO040001

et "Fault under Tsuruga nuclear plant could trigger M7.4 quake: Research." *Mainichi Shimbun*, 6 March 2012. Archive link: https://web.archive.org/web/20120307172308/http://mdn.mainichi.jp/mdnnews/news/20120306p2g00m0dm081000c.html

eu Suzuki, Yasuhiro, *Active Faults and Nuclear Regulation: Background to Requirement Enforcement in Japan*, pp. 42-43. Springer, January 2020.

ev "Tsuruga nuke plant operator disputes NRA verdict on fault." *Japan Times*, 12 December 2012. https://www.japantimes.co.jp/news/2012/12/12/national/tsuruga-nuke-plant-operator-disputes-nra-verdict-on-fault/

ew "Second Japan nuclear plant at fault risk." Phys.org, 14 December 2012. https://phys.org/news/2012-12-japan-nuclear-fault.html

ex "Faults under Japan's Shika NPP said to be active." *Nuclear Engineering International*, 4 March 2016. https://www.neimagazine.com/news/newsfaults-under-japans-shika-npp-said-to-be-active-4830171

ey Higashiyama, Masanobu, "Fault beneath Monju nuclear reactor likely inactive." *Asahi Shimbun*, 8 October 2015. Archive link: https://web.archive.org/web/20151213012057/http://ajw.asahi.com/article/behind_news/social_affairs/AJ201510080072

ez "N-safety official fired over leak." *Yomiuri Shimbun*, 3 February 2013. http://www.yomiuri.co.jp/dy/national/T130202003272.htm)

fa "NRA secretariat must halt collusion." *Japan Times*, 19 February 2013. https://www.japantimes.co.jp/opinion/2013/02/19/editorials/nra-secretariat-must-halt-collusion/

fb Nuclear Regulatory Authority, "Enforcement of the new regulatory requirements for commercial nuclear power reactors," 8 July 2013. http://www.nsr.go.jp/data/000067212.pdf

fc "Kyushu misses deadline for Sendai 1 emergency facilities." *World Nuclear News*, 16 March 2020. https://www.world-nuclear-news.org/Articles/Kyushu-misses-deadline-for-Sendai-1-emergency-faci

fd "TEPCO declares Fukushima Daini for decommissioning." *World Nuclear News*, 31 July 2019. https://world-nuclear-news.org/Articles/Tepco-declares-Fukushima-Daini-for-decommissioning

fe "TEPCO aims for nuclear power business realignment in 2020s." *Mainichi Shimbun*, 12 May 2017. https://mainichi.jp/english/articles/20170512/p2a/00m/0na/012000c

ff "Niigata puts TEPCO's reactor inspection request on hold." *Japan Times*, 25 September 2013. https://www.japantimes.co.jp/news/2013/09/25/national/niigata-puts-tepcos-reactor-inspection-request-on-hold/

fg Okada, Yuji, and Suga, Masumi, "TEPCO vows to push ahead with restart of Niigata Nuclear Station." Bloomberg, 25 September 2013. https://www.bloomberg.com/news/articles/2013-09-25/tepco-vows-to-push-ahead-with-restart-of-niigata-nuclear-station

fh "Fukushima No. 1 mishandling may foreclose on TEPCO reactor restarts: NRA." *Japan Times*, 2 October 2013. https://www.japantimes.co.jp/news/2013/10/02/national/fukushima-no-1-mishandling-may-foreclose-on-tepco-reactor-restarts-nra/

fi "Onagawa 2 upgrade faces further delay." *World Nuclear News*, 4 May 2020. https://www.world-nuclear-news.org/Articles/Further-delay-in-completion-of-Onagawa-2-safety-up

fj "Governor of Japan's Niigata resigns to avoid 'turmoil' over magazine article." Reuters, 18 April 2018. https://www.reuters.com/article/us-japan-nuclear-governor/governor-of-japans-niigata-resigns-to-avoid-turmoil-over-magazine-article-idUSKBN1HP1EP.

fk "Mayor approves TEPCO's N-reactor decommissioning plan." *Japan News*, 19 November 2019. https://the-japan-news.com/news/article/0006171647

fl Ryall, Julian, "Japanese nuclear plant sparks panic after ticking wrong box on fax in wake of earthquake." *Telegraph*, 20 June 2019. https://www.telegraph.co.uk/news/2019/06/20/japanese-nuclear-plant-sparks-panic-ticking-wrong-box-fax-wake/

fm Kiyo, Dörrer, "Japan's TEPCO fights for return to nuclear power after Fukushima." *Deutsche Welle*, 11 March 2019. https://www.dw.com/en/japans-tepco-fights-for-return-to-nuclear-power-after-fukushima/a-47836968

fn Stapczynski, Stephen, "World's biggest nuclear plant is center stage in rural election." Bloomberg, 7 June 2018. https://www.bloomberg.com/news/articles/2018-06-07/world-s-biggest-nuclear-plant-is-center-stage-in-rural-election.

fo "Review of the expansion of nuclear power plants 'prematurely' requested" (in Japanese). *Fukui Shimbun*, 5 April 2011. Archive link: https://web.archive.org/web/20110422220243/https://www.fukuishimbun.co.jp/localnews/earthquake/27359.html

fp Naylor, Brian, "Japan backs off of nuclear power after public outcry." NPR, 11 May 2011. https://www.npr.org/2011/05/11/136209502/japan-backs-off-of-nuclear-power-after-public-outcry

fq "Japan to scrap nuclear power in favour of renewables." *Guardian* (Associated Press), 10 May 2011. https://www.theguardian.com/environment/2011/may/10/japan-nuclear-renewables

fr According to the Federation of Electric Power Companies (FEPC).

fs Abe, Yuki, "The nuclear power debate after Fukushima: A text-mining analysis of Japanese newspapers." *Contemporary Japan*, vol. 27, issue 2, 2015. https://www.degruyter.com/view/journals/

cj/27/2/article-p89.xml

ft   IEEJ, "Short-term energy supply."

fu   "Japan switches on Ohi nuclear reactor amid protests." BBC, 2 July 2012. https://www.bbc.co.uk/news/world-asia-18662892

fv   Kingston, Jeff, "Japan's Nuclear Village." *Asia-Pacific Journal*, vol. 10, issue 37, no. 1, 9 September 2012. https://apjjf.org/2012/10/37/Jeff-Kingston/3822/article.html

fw   McCurry, Justin, "Japan plans to end reliance on nuclear power within 30 years." *Guardian*, 14 September 2012. https://www.theguardian.com/world/2012/sep/14/japan-end-nuclear-power

fx   Tabuchi, Hiroko, "Japan, under pressure, backs off goal to phase out nuclear power by 2040." *New York Times*, 19 September 2012. https://www.nytimes.com/2012/09/20/world/asia/japan-backs-off-of-goal-to-phase-out-nuclear-power-by-2040.html

fy   "90 percent of public submissions favor zero nuclear power plants." *Mainichi Shimbun*, 23 August 2012. Archive link: https://web.archive.org/web/20120919173558/http://mainichi.jp/english/english/newsselect/news/20120823p2a00m0na010000c.html

fz   Tabuchi, Hiroko, "Japan sets policy to phase out nuclear power plants by 2040." *New York Times*, 14 September 2012. https://www.nytimes.com/2012/09/15/world/asia/japan-will-try-to-halt-nuclear-power-by-the-end-of-the-2030s.html

ga   "Monju operator skipped inspections of another 2,300 devices." *Mainichi Shimbun*, 22 June 2013. Archive link: https://web.archive.org/web/20130630013612/http://mainichi.jp/english/english/newsselect/news/20130622p2g00m0dm004000c.html

gb   "Regulators warn Monju operator over breach of nuclear security rules." *Mainichi Shimbun*, 6 November 2013. Archive link: https://web.archive.org/web/20131124154101/http://mainichi.jp/english/english/newsselect/news/20131106p2g00m0dm056000c.html

gc   "Monju reactor project failed to pay off after swallowing ¥1.13 trillion of taxpayers' money: auditors." *Japan Times*, 11 May 2018. https://www.japantimes.co.jp/news/2018/05/11/national/monju-reactor-project-failed-pay-off-swallowing-%C2%A51-13-trillion-taxpayers-money-auditors/

gd   Ishii, Noriyuki. "NRA deems JAEA unfit to operate FBR Monju." *Japan Atomic Industrial Forum*, 5 November 2015. https://www.jaif.or.jp/en/nra-deems-jaea-unfit-to-operate-fbr-monju/

ge   "The stalled restart of idled reactors." *Japan Times*, 20 December 2019. https://www.japantimes.co.jp/opinion/2019/12/20/editorials/stalled-restart-idled-reactors/

gf   "Onagawa 2 upgrade faces further delay." *World Nuclear News*, 4 May 2020. https://world-nuclear-news.org/Articles/Further-delay-in-completion-of-Onagawa-2-safety-up

gg   "Despite risk from the 'Big One,' are Shizuoka and the rest of Japan becoming complacent about earthquakes?" *Japan Times*, 3 January 2020. https://www.japantimes.co.jp/news/2020/01/03/national/despite-risk-big-one-shizuoka-rest-japan-becoming-complacent-earthquakes/

gh   Clark, Aaron, Stapczynski, Stephen, and Takezawa, Shiho, "Executives in Japan nuclear scandal blame dead local official." *Bloomberg*, 2 October 2019. https://www.bloomberg.com/amp/news/articles/2019-10-02/execs-in-japan-nuclear-scandal-vow-to-stay-on-blame-dead-local

gi   KEPCO, "Investigation report" (in Japanese), March 2020. https://www.kepco.co.jp/corporate/pr/2020/pdf/0314_2j_01.pdf

gj   Ljubas, Zdravko, "Japan: KEPCO's officials apologize for taking gifts." Organized Crime and Corruption Reporting Project (OCCRP), 30 September 2019. https://www.occrp.org/en/daily/10755-japan-kepco-s-officials-apologize-for-taking-gifts

gk   "Gift scandal extends to KEPCO's Kyoto branch, utility says." *Japan Times*, 6 October 2019. https://www.japantimes.co.jp/news/2019/10/06/national/bribery-scandal-kansai-electric-kyoto-branch/

gl   "Kansai Electric admits execs' acceptance of gifts; president won't resign." *Japan Today*, 2 October 2019. https://japantoday.com/category/national/Kansai-Electric-admits-execs'-acceptance-of-gifts-president-won't-resign; archived link: https://web.archive.org/web/20191004022943/https://japantoday.com/category/national/Kansai-Electric-admits-execs'-acceptance-of-gifts-president-won't-resign

gm   "Japan's Kansai Electric vows to improve governance after bribery scandal." *Mainichi Shimbun*, 30 March 2020. https://mainichi.jp/english/articles/20200330/p2g/00m/0bu/052000c

gn   "Japan nuclear regulator effectively made safety measure decision behind closed doors." *Mainichi Shimbun*, 4 January 2020. https://mainichi.jp/english/articles/20200104/p2a/00m/0na/013000c

go   "Japan Atomic accused of altering data about fault under Tsuruga reactor." *Japan Times*, 15 February 2020. https://www.japantimes.co.jp/news/2020/02/15/national/nra-accuses-japan-atomic-power-altering-data-used-evaluate-fault-tsuruga-plants-no-2-reactor/

gp   "Tokai reprocessing plant to shut." *World Nuclear News*, 29 September 2014. https://www.world-nuclear-news.org/WR-Tokai-reprocessing-plant-to-shut-2909144.html. Statistic from here: http://www.jaea.go.jp/about_JAEA/reorganization/jaea_kaikaku03/shiryo3-1.pdf

gq "Tokai reprocessing facility resumes vitrification of highly radioactive liquid waste, reducing risk early" (in Japanese). *Japan Atomic Industrial Forum*, 9 July 2019. https://www.jaif.or.jp/190709-1

gr "Editorial: Japan should end its nonsensical effort to recycle nuclear fuel." *Asahi Shimbun*, 14 May 2020. http://www.asahi.com/ajw/articles/13372798

gs Sekiguchi, Yakari, "Politics and Japan's Rokkasho Reprocessing Plant." Center for Strategic & International Studies, 12 September 2014. https://www.csis.org/analysis/politics-and-japans-rokkasho-reprocessing-plant

gt International Panel on Fissile Materials (IPFM), "Plutonium separation in nuclear power programs: Status, problems, and prospects of civilian reprocessing around the world," p. 108, 2015. http://fissilematerials.org/library/rr14.pdf

gu Takubo, Masafumi, and von Hippel, Frank, "Forty years of impasse: The United States, Japan, and the plutonium problem." *Bulletin of the Atomic Scientists*, 23 August 2017, pp. 337-343. https://www.tandfonline.com/doi/full/10.1080/00963402.2017.1364007

gv Nuclear Engineering International, "France cancels ASTRID fast reactor project," 2 September 2019. https://www.neimagazine.com/news/newsfrance-cancels-astrid-fast-reactor-project-7394432

gw Tsukimori, Osamu, "With coal under fire, 2020 could be a big year for wind power in Japan." *Japan Times*, 2 January 2020. https://www.japantimes.co.jp/news/2020/01/02/business/wind-power-2020-japan/

gx "With coal under fire, 2020 could be a big year for wind power in Japan." *Japan Times*, 2 January 2020. https://www.japantimes.co.jp/news/2020/01/02/business/wind-power-2020-japan/

gy Ueda, Shiko, and Kurimoto, Suguru, "Fukushima to be reborn as $2.7bn wind and solar power hub." Nikkei, 10 November 2019. https://asia.nikkei.com/Business/Energy/Fukushima-to-be-reborn-as-2.7bn-wind-and-solar-power-hub

gz International Energy Agency data.

ha US Energy Information Administration, "Japan is the world's third-largest coal-importing country," 14 June 2019. https://www.eia.gov/todayinenergy/detail.php?id=39853

hb Shrestha, Priyanka, "Japan 'to build 20GW of coal-fired capacity over next decade.'" *Energy Live News*, 20 June 2019. https://www.energylivenews.com/2019/06/20/japan-to-build-20gw-of-coal-fired-capacity-over-next-decade/

hc Sato, Makiko, and Kharecha, Pushker A., "Implications of energy and CO_2 emission changes in Japan and Germany after the Fukushima accident." *Energy Policy*, vol. 132, pp. 647-653, September 2019. https://www.sciencedirect.com/science/article/pii/S0301421519303611

hd "Japan: Energy imports, net (% of energy use)," n.d. Trading Economics. https://tradingeconomics.com/japan/energy-imports-net-percent-of-energy-use-wb-data.html

he "Orano wins further MOX order from Japan." *World Nuclear News*, 6 February 2020. https://www.world-nuclear-news.org/Articles/Orano-wins-further-MOX-order-from-Japan

hf "France and Japan collaborate in fast reactor development." *World Nuclear News*, 9 December 2019. https://world-nuclear-news.org/Articles/French-Japanese-collaboration-on-fast-reactor-deve

hg "Ohma start-up delayed by a further two years." *World Nuclear News*, 5 September 2018. https://www.world-nuclear-news.org/Articles/Ohma-start-up-delayed-by-a-further-two-years

hh "Cabinet minister rules out new nuclear reactors for 10 years." *Asahi Shimbun*, 12 November 2020. http://www.asahi.com/ajw/articles/13924315

hi Piller, Frank, "Energy regulations and electricity deregulation in Japan." *Japan Industry News*, 4 April 2016. https://www.japanindustrynews.com/2016/04/energy-regulations-electricity-deregulation-japan/

hj Neidell, Matthew, Uchida, Shinsuke, and Veronesi, Marcella, "Be cautious with the precautionary principle: Evidence from Fukushima Daiichi nuclear accident." IZA discussion paper no. 12687, 28 October 2019. https://ssrn.com/abstract=3475793

hk International Energy Agency. "Japan." https://www.iea.org/countries/japan. This page has all sorts of data on Japan's energy usage.

hl Evangeliou, Nikolaos, Balkanski, Yves, Cozic, Anne, and Møller, Anders Pape, "How 'lucky' we are that the Fukushima disaster occurred in early spring: Predictions on the contamination levels from various fission products released from the accident and updates on the risk assessment for solid and thyroid cancers." *Science of the Total Environment*, vol. 500-501, pp. 155-172, December 2014. https://www.sciencedirect.com/science/article/pii/S0048969714012819

hm Investigation Committee, "Appendices: Commission meeting reports," p. 82. *Final Report*. Government of Japan, July 2012. https://www.cas.go.jp/jp/seisaku/icanps/eng/final-report.html

hn USNRC, "Reflections on Fukushima: NRC senior leadership visit to Japan," p. 16, 2014. https://www.nrc.gov/docs/ML1435/ML14353A089.pdf

ho Japan Nuclear Fuel Limited, "Leadership/profile," 2019. https://www.jnfl.co.jp/en/about/company/

FINAL THOUGHTS

This book was hard work, but I was helped by many Japanese and non-Japanese people who contributed in big and small ways – not least my father, who has endured me wittering on about Japan for the last several years. I tried to compile a list of them all, but I know that I've missed loads (sifting through four years of emails isn't easy), so I have decided that a blanket 'thank you' is more appropriate. Those of you who contributed in some way know who you are, and I cannot thank you enough.

One final thing. So many people have contacted me since I released *01:23:40* with some variation of "I'd love to write my own non-fiction book, but..." To those people I say: just do it. Pick a topic that interests you and start Googling, you never know where it might take you. Everyone has experienced falling down the Wikipedia rabbit hole, this is just like that on steroids. It's lonely, difficult, time-consuming work, and the overwhelming majority of people you contact for advice or information will either ignore you or offer minimal assistance. But it is also fun and intellectually rewarding, some people are generous with their time, and you may form friendships in the most unexpected of places. Self-publishing is straightforward these days, and if you can afford to not need to make your fortune from royalties and can work at your own pace in your spare time, then there's nothing to hold you back. I can't wait to read what you come up with.